EPITHELIA
Advances in Cell Physiology
and Cell Culture

EPITHELIA
Advances in Cell Physiology and Cell Culture

Edited by
Christopher J. Jones
formerly Division of Biomedical Sciences,
University of California,
Riverside,
California,
USA

KLUWER ACADEMIC PUBLISHERS
DORDRECHT / BOSTON / LONDON

Correspondence to the Editor should be sent via
the publishers, Kluwer Academic Publishers, PO Box 55,
Lancaster, LA1 1PE, UK

Distributors

for the United States and Canada: Kluwer Academic Publishers, PO Box 358,
Accord Station, Hingham, MA 02018-0358, USA
for all other countries: Kluwer Academic Publishers Group, Distribution
Center, PO Box 322, 3300 AH Dordrecht, The Netherlands

British Library Cataloguing in Publication Data

Epithelia.
   1. Mammals. Epithelia. Cells
   I. Jones, Christopher J.
   599.087

ISBN 0-7923-8948-4

Library of Congress Cataloging in Publication Data

Epithelia: advances in cell physiology and cell culture/edited by
   Christopher J. Jones.
      p.   cm.
   Includes bibliographical references and index.
   ISBN 0-7923-8948-4 (casebound)
   1. Epithelium—Cytology.   2. Epithelium—Cultures and culture
media. I. Jones, Christopher J.
   [DNLM: 1. Biological Transport. 2. Cells, Cultured—physiology.
3. Epithelium—cytology. 4. Epithelium—physiology. QS 532.5.E7
E635]
QP88.4.E6   1990
611'.0187—dc20
DNML/DCL
for Library of Congress   90-5342
                                               CIP

Copyright

Published in the United Kingdom by Kluwer Academic Publishers,
PO Box 55, Lancaster, UK

Kluwer Academic Publishers BV incorporates the publishing programmes of
D. Reidel, Martinus Nijhoff, Dr W. Junk and MTP Press.

Printed in Great Britain by Butler and Tanner Ltd., Frome and London

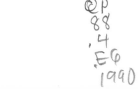

Contents

CONTENTS

List of Contributors

B. E. ARGENT
Department of Physiological Sciences
The Medical School
University of Newcastle upon Tyne
Newcastle upon Tyne NE2 4HH
UK

S. T. BALLARD
Department of Physiology
University of South Alabama
School of Medicine, MSB3024
Mobile, AL 36688
USA

R. C. BOUCHER
Department of Medicine, Division of
  Pulmonary Diseases
CB#7020, 724 Burnett-Womack Building
University of North Carolina at Chapel Hill
Chapel Hill, NC 27599–7020
USA

A. W. CUTHBERT
Department of Pharmacology
University of Cambridge
Tennis Court Road
Cambridge CB2 1QJ
UK

J. T. GATZY
Department of Pharmacology
Medical School, CB#7365 FLOB
University of North Carolina at Chapel Hill
Chapel Hill, NC 27599
USA

M. A. GRAY
Department of Physiological Sciences
The Medical School
University of Newcastle upon Tyne
Newcastle upon Tyne NE2 4HH
UK

A. HARRIS
Paediatric Research Unit
Division of Medical and Molecular Genetics
UMDS Guy's Campus, Guy's Tower
London Bridge,
London SE1 9RT
UK

B. H. HIRST
Department of Physiological Sciences
The Medical School
University of Newcastle upon Tyne
Newcastle upon Tyne NE2 4HH
UK

M. HUNTER
Department of Physiology
Worsley Medical and Dental Building
University of Leeds
Leeds LS2 9NQ
UK

P. K. JEFFERY
Department of Lung Pathology
National Heart and Lung Institute
Brompton Hospital
London SW3 6HP
UK

C. J. JONES
formerly of the
Department of Zoology
University of Durham
Durham DH1 3LE
UK
and the
Division of Biomedical Sciences
University of California
Riverside
CA 92521–0121
USA

LIST OF CONTRIBUTORS

T. KEALEY
Department of Clinical Biochemistry
University of Cambridge
Addenbrooke's Hospital
Hills Road, Cambridge CB2 2QR
UK

S. C. KIRLAND
Imperial Cancer Research Fund
  Histopathology Unit
Department of Histopathology
Royal Postgraduate Medical School
  Hammersmith Hospital, Ducane Road
London W12 OHS
UK

E. M. KROCHMAL
Department of Pediatrics
Division of Lung Biology
University of Utah, 50 North Medical Drive
Salt Lake City, UT 84132
USA

C. M. LEE
Pharmacogenetics Research Unit
Department of Pharmacological Science
The Medical School,
University of Newcastle upon Tyne
Newcastle upon Tyne NE2 4HH
UK

M. H. MONTROSE
Department of Medicine
Johns Hopkins University
Hunterian 515
725 N Wolfe St,
Baltimore, MD 21205
USA

N. L. SIMMONS
Department of Physiological Sciences
The Medical School
University of Newcastle upon Tyne
Newcastle upon Tyne NE2 4HH
UK

J. J. SMITH
Department of Pediatrics
University of Iowa College of Medicine
2542 Colloton Pavilion
Iowa City, IA 52242
USA

M. R. VAN SCOTT
Department of Medicine,
  Division of Pulmonary Diseases
CB#7020, 724 Burnett-Womack Building
University of North Carolina at Chapel Hill
Chapel Hill, NC 27599–7020
USA

J. R. YANKASKAS
Department of Medicine,
  Division of Pulmonary Diseases
CB#7020, 724 Burnett-Womack Building
University of North Carolina at Chapel Hill
Chapel Hill, NC 27599—7020
USA

Preface

Epithelial cells probably constitute the most diverse group of cells found in the body. In addition to serving as interfaces between external and internal environments, their functions include ion and fluid secretion and reabsorption, protein exocytosis, hormone secretion, recognition, surface protection and the control of ciliary movement. By their very exposure on the surfaces of the body, epithelial cells are subjected to wide-ranging assault, by micro-organisms and by chemical and physical forces. They are the targets for abrasion, infection and malignant transformation. Some epithelial cells show altered behaviour in inherited syndromes, such as cystic fibrosis, characterized by serious pancreatic and pulmonary disease.

In view of the importance of epithelia and the fact that their function can be altered by environmental and inherited factors, they are the subject of intensive research, particularly so in the case of cancer where most tumours are of epithelial origin. The use of animal tissues in epithelial research continues to provide important advances and this, coupled with the need to focus more on human tissues, has prompted a greater research emphasis on accessible human epithelia and on the establishment of cell cultures from animal and human sources. For primary cell cultures and cell lines to be of value, they need to express properties appropriate to their progenitors and relevant to the study in progress.

The purpose of this book is to consider some of the advances which have been made in the study of native epithelia, both animal and human, and in the development of cell cultures and cell lines from these sources. To facilitate this presentation, the text has been divided into five sections: Gastrointestinal Epithelia, Pancreas, Kidney, Respiratory Epithelia, and Skin and Skin Glands.

Acknowledgements

I should like to thank all of the authors for contributing such excellent chapters in their specialist fields and for adhering as much as possible to the desired format, particularly with respect to a consideration of techniques, justification of the use of cultured cells and a look ahead to future developments. I also wish to thank Dr Barry Argent and Dr M. M. Reddy for giving second opinions on certain chapters. This book was developed in response to an invitation from Kluwer Academic Publishers and I should especially like to thank Dr Michael Brewis, Dr Peter Clarke and Mr Philip Johnstone for their assistance during the production and publication.

Section I
GASTROINTESTINAL EPITHELIA

Introduction

There are three chapters in the gastrointestinal section of the book. The first, by Barry Hirst, considers the crucial, but often overlooked, barrier property of the upper gastrointestinal epithelium to acid and how this can be compromised by exposure of the epithelium to alcohols, aspirin and bile salts. After briefly reviewing studies conducted with the native epithelium, Dr Hirst considers potential model systems, in particular isolated membrane vesicles and cultured epithelial cells. The effects of alcohols and bile salts on the proton permeability and fluidity of apical membranes are described as is the resistance of epithelial cell cultures to acid. Both systems are judged to have provided important insights into the nature of the epithelial barrier to acid.

The chapter by Susan Kirkland considers the establishment and properties of cell lines derived from human colorectal carcinoma. In addition to their use in the study of colorectal cancer, some cell lines retain differentiated properties characteristic of cells in the normal epithelium and so are being used as models with which to study aspects of normal epithelial function. This is especially valuable in view of the current inability to establish differentiating cultures from the normal epithelium itself. Features expressed by the cell lines include absorptive and mucin-producing capabilities and the ability to transport ions, and possibly fluid, vectorially. Some cell lines have been found to be multipotential, having the ability to differentiate into two or three of the differentiated cell types found in the normal epithelium. Such cell lines are proving useful in studying the control of differentiation in colorectal epithelium.

Alan Cuthbert assesses the advantages and disadvantages of working with cultured epithelial cells and reviews the development of systems designed to prepare cultured monolayers for transport studies. Once the monolayers have become electrically 'tight' they are transferred to modified Ussing chambers where the biophysics and regulation of ion transport can be studied. In cell lines established from colorectal carcinomas and their metastases, electrogenic chloride secretion accounts for a major fraction of the responses to many secretagogues, including VIP and PGE_1. The absence

3

of a convincing demonstration of neutral NaCl absorption or electrogenic Na^+ absorption and lack of an aldosterone effect suggests that the mono-layer cells may be more like crypt cells in their properties rather than surface or villous cells. The mechanisms whereby the secretagogues act and the ion-transporting processes themselves are considered more fully.

In the renal section of the book, Marshall Montrose assesses the contribution of both renal and intestinal epithelial cell lines to the study of transport physiology. There is a detailed compilation of cell lines, and sub-clones where appropriate, their tissue of origin, expressed membrane enzymes and transport functions.

1
Gastrointestinal Epithelial Barrier to Acid: Studies with Isolated Membrane Vesicles and Cultured Epithelial Cells

B. H. HIRST

INTRODUCTION

All epithelia share a common function in acting as selective barriers between two compartments. In the case of the gastrointestinal tract, the epithelial cells separate the luminal contents from the interstitial fluid. The gut epithelium has several barrier functions. It has an important immunological role in protecting the body from micro-organisms ingested with food or those colonizing the gut lumen. The barrier mechanisms of the gut against micro-organisms are many and varied. They range from the production of a highly acidic environment within the stomach, to kill bacteria, to the selective sampling of luminal antigens (bacteria) by the intestinal M cells overlying the gut-associated lymphoid tissue, resulting in a mucosal immunological response. The digestive environment of the gut lumen is also a formidable barrier in itself.

The gut is, however, a very selective barrier. It must selectively allow essential nutrients to be absorbed into the body. Some small molecules are able to diffuse passively through the gut mucosa. Most molecules, however, are aided in their absorption by specific transport processes. Many such processes have been defined in gut epithelial cells, including those for sugars and amino acids (usually Na^+-coupled), dipeptides (usually H^+-coupled), water-soluble vitamins, such as folic acid and vitamin B_{12}, ions such as iron and calcium, and other nutrients essential to the body.

The acid secreted by the stomach acts as barrier to ingested microbes. But the upper gastrointestinal tract mucosa must also act as a barrier to prevent the back-diffusion of its acidic contents. This barrier to acid also, if only to a lesser degree, extends into the upper intestine, and upward into the oesophagus. The gastric mucosal barrier to luminal acid has been the subject of continued and considerable study. The impetus for such sustained

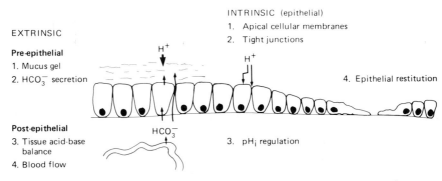

Figure 1.1 Components of the upper gastrointestinal barrier to acid. (From Hirst[2])

scientific interest has been the fascination, at least since the time of John Hunter[1], in understanding how the stomach is able to carry out its normal digestive functions without digesting itself. The integrity of the stomach has to be maintained in the face of a luminal proton concentration of up to 160 mmol L^{-1}, i.e. a pH of around 0.8. This represents a proton concentration gradient across the gastric epithelium of $>10^6$. The maintenance of an upper gastrointestinal barrier to protons has clear clinical importance; breach of the barrier is associated with peptic ulceration. Components of the barrier to acid are illustrated in Figure 1.1. These may be conveniently divided into extrinsic and intrinsic mechanisms. Intrinsic mechanisms are native to the epithelia *per se* and include the role of the apical cell membranes and intercellular junctional complexes in resisting proton permeation, and mechanisms for controlling intracellular pH. Extrinsic mechanisms include the secretion of an alkaline mucus gel onto the apical surface of the cells and, on the serosal side, an adequate supply of blood and the overall tissue acid–base balance[2].

Until recently, most of the studies on upper gastrointestinal resistance to damage by acid and other agents likely to be found in the gut lumen, such as pepsin, bile salts, aspirin and ethanol, have employed animal or human studies *in vivo* or pieces of whole mucosa mounted *in vitro*. In particular, the earlier work has concentrated on the use of mucosal damaging agents, including ethanol, bile salts and aspirin, to define the barrier. Such studies have yielded valuable information. Hollander was early to recognize the epithelial cell layer as an important component of the mucosal barrier[3]. The studies of Davenport, however, must be acknowledged as giving the essential insights into describing the components of the barrier[4-15]. In particular, Davenport's experiments, together with those of other scientists, have pointed to the epithelial cell layer as being the principal structure conferring physiological barrier function to acid in the upper gastrointestinal tract[16]. The earlier work on the gastric mucosal barrier has been the subject of a recent review[2], and is also reviewed elsewhere[17-21]. In contrast, the present discourse will review recent studies using two modern cell biology techniques, isolated membrane vesicles and cultured epithelial cells, which have helped to better define the nature of the epithelial barrier to acid.

ISOLATED APICAL (LUMINAL) MEMBRANE VESICLES

Preparations of isolated membrane vesicles have been used extensively for transport studies in epithelial tissues. These studies have yielded important and novel advances in several fields[22], not least in solute and ion transport in the small intestine[23] and the stomach[24]. By extending the techniques, the same vesicle preparations have been used to study the passive permeability properties of the membranes and their modulation (e.g. membranes from kidney[24,25], intestine[26], placenta[27] and stomach[28]). Work in the author's laboratory has concentrated on quantifying the proton permeability of apical membrane vesicles isolated from the upper gastrointestinal tract. Such preparations offer a number of advantages. The vesicles can be purified from one plasma membrane domain, i.e. apical, thus allowing study of properties of this domain alone; different isolation/purification techniques allow isolation of the basolateral membrane. The influence of other cellular events on membrane function are minimized and the intra- and extravesicular environments can be readily controlled and manipulated.

One must also be aware of the potential disadvantages of these membrane vesicles for quantifying membrane functions. The normal interaction between other cellular functions and the membrane are minimized or lost. The structure, both chemical and physical, of the membranes may be perturbed during the preparative procedures. Such perturbations may arise, for example, due to shear stress stripping out integral components, other cellular degradation products, particularly lipids, becoming incorporated into the membrane preparations, or damage to membrane components due to enzymic processes activated during the isolation procedures. Practical examples of such problems are illustrated below. However, on balance, the information gained from the introduction of membrane vesicle techniques has far outweighed any potential disadvantages in their use.

Parietal cell apical membranes

Gastric acid is secreted as a result of the (H^++K^+)-ATPase located in the apical membrane of the parietal cell. As the site of acid secretion, these apical membranes must, *in situ*, be able to withstand the highest concentrations of acid found in the stomach, and therefore the whole body. Parietal cell apical membrane vesicles are prepared by differential and density-gradient centrifugation from homogenates of rabbit fundic mucosa stimulated to secrete acid[29-31]. These membrane vesicles are rich in (H^++K^+)-ATPase and associated ouabain-insensitive K^+-stimulated p-nitrophenyl-phosphatase (pNPPase) activity. From 55–85% of these activities are latent, that is the enzyme activities are only expressed after pretreatment of the vesicles with a mild detergent such as 9–15 mmol L^{-1} octyl glucoside[31],

7

5–7 mmol L^{-1} octyl thioglucoside or 1.5 mmol L^{-1} 3-[(3-cholamidopropyl)-dimethylammonio]-1-propanesulphonate (CHAPS)[32]. Latent enzyme activity can also be expressed by mild hypotonic shock activity[31]. These latency studies indicate that the majority of the vesicles form with a right-side-out, or extracellular face outwards, orientation. This interpretation was confirmed by analysis of freeze-fracture replicates of the membrane vesicles by electron microscopy. In such replicates, the majority of the membrane-associated particles visible by electron microscopy are associated with the convex surface, consistent with a right-side-out orientation of the vesicles[33]. Thus, these parietal cell apical membrane vesicles are an appropriate model for studying the effect of luminal damaging agents on gastric membrane barrier function.

Intestinal brush-border membrane vesicles

Brush-border membrane vesicles (BBM) are prepared from discrete areas of the small intestine by the divalent cation precipitation and differential centrifugation technique described by Kessler *et al.*[34], but instead using $MgCl_2$ (Figure 1.2). The original technique for preparing BBM used 10 mmol L^{-1} $CaCl_2$ as the precipitating agent, but this was reported to be associated with activation of Ca^{2+}-dependent phospholipase A activity, with subsequent lipid decomposition[35]. $MgCl_2$ was thought to have a lesser effect on lipid decomposition[35], but recent studies[36] questioned the benefits of $MgCl_2$. Intestinal BBM are predominantly (>90–95%) in the right-side-out orientation[37,38].

PROTON PERMEABILITY OF APICAL MEMBRANES

The net proton permeability of upper gastrointestinal apical membranes has been quantified using membrane vesicles and the acridine orange fluorescence quenching technique[31,39]. Membrane vesicles are pre-incubated in an acidic solution (e.g. pH 6.5) to equilibrate the intravesicular pH. The vesicles are then added to a similar solution at higher pH (e.g. pH 8.0) containing acridine orange and the fluorescence monitored continuously. In addition, the vesicles have to be voltage-clamped to eliminate H^+–OH^- diffusion potentials. This may be achieved by carrying out the experiments with 150 mmol L^{-1} K^+ on both sides of the vesicle membranes, and by adding valinomycin, a K^+-ionophore. Proton permeation, as measured by the recovery of acridine orange fluorescence, is indeterminably slow when vesicles are not voltage-clamped[39]. Representative traces from experiments to determine proton permeability are illustrated in Figure 1.3. It can be seen that the original acridine orange fluorescence is quenched upon addition of the acidic vesicles. This is a result of the intravesicular accumulation of the acridine orange, and the subsequent proton-quenching of its fluorescence signal. As the transvesicular pH gradient decays, due to the net permeation of protons, the fluorescence signal recovers. This fluorescence recovery follows simple first-order kinetics[31], and enables determination of the rate

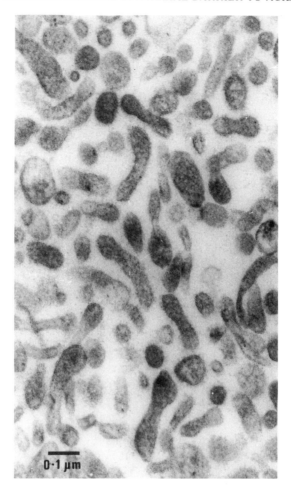

Figure 1.2 Electron micrograph of rabbit duodenal brush-border membrane vesicles prepared by the $MgCl_2$-precipitation method

constant for the recovery of the fluorescence, equivalent to proton permeation; k_H^+. The exponential time constant (τ) for this process is the inverse of k_H^+. From these values, and the radius (r) of the vesicles, the intracellular concentration of protons ($[H^+]_i$), and the intravesicular (β_{end}) and extravesicular buffer capacities (β_{ex}), the net proton permeability coefficient (P_{net}), which is the sum of the flux of protons (P_H) and hydroxide (P_{OH}), may be calculated:

$$P_{net} = P_H + P_{OH} = \frac{r \cdot \beta_{end} + \beta_{ex}}{3\tau[H^+]_i \cdot \quad \ln 10}$$

9

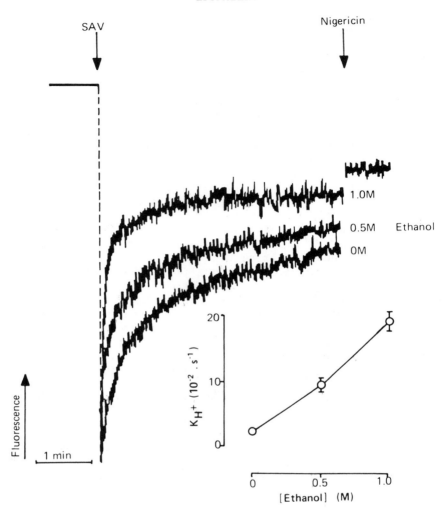

Figure 1.3 Effects of ethanol on proton permeation in parietal cell apical membrane vesicles. Vesicles (stimulation-associated; SAV) were voltage-clamped with K⁺-valinomycin, and equilibrated at pH 6.5, and then diluted into a pH 8.0 solution containing acridine orange (*first arrow*). Addition of vesicles leads to a quenching of the fluorescence signal as the dye accumulates in the acidic intravesicular space. Recovery of fluorescence gives the rate of H⁺ permeation. Nigericin, a H⁺-K⁺ ionophore, was added at the *second arrow* and fully dissipates the proton gradient. Increasing concentrations of ethanol accelerate the rate of fluorescence recovery, i.e. increase proton permeation. *Inset*: apparent rate constant for the recovery of fluorescence plotted against concentration of ethanol. (From Hirst[2])

Such experiments yield a value for P_{net} in parietal cell membranes, at 20°C, of 4×10^{-4} cm/s[31]. This value is slightly lower than P_{net} in duodenal

$(6-10 \times 10^{-4} \text{ cm/s})^{31,40}$ and jejunal BBM (6×10^{-4} cm/s; D. Zhao and B. H. Hirst, unpublished), but considerably lower than in renal cortical BBM $(5-10 \times 10^{-3} \text{ cm/s})^{25,31}$.

The rank order of proton permeability for these membrane types (parietal < duodenal = jejunal << renal) is in general agreement with their known physiological environments. Parietal cell membranes are likely to be exposed to pH < 1, with duodenal pH 3–4, and renal pH 5–6. Quantitatively, however, these permeabilities appear inconsistent with the properties of the membranes *in situ*. The estimated P_H for intact gastric mucosa[41] is around $0.4-4.5 \times 10^{-5}$ cm/s; one to two orders of magnitude lower than the value for P_{net} estimated for isolated parietal cell membranes. Moreover, the estimated maximum value for P_H if the observed pH gradients (intracellular pH ~ 7.2; intragastric pH ~ 0.8) across the parietal cell apical membrane during secretion are to be maintained, is 10^{-7} cm/s[42]. Although P_{net} determined for the isolated membranes is the sum of P_H and P_{OH}, P_H is the dominating variable[43]. A potential explanation for the greater values for P_{net} estimated in isolated membranes is their contamination with endogenous protonophores, i.e. agents which can carry protons across the lipid bilayer. These might include fatty acids generated during the homogenization and isolation procedures[44]. Bovine serum albumin added to several membranes reduced P_{net}[44], arguably by binding these contaminants. The reduction in parietal cell membranes, though, was minimal, about 30%[31], and the reduction in P_{net} caused by the albumin might better be explained by interaction with valinomycin, so reducing the effectiveness of the voltage-clamp[43]. Other factors resulting in differences in estimates of P_{net} *in vitro* as compared with *in situ*, may include the greater pH gradient *in vivo* and thus pH-dependent factors[31,43], cellular transmembrane potentials[31], and the small radius of curvature for membrane vesicles as compared with intact cells[31].

The routes of proton permeation through biological membranes have been recently reviewed[43,44]. Possibilities include fatty acids and other contaminants acting as protonophores (*vide supra*), permeation as neutral molecules such as HCl at low pH via intramembrane water, or by thermally activated permeation of the lipid bilayer. Permeation through protein channels, or via transmembrane proteins, is an attractive explanation in several epithelial apical membrane systems. In duodenal BBM, we have reported an amiloride-sensitive proton-conductive pathway, the properties of which are consistent with a role for Na^+–H^+ exchangers as a leak pathway for protons[45]. We have also observed amiloride-sensitive proton permeation in renal BBM[31], although others failed to detect such an amiloride-sensitivity, but did report sensitivity to *p*-chloromercuribenzenesulphonate, suggesting a role for sulphydryl groups in the pathway[46]. Placental, but not renal, BBM proton permeation is reported to be reduced by the anion exchange and chloride conductance inhibitor, DIDS[27]. Proton permeation in gastric parietal cell apical membranes is insensitive to amiloride, arguably because these membranes do not possess a Na^+–H^+ exchanger (J. M. Wilkes and B.H. Hirst, unpublished). It may be postulated that proton permeation in parietal cell apical membranes might occur via leakage through the

11

$(H^+ + K^+)$-ATPase, or the apical conductances[24] for Cl^- or, more likely, K^+. Thus, integral membrane proteins are likely to act as an important, and cell specific, route for proton permeation.

The dynamics of the lipid phase of the apical membranes are likely to play a critical role in determining proton permeability[25,31]. The lipid dynamics of membranes may be investigated by several techniques, each with their own benefits and shortcomings. In conjunction with studies of proton permeation in apical membrane vesicles, we and others have applied the technique of diphenylhexatriene (DPH) fluorescence anisotropy, under steady-state and time-resolved conditions[25,31]. The term *membrane fluidity* may be used to describe the constraints on the free rotation of DPH within the lipid environment of the membrane bilayer, and is quantified by the parameter $(r_0/r)-1$ which is directly related to viscosity in an isotropic system[31]. The proton permeability of gastric, duodenal and jejunal membranes was not simply related to their native membrane fluidity. If any relationship was apparent, it was an inverse one; $(r_0/r)-1$ for renal BBM = duodenal BBM < parietal cell membranes[31]. In contrast to this lack of relationship between native membrane fluidity and P_{net}, perturbations in membrane fluidity are correlated with changes in P_{net} (*vide infra*).

STUDIES WITH BARRIER-BREAKING AGENTS

Studies on the nature of the gastric mucosal barrier and, to a lesser extent, intestinal barrier function, have concentrated on the effect of a variety of agents which damage or break the barrier. The physiological breaking of the gastric mucosal barrier in intact mucosa, whether *in vivo* or *in vitro*, is indicated by the increased flux of ions across the gastric mucosa, particularly H^+ back-diffusion and Na^+ and K^+ flux into the stomach, and an associated fall in transmucosal potential difference and electrical resistance. Agents widely associated with such barrier-breaking actions include aspirin and other weak acids, bile acids and other detergents, alcohols, and proteolytic enzymes, especially pepsin. The action of these agents in whole mucosa have recently been reviewed[2,20]. These earlier studies have indicated that the major action of these barrier-breaking agents is on the plasma membrane[2,16,21]. Here, we concentrate on data concerning the direct action of selected agents on gastrointestinal isolated apical membranes.

Alcohols

Alcohols, including ethanol, have been the agents most extensively studied with isolated membrane vesicles. *In vivo*, alcohols break the gastric mucosal barrier, and the potency of individual alcohols is related to their oil–water partition coefficients[47]. Acute and chronic administration of ethanol impairs intestinal absorption of glucose and amino acids *in vivo* and *in vitro*[48,49]. In the mouse gastric mucosa, ethanol (25% or approximately 5 mol L^{-1} in

100 mmol L^{-1} HCl) after 5 min caused distortion of the apical membrane, with swelling of mitochondria and chromatin clumping[50]. The tight junctions, however, appeared normal. In contrast, in isolated *Necturus* antral mucosa, ethanol (20%) at pH 3.0 caused an immediate reduction in transepithelial electrical resistance, suggesting that paracellular conductance primarily contributes to acid back-diffusion. At a slightly later time (4–6 min), the ethanol reduced cell membrane resistances and intracellular pH[51]. Thus, in intact gastric mucosa, high concentrations of ethanol increase apical membrane proton conductance, but gross acid back-diffusion may also be contributed to by increased paracellular conductance.

Alcohols are well-recognized membrane perturbing agents. In several systems, including isolated gastric surface[52] and parietal cells[53], parietal cell apical[31] and microsomal vesicles[54], and duodenal and jejunal BBM[31,55], alcohols perturb membrane phospholipid dynamics, resulting in a general fluidization of the plasma membrane. These increases in membrane fluidity are associated with generalized increases in membrane permeability. For example, low concentrations of ethanol (0.3–0.5 mmol L^{-1}), equivalent to those likely to be found in the human gut after what might be considered as moderate alcohol ingestion[47,56], indirectly reduce alanine and glucose uptake into jejunal BBM vesicles[57,58] by increasing membrane conductance to Na^+. Glucose and alanine transport are coupled to, and driven by, the Na^+ gradient; hence increased membrane Na^+ conductance allows uncoupled Na^+ uptake with reduced solute uptake. Similarly, we have reported that benzyl alcohol increases the passive permeability of duodenal BBM to NaCl[39].

In duodenal BBM and gastric parietal cell membranes, alcohols increase P_{net}[31]. Similar results have been reported for renal BBM[25]. Typical experiments illustrating increased rate of proton permeation in gastric parietal cell apical membrane vesicles induced by ethanol are illustrated in Figure 1.3 P_{net} is increased by similar low concentrations of ethanol reported to increase other ionic permeabilities. The increases in P_{net} in parietal cell and duodenal BBM vesicles are correlated with the alcohol-induced increases in membrane fluidity (Figure 1.4), suggesting a simple causal relationship. The gastric membranes are more sensitive than the duodenal BBM. The concentrations of ethanol required to increase P_{net} are not only equivalent to those likely to be found in the gut after moderate social drinking[47,56], but, moreover, considerably lower than the concentrations required to elicit immediate gross damage to the whole mucosa. The more subtle increases in P_{net} observed with these lower concentrations of ethanol are likely to be important early events in pathophysiological damage to the mucosa *in situ*.

Slightly greater concentrations of ethanol, as well as other alcohols, cause more extensive damage to membrane vesicles, resulting in increased fragility and disruption (Figure 1.5). This increased membrane fragility, as assessed by decreases in the vesicle volume, are also correlated with increased membrane fluidity (Figure 1.6). Of interest is the greater resistance of duodenal, as compared with jejunal and gastric, membranes to disruption by alcohols (Figure 1.6). The difference cannot be explained by differences in native membrane fluidity as these are similar in duodenal and jejunal BBM from the rat[59,60] and rabbit[55]. Duodenal BBM are also less sensitive to

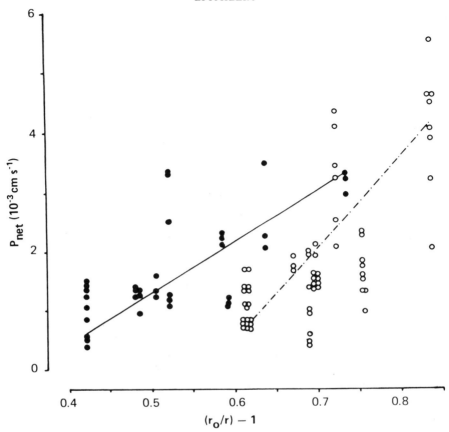

Figure 1.4 Correlation between membrane fluidity $(r_o/r)-1)$ and proton permeability (P_{net}) in gastric parietal cell (○) and duodenal BBM (●) vesicles treated with alcohols. (From Wilkes *et al.*[31])

alcohol-induced increases in P_{net} (Figure 1.4). The mechanism of greater resistance of duodenal membranes has yet to be explained.

Bile salts

Bile salts are another group of agents which cause increased ionic, including H^+, fluxes and reduced transmucosal electrical resistance upon luminal addition. Ultrastructural studies have localized the initial site of damage to the apical plasma membranes[61]. Taurocholate (10 mmol L^{-1}) at pH 3.0 resulted in rapid acidification of *Necturus* antral surface epithelial cells, concomitant with an increase in apical cell conductance[51]. Bile salts have important detergent qualities[62], which would appear to provide a likely explanation for their ability to damage the gastrointestinal tract[2].

14

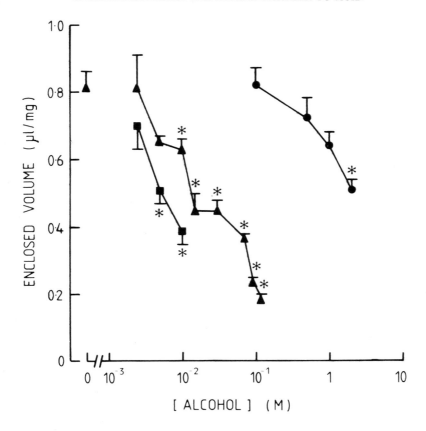

Figure 1.5 Effect of octyl (■), benzyl (▲) and ethyl (●) alcohols on the enclosed volume of duodenal BBM vesicles. Vesicles were equilibrated for 45-min in the presence of [^{14}C]glucose, and various concentrations of one of the alcohols. Values are mean enclosed volumes calculated from the [^{14}C]glucose retained at equilibrium with error bars of ±1sem.*, $p < 0.05$ compared with control without alcohol. (From Ballard *et al.*[55])

Low concentrations of bile salts (e.g. 0.1–1.0 mmol L^{-1} deoxycholate and its conjugates), likely to be below their critical micellar concentration (the concentration above which amphiphile molecules, such as bile salts, associate to form thermodynamically stable colloidal aggregates[62]), increase duodenal BBM membrane fragility and P_{net}[40]. This is consistent with the initial site of action described from ultrastructural and electrophysiological studies in intact mucosa. In contrast to the effects of the alcohols, however, these bile salt-induced increases in membrane fragility and P_{net} are not associated with significant changes in membrane fluidity[40]. Thus, small changes in membrane fluidity may produce a disproportionate increase in P_{net} and fragility. Alternatively, changes in membrane fluidity may be only a secondary factor involved in bile salt-induced increases in P_{net} and fragility. As was noted earlier for the action of alcohols, duodenal BBM are less

15

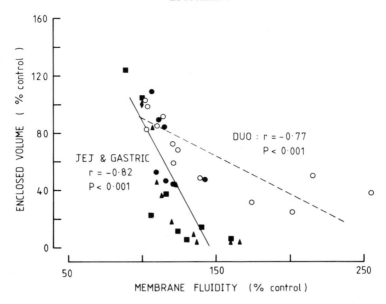

Figure 1.6 Relationship between enclosed volume and membrane fluidity during treatment with alcohols, for duodenal BBM (○, *dashed* line), and for jejunal BBM (●), parietal cell apical (▲) and gastric microsomal (■) membrane vesicles. The latter three membrane types fell on the same correlation (*solid*) line. Enclosed volume and membrane fluidity have been normalized to the value for untreated vesicles (◆). (From Ballard *et al.*[55])

sensitive than jejunal BBM to bile salts, and in jejunal BBM, bile salts induce a significant increase in membrane fluidity (D. Zhao and B. H. Hirst, unpublished). The concentrations of bile salts required to induce increases in P_{net} are within the physiological range, even for that found in the normal human stomach [63]. Bile salts, therefore, might be considered as a normal luminal factor likely to influence intestinal permeability to acid.

RESISTANCE OF CULTURED EPITHELIAL CELLS TO ACID

The major anatomical site conferring physiological barrier function *in vivo* has been localized to the epithelial cell layer, and, in particular, to the apical cell membranes and the occluding tight junctions[2]. Studies with intact gastrointestinal mucosal tissue, *in vitro* as well as *in vivo*, are complicated by other components with barrier functions. These will include the mucous layer overlying the epithelium, and an adequate supply of blood providing not only nutrients, but also, in the case of resistance to acid, buffer capacity.

Gastric mucosal cells, both primary cultures and established cell lines, have been used for studies of gastric damaging agents such as bile salts and non-steroidal anti-inflammatory drugs (*vide infra*). Long-term primary culture of gastrointestinal cells has not proved successful to date, often due

16

to overgrowth by fibroblasts. In the majority of studies with damaging agents, the cells have been grown on an impermeable support, such as plastic multi-well dishes, which does not allow the study of integrated physiological functions, such as the vectorial transport of ions or epithelial barrier functions. In contrast, monolayers of functional epithelial cells grown on permeable supports, such as cellulose esters, polycarbonate or anodized aluminium, in culture provide a simplified system in which the barrier properties of the isolated epithelial cell layer may be studied. In such a system, the first barrier to acid will be presented by the apical cell membranes in parallel with the tight junctions. Subsequent cellular functions, including regulation of intracellular pH, will also play a role in resistance to acid in these simplified epithelial systems. Some of the advantages of these cell culture systems include the relative homogeneity of the cells under study and the absence of other complicating factors, such as blood supply, smooth muscle tissue and a mucous layer. This allows epithelial cell barrier function to be easily studied in isolation. However, it should be noted that the mucosa *in vivo* is not such a simple system, with its heterogeneous cell population and all the other components aside from the epithelial cells. Nevertheless, culture systems have provided a wealth of important information on epithelial cell function, and, in many cases, are the only appropriate system for studying some aspects of epithelial cell biology (see other contributions in this volume, and Matlin and Valentich[64]).

Choice of the cell model is an obvious consideration, although not a simple one. A prerequisite for studying the function of an epithelium is that the chosen cell model can reconstitute as such an epithelium. The electrical properties of the cell monolayers may then be measured in a modified Ussing chamber system. The cell model should, in addition, simulate those features of the normal gastrointestinal tract one wishes to investigate. Ideally, therefore, the cell culture systems should reconstitute as an epithelium with histotypic and physiological features of the part of the gastrointestinal tract to be represented. In practice, however, these ideals often have to be compromised to experimental limitations of the cell systems available. One successful approach has been the primary culture of canine peptic cells. The approach chosen by our own laboratory has been to select established epithelial cell lines which are able to reconstitute as high-resistance monolayers and, as such, mimic one aspect of the gastric mucosal barrier to acid: the high electrical resistance of the gastric epiethelium.

CELLS GROWN ON PLASTIC

Several studies on the effects of barrier-breaking agents on gastric epithelial cells in culture have been reported. The disadvantages of these studies using cells grown on an impermeable support, such as plastic, is that only the apical surface of the cells is exposed to the damaging agent, and that the variables measured, usually related to cell viability, are individual cell, rather than epithelial, responses.

Damage to primary cultures of rat gastric mucosal cells has been assessed in terms of trypan blue dye exclusion, and release of ^{51}Cr. Damage was observed with saline at pH 5.0, but not pH 6.0, after 1 or 2 h incubation. With a pH 4.0 solution, ^{51}Cr release was maximal[65]. Aspirin, 10 mmol L^{-1}, when added at pH 5.0, increased cellular damage but was without effect at pH 7.4 or 6.0[65]. Indomethacin, 2.5–10 mmol L^{-1}, at neutral pH also increased ^{51}Cr release and reduced dye exclusion[66]. Bile salt-induced cellular damage was also observed in these rat gastric cells. Taurocholate, 5–20 mmol L^{-1}, resulted in a dose-related increase in ^{51}Cr release and decrease in dye exclusion[67]. The damaging effects of aspirin, indomethacin and taurocholate could be ameliorated by addition of 16,16-dimethyl prostaglandin E_2[65-67].

Similar results to those described with primary cultures of rat gastric mucosal cells have been obtained with a human gastric cell line, MKN 28. MKN 28 is a cell line with epithelial characteristics, including the presence of junctional complexes, microvilli, and positive staining with periodic acid–Schiff but not Alcian blue[68]. These characteristics are similar to primary cultures of human gastric epithelial cells[68,69]. In MKN 28 cells, taurocholate, ethanol and indomethacin increased ^{51}Cr release[68,70,71]. The concentrations of these agents required to elicit damage are relatively high, as compared with those described above to alter the isolated apical membranes; e.g. ethanol concentrations of 15% are required to increase ^{51}Cr release from MKN 28 cells, comparable to those of around 10% required to cause damage *in vivo*, but greater than those required to increase proton permeability in apical membrane vesicles, ~ 2% (*vide supra*). We have reported similar damaging effects of a variety of non-steroidal anti-inflammatory drugs on an ileocaecal cell line, HCT-8, grown on plastic. Indomethacin was the most potent compound, as assessed by release of the dye neutral red (monitoring plasma membrane integrity similar to trypan blue exclusion or ^{51}Cr release) or decreased tetrazolium dye reduction (MTT assay), an indication of inhibition of mitochondrial enzyme activity[72].

RECONSTITUTED EPITHELIAL MONOLAYERS OF GASTRIC CELLS IN PRIMARY CULTURE

The cells of the gastric gland, as part of their normal function, have to be able to withstand acid concentrations of ~ 150 mmol L^{-1}. The two major cell types within the gland are the parietal cell, which produces this acid, and the peptic cell, responsible for pepsinogen synthesis and secretion. Canine isolated peptic cells have been grown in short-term culture to form reconstituted epithelial monolayers[73]. These cells retain functional differentiation, including pepsinogen granules which may be stimulated to release their contents by agents, such as carbachol and secretin, recognized as stimulants of pepsinogen secretion in the intact mucosa[74,75]. The monolayers grown on permeable supports are electrically tight, forming tight junctions, and the transepithelial electrical resistance[76] (R_e) is around 1500 $\Omega \cdot cm^2$. Thus, these peptic cell monolayers may be considered as useful model systems mimicking at least some features of the gastric epithelium and particularly gastric glands.

18

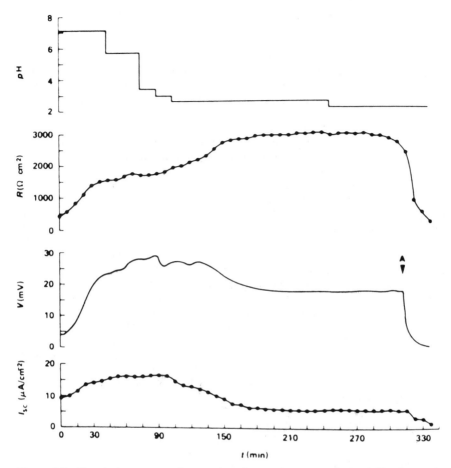

Figure 1.7 Electrical response of a peptic cell monolayer to apical acidification. With reduction of apical pH to <2.5, transepithelial resistance (R) increased, and remained stable for >3 h. Addition of aspirin, 4 mmol l^{-1} (*A, arrow head*), caused a rapid decay in electrical resistance, potential difference (V) and short-circuit current (*Isc*). (From Sanders *et al.*[73])

Sanders *et al.*[73] have demonstrated that these peptic cell monolayers are able to maintain epithelial electrical integrity despite the imposition of a large pH gradient (Figure 1.7). R_e more than doubled during acidification of the apical solution to pH 2. The monolayers were able to maintain this pH gradient for more than 4 h. The peptic cell barrier to acid was overcome by a critical pH between 2.5 and 2.0 for 70% of monolayers, and < 2.0 in 20%. Aspirin, 4 mmol L^{-1}, caused a rapid decline in R_e when added to the acidified solution (Figure 1.7), but not when added to a neutral apical solution. Cellular integrity was maintained during acidification, as indicated by the changes in R_e being fully reversible. The peptic cell monolayers showed marked polarity in their resistance to acid. Acidification of the

19

basolateral solution to pH $<$ 5.5 resulted in a rapid and irreversible decay in R_e.

Resistance of the peptic cell monolayers to acid was suggested not to be dependent upon Na^+ or Cl^- transport[73]. Barrier function was not prevented by treatment with ouabain, nor by absence of basolateral HCO_3^-. The monolayers were also resistant to apical addition of acidified pepsin, consistent with the normal function of peptic cells: secretion of pepsinogen into the gastric gland lumen, where it will be activated by the acid secreted by the adjacent parietal cells to give the active enzyme, pepsin. In another study, apical acid, up to 150 mmol L^{-1} for 4 h, did not affect the synthesis of pepsinogen or its secretion in response to stimulants in these peptic cell monolayers[75].

The lowest pH the peptic cell monolayers were able to withstand[73], as measured by R_e, was around 2.0. *In situ*, however, the intraluminal pH in gastric glands falls below pH 1. The failure of the monolayers to be able to resist such a low pH suggests that other factors, including intracellular and interstitial buffering, may contribute to acid resistance *in vivo*. Alternatively, during adaptation to the culture conditions, the peptic cells may have, to some degree, lost some of their ability to sustain pH gradients. More recent studies from another laboratory suggest that the monolayers are able to withstand physiologically lower pH (150 mmol L^{-1} HCl; presumably pH $<$ 1.0), although R_e was not monitored in this study[75].

Primary cultures of gastric surface epithelial cells from the guinea pig[77] and foetal rabbit gastric epithelial cells[78] have also been reported to form monolayers with significant, although lower ($\sim$ 280 $\Omega \cdot cm^2$), transepithelial resistance. These monolayers have not been studied to investigate whether they are able to maintain a pH gradient.

RESISTANCE OF ESTABLISHED EPITHELIAL CELL LINES TO ACID

Studies in our laboratory have used established epithelial cell lines as a simpler system, compared with primary cultures, for studying resistance to acid. In addition, it has allowed us to address the question of whether resistance to acid is a unique property of gastric epithelia. We have investigated three cell lines, each of which is able to reconstitute as high resistance epithelial monolayers when grown on permeable supports. Madin–Darby canine kidney (MDCK) cells form high-resistance monolayers[79], e.g. mean R_e approximately 2500 $\Omega \cdot cm^2$. Acidification of the apical solution to pH 4.5 was associated with a gradual increase in R_e. Further reduction in apical pH resulted in a decrease in R_e back to control values (Figure 1.8). Basolateral acidification was also associated with a gradual increase in R_e, with maximum R_e observed at pH 3.5, and R_e still elevated at pH 3.0 (Table 1.1). The resistance of MDCK monolayers to basolateral acid contrasts with the results with peptic cell monolayers, where marked difference in apical and basolateral susceptibility to acid was observed. More recent studies (A. B. Chan and B. H. Hirst; unpublished) have shown some differentiation between the apical and basolateral sides;

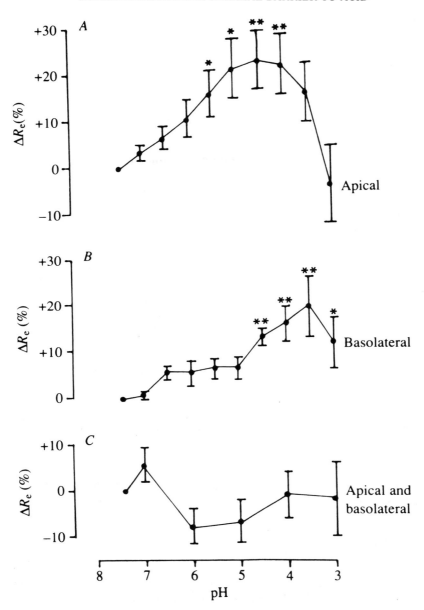

Figure 1.8 Transepithelial electrical resistance of MDCK monolayers as a function of pH of apical and basolateral solutions. The pH of the apical (*A*) or basolateral (*B*) solution was acidified while the contralateral solution was maintained at pH 7.4. In (*C*), both solutions were acidified simultaneously. Even at pH 3.0, electrical resistance of the monolayers is maintained. (From Chan *et al.*[82])

Table 1.1 Transepithelial electrical resistance and potential difference in MDCK, HCT-8 and T84 monolayers in response to low pH

Cell line	n	Basolateral pH	Apical pH	R_e^a ($\Omega \cdot cm^2$)	$\hat{R}_e^b$ (%)	V_t^c (mV)
MDCK	11	7.4	7.4	3404±313	+2.9±1.7	+0.7±0.1
	6	7.4	3.0	2322±250	−3.6±8.5	+5.2±1.0
	6	3.0	7.4	3291±391	+12.0±5.2	−3.0±0.3
	6	3.0	3.0	1475±207	−1.5±7.3	+0.6+±0.1
HCT-8	6	7.4	7.4	772±62	+12.3±8.6	+1.0±0.0
	6	7.4	3.0	1063±70	+32.1±8.5	+3.5±0.0
	6	3.0	7.4	1331±96	+39.7±10.0	−1.3±0.0
T84	5	7.4	7.4	1103±93	+15.5±7.7	+0.4±0.1
	4	7.4	3.0	1017±88	−14.6±10.4	+6.2±0.8
	5	3.0	7.4	906±90	−12.3±9.1	−1.3±0.4

[a] Measured at 95 min (see Figure 1.8)
[b] R_e at 95 min expressed as the per cent change from the original R_e with both apical and basolateral solutions at pH 7.4
[c] Basal positive; measured at the same time as R_e

R_e was maintained with apical pH 3.0, but not basolateral pH 3.0 for 1.5 h, although pH 2.5 on either side resulted in irreversible decreases in R_e. MDCK cell monolayers were also able to withstand simultaneous apical and basolateral acidification (Figure 1.8). The lesser expression of polarization with respect to resistance to acid in the established MDCK cell line, as compared with primary cultures of peptic cells, might be indicative of some loss of polarity in the cell line, perhaps reflecting selection pressures which have allowed the cell line to survive in culture.

Similar results were observed with two human intestinal cell lines, T84 derived from a colonic adenocarcinoma, and HCT-8 derived from an ileocaecal adenocarcinoma. Numerous reports are available describing the use of T84 monolayers as a model system for colonic ion transport (e.g. see McRoberts and Barrett[80]). HCT-8 cells show several epithelial characteristics similar to T84 cells[72,81]. Both these human gut cell lines reconstitute as high-resistance monolayers ($R_e \sim 1000\ \Omega \cdot cm^2$). HCT-8 and T84 monolayers, in an analogous manner to MDCK cells, are able to maintain R_e with apical and/or basolateral acidification to pH 3.0 (Table 1.1). The increases in R_e with at least mild acidification are consistent with acid-induced reductions in the paracellular shunt pathway[2].

Acidification of either the basolateral or apical solutions in MDCK, HCT-8 and T84 monolayers resulted in pH-related increases in potential difference, consistent with development of H^+ diffusion potentials (Table 1.1). Apical, but not basolateral, amiloride reduced the ability of MDCK monolayers to resist acidic challenges[82]. This suggests a role for $Na^+–H^+$ exchangers in the normal maintenance of epithelial integrity in the face of acidic challenges.

We have also investigated the effect of combinations of acid and aspirin on transepithelial resistance. In HCT-8 cell monolayers, addition of aspirin,

5 mmol L^{-1}, to a pH 3.0 solution resulted in a marked decline in R_e[72]. At neutral pH, this concentration of aspirin is relatively innocuous. Similarly, in MDCK cell monolayers, acute or chronic (12 h) application of aspirin, >1 mmol L^{-1}, caused a decline in R_e associated with increased trans-epithelial flux of ^{14}C-thiourea[83]. These results are consistent with the pH-partition hypothesis; the weak acid aspirin may more readily diffuse across the cell membranes in an unionized form from acidic solutions[2,84].

These studies with established intestinal and renal cell lines illustrate that they are useful models for studying epithelial barrier function. The demonstration that reconstituted epithelial monolayers of such cells are resistant to basolateral, in addition to apical, acidification might suggest some loss of polarity compared with the peptic cell monolayers.

FUTURE DIRECTIONS

The study of upper gastrointestinal resistance to acid has been advanced by the use of modern cell biology techniques. In particular, the use of isolated membrane vesicles has enabled quantitation of apical membrane permeability to protons. These vesicles have also illustrated that the apical membranes are likely to be the initial site of damage caused by aggressive agents. Studies with membrane vesicles should allow the definition of the role of specific membrane components, such as the membrane lipid composition, intramembranous proteins and the role of the glycocalyx, as pathways for proton permeation, and mechanisms by which it is modulated.

Epithelial cells in culture, particularly in the form of reconstituted epithelial monolayers on permeable supports, offer a novel model system for investigating barrier functions, similar to the advances they have provided for studying transport functions. Such reconstituted monolayers have already demonstrated that resistance to acid is a common property of several 'tight' epithelia. The continued use of these cells will allow definition of cellular processes involved in such resistance. In particular, the bio-physical measurements of these simple epithelia, coupled with intracellular measurements of H^+, and other ions, using microspectrofluorimetric techniques will allow mechanisms of resistance to acid to be elucidated.

Acknowledgements

The work reported from the author's laboratory was supported by grants from the Medical Research Council (G8418056CA), Smith Kline Foundation, University of Newcastle upon Tyne Research Committee (563022 and 563262), and Science and Engineering Research Council CASE awards with Smith Kline and French Research, Welwyn, and Sterling–Winthrop Research Centre, Alnwick. The author is pleased to record the scientific collaboration of Dr J. M. Wilkes, Dr D. Zhao, Dr H. J. Ballard, A. B. Chan, C. N. Allen and Dr N. L. Simmons.

References

1. Hunter, J. (1772). On the digestion of the stomach after death. *Philos Trans.*, **62**, 447–454
2. Hirst, B. H. (1989). The gastric mucosal barrier. In Forte, J. G. (ed.) *Handbook of Physiology. The Gastrointestinal System. Vol. III, Gastrointestinal Secretion*, 2nd Edn., pp. 279–308. (Bethesda, MD: American Physiological Society)
3. Hollander, F. (1952). The two-component mucous barrier: its activity in protecting the gastroduodenal mucosa against peptic ulceration. *Arch. Intern Med.*, **93**, 107–120
4. Davenport, H. W. (1964). Gastric mucosal injury by fatty and acetylsalicylic acids. *Gastroenterology*, **46**, 245–253
5. Davenport, H. W. (1965). Damage to the gastric mucosa: effects of salicylates and stimulation. *Gastroenterology*, **49**, 189–196
6. Davenport, H. W. (1965). Potassium fluxes across the resting and stimulated gastric mucosa: injury by salicylic and acetic acids. *Gastroenterology*, **49**, 238–245
7. Davenport, H. W. (1966). Fluid produced by the gastric mucosa during damage by acetic and salicylic acids. *Gastroenterology*, **50**, 487–499
8. Davenport, H. W. (1967). Absorption of taurocholate-24-^{14}C through the canine gastric mucosa. *Proc. Soc. Exp. Biol. Med.*, **125**, 670–673
9. Davenport, H. W. (1967). Ethanol damage to canine oxyntic glandular mucosa. *Proc. Soc. Exp. Biol. Med.*, **126**, 657–662
10. Davenport, H. W. (1967). Salicylate damage to the gastric mucosal barrier. *N. Engl. J. Med.*, **276**, 1307–1312
11. Davenport, H. W. (1968). Destruction of the gastric mucosal barrier by detergents and urea. *Gastroenterology*, **54**, 175–181
12. Davenport, H. W. (1969). Gastric mucosal hemorrhage in dogs: effects of acid, aspirin, and alcohol. *Gastroenterology*, **59**, 439–449
13. Davenport, H. W. (1970). Effect of lysolecithin, digitonin, and phospholipase A upon the dog's gastric mucosal barrier. *Gastroenterology*, **59**, 505–509
14. Davenport, H. W. (1971). Protein-losing gastropathy produced by sulfhydryl reagents. *Gastroenterology*, **60**, 870–879
15. Davenport, H. W. (1974). Plasma protein shedding by the canine oxyntic glandular mucosa induced by topical application of snake venoms and ethanol. *Gastroenterology*, **67**, 264–270
16. Davenport, H. W. (1975). The gastric mucosal barrier: past, present and future. *Mayo Clin. Proc.*, **50**, 507–514
17. Fromm, D. (1981). Gastric mucosal barrier. In Johnson, L. R. (ed.) *Physiology of the Gastrointestinal Tract*, pp. 733–748. (New York: Raven Press)
18. Powell, D. W. (1981). Barrier function of epithelia. *Am. J. Physiol.*, **241**, G275–G288
19. Powell, D. W. (1984). Physiological concepts of epithelial barriers. In Allen, A., Flemström, G., Garner, A., Silen, W. and Turnberg, L. A. (eds.) *Mechanisms of Mucosal Protection in the Upper Gastrointestinal Tract*, pp. 1–5. (New York: Raven Press)
20. Kauffman, G. L. (1985). The gastric mucosal barrier. Component control. *Dig. Dis. Sci.*, **30**, 69S–76S
21. Lichtenberger, L. M. (1987). Membranes and barriers: with a focus on the gastric mucosal barrier. *Clin. Invest. Med.*, **3**, 181–188
22. Sachs, G., Jackson, R. J. and Rabon, E. C. (1980). Use of plasma membrane vesicles. *Am. J. Physiol.*, **238**, G151–G164
23. Murer, H. and Kinne, R. (1980). The use of isolated membrane vesicles to study epithelial transport processes. *J. Membr. Biol.*, **55**, 81–95
24. Wolosin, J. M. (1985). Ion transport studies with H^+-K^+-ATPase-rich vesicles: implications for HCl secretion and parietal cell physiology. *Am. J. Physiol.*, **248**, G595–G607
25. Ives, H. E. and Verkman, A. S. (1985). Effects of membrane fluidizing agents on renal brush border proton permeability. *Am. J. Physiol.*, **249**, F933–F940
26. Gunther, R. D., Schell, R. E. and Wright, E. M. (1984). Ion permeability of rabbit intestinal brush border membrane vesicles. *J. Membr. Biol.*, **78**, 119–127
27. Verkman, A. S. (1987). Passive H^+/OH^- permeability in epithelial brush border membranes. *J. Bioenerg. Biomembr.*, **19**, 481–493
28. Rabon, E., Takeguchi, N. and Sachs, G. (1980). Water and salt permeability of gastric vesicles. *J. Membr. Biol.*, **53**, 109–117

29. Wolosin, J. M. and Forte, J. G. (1981). Changes in the membrane environment of the $(K^+ + H^+)$ -ATPase following stimulation of gastric oxyntic cells. *J. Biol. Chem.*, **256**, 3149–3152

30. Hirst, B. H. and Forte, J. G. (1985). Redistribution and characterization of $(H^+ + K^+)$-ATPase membranes from resting and stimulated gastric parietal cells. *Biochem. J.*, **231**, 641–649

31. Wilkes, J. M., Ballard, H. J., Dryden, D. T. F. and Hirst, B. H. (1989). Proton permeability and lipid dynamics of gastric and duodenal apical membrane vesicles. *Am. J. Physiol.*, **256**, G553–G562

32. Hirst, B. H., Ballard, H. J., Wilkes, J. M. and Forte, J. G. (1989). Gastric and small intestinal membrane vesicle resistance to trypsin: implications for mucosal protection. In Davison, J. S. and Schaffer, E. A. (eds.) *Gastrointestinal and Hepatic Secretions: Mechanisms and Control*, pp. 111–114. (Calgary: University of Calgary Press)

33. Black, J. A., Forte, J. G. and Hirst, B. H. (1985). Orientation of $(H^+ + K^+)$-ATPase membrane vesicles from rabbit resting and stimulated gastric mucosa. *J. Physiol. London*, **371**, 136P

34. Kessler, M., Acuto, O., Murer, H., Muller, M. and Semenza, G. (1978). A modified procedure for the rapid preparation of efficiency transporting vesicles from small intestinal brush border membranes. Their use of investigating some properties of D-glucose and choline transport systems. *Biochim. Biophys. Acta*, **506**, 136–154

35. Hauser, H., Howell, K., Dawson, R. M. C. and Bowyer, D. E. (1980). Rabbit intestinal brush border membrane preparation and lipid composition. *Biochim Biophys. Acta*, **602**, 567–577

36. Aubry, H., Merrill, A. R. and Proulx, P. (1986). A comparison of brush-border membranes prepared from rabbit small intestine by procedures involving Ca^{2+} and Mg^{2+} precipitation. *Biochim. Biophys. Acta*, **856**, 610–614

37. Haase, W., Schäfer, A., Murer, H. and Kinne, R. (1978). Studies on the orientation of brush-border membrane vesicles. *Biochem. J.*, **172**, 57–62

38. Hearn, P. R., Russell, R. G. G. and Farmer, J. (1981). The formation and orientation of brush border vesicles from rat duodenal mucosa. *J. Cell Sci.*, **47**, 227–236

39. Wilkes, J. M. Ballard, H. J., Latham, J. A. E. and Hirst, B. H. (1987). Gastroduodenal epithelial cells: the role of the apical membrane in mucosal protection. In Reid, E., Cook, G. M. W. and Luzio, J. P. (eds.) *Cells, Membranes, and Disease, Including Renal*, pp. 243–254. (New York: Plenum Press)

40. Zhao, D. and Hirst, B. H. (1990). Bile salt induced increases in duodenal brush-border membrane proton permeability, fluidity and fragility. *Dig. Dis. Sci.*, **35**, 589–595

41. Durbin, R. P. (1984). Backdiffusion of H^+ in isolated frog gastric mucosa. *Am. J. Physiol.*, **246**, G114–G119

42. Deamer, D. W. and Nichols, J. W. (1983). Proton-hydroxide permeability of liposomes. *Proc. Natl. Acad. Sci. USA*, **80**, 165–168

43. Deamer, D. W. (1987). Proton permeation of lipid bilayers. *J. Bioenerg. Biomembr.*, **19**, 457–479

44. Gutknecht, J. (1987). Proton conductance through phospholipid bilayers: water wires or weak acids? *J. Bioenerg. Biomembr.*, **19**, 427–442.

45. Briggs, R., Hirst, B. H. and Wilkes, J. M. (1988). Amiloride sensitivity of proton conductance in rabbit duodenal brush-border membrane vesicles *in vitro*. *J. Physiol. (London)*, **406**, 117P

46. Ives, H. E. (1985). Proton/hydroxyl permeability of proximal tubule brush border vesicles. *Am. J. Physiol.*, **248**, F78–F86

47. Weisbrodt, N. M., Kienzle, M. and Cooke, A. R. (1973). Comparative effects of aliphatic alcohols on the gastric mucosa. *Proc. Soc. Exp. Biol. Med.*, **142**, 450–454

48. Israel, Y., Salazar, I. and Rosenmann, E. (1968). Inhibitory effects of alcohol on intestinal amino acid transport in vivo and in vitro. *J. Nutr.*, **96**, 499–504

49. Chang, T., Lewis, J. and Glazko, A. J. (1967). Effect of ethanol and other alcohols on the transport of amino acids and glucose by everted gut sacs of rat small intestine. *Biochim. Biophys. Acta*, **135**, 1000–1007

50. Eastwood, G. L. and Kirchner, J. P. (1974). Changes in the fine structure of mouse gastric epithelium produced by ethanol and urea. *Gastroenterology*, **67**, 71–84

on intracellular pH and epithelial membrane resistances: studies in isolated Necturus antral mucosa exposed to acid. *Gastroenterology*, **96**, 1410–1418

52. Bailey, R. E., Levine, R. A., Nandi, J., Schwartzel, E. H., Beach, P. N. Jr., Borer, P. N. and Levy, G. C. (1987). Effects of ethanol on gastric epithelial cell phospholipid dynamics and cellular function. *Am. J. Physiol.*, **252**, G237–G243

53. Mazzeo, A. R., Nandi, J. and Levine, R. A. (1988). Effects of ethanol on parietal cell membrane phospholipids and proton pump function. *Am. J. Physiol.*, **254**, G57–G64

54. Bailey, R. E., Nandi, J., Levine, R. A., Ray, T. K., Borer, P. N. and Levy, G. C. (1986). NMR studies of pig gastric microsomal H^+, K^+-ATPase and phopholipid dynamics: effects of ethanol perturbation. *J. Biol. Chem.*, **261**, 11086–11090

55. Ballard, H. J., Wilkes, J. M. and Hirst, B. H. (1988). Effect of alcohols on gastric and small intestinal apical membrane integrity and fluidity. *Gut*, **29**, 1648–1655

56. Halstead, C. H., Robles, E. A. and Mezey, E. (1973). Distribution of ethanol in the human gastrointestinal tract. *Am. J. Clin. Nutr.*, **26**, 831–834

57. Hunter, C. K., Treanor, L. L., Gray, J. P., Halter, S. A.. Hoyumpa, A. and Wilson, F. A. (1983). Effects of ethanol in vitro on rat intestinal brush-border membranes. *Biochim. Biophys. Acta*, **732**, 256–265

58. Beesley, R. C. (1986). Ethanol inhibits Na^+-gradient-dependent uptake of L-amino acids into intestinal brush border membrane vesicles. *Dig. Dis. Sci.*, **31**, 987–992

59. Schachter, D., Cogan, U. and Shinitzky, M. (1976). Interaction of retinol and intestinal microvillus membranes studied by fluorescence polarization. *Biochim. Biophys. Acta*, **448**, 620–624

60. Schachter, D. and Shinitzky, M. (1977). Fluorescence polarization studies of rat intestinal microvillus membranes. *J. Clin. Invest.*, **59**, 536–548

61. Forte, T. M., Silen, W. and Forte, J. G. (1976). Ultrastructural lesions in gastric mucosa exposed to deoxycholate: implications toward the barrier concept. In Kasbekar, D. K., Sachs, G. and Rehm, W. S. (eds.) *Gastric Hydrogen Ion Secretion*, pp. 1–28. (New York: Dekker)

62. Helenius, A. and Simons, K. (1975). Solubilization of membranes by detergents. *Biochim. Biophys. Acta*, **415**, 29–79

63. Muller-Lessner, S. A., Bauerfeind, P. and Blum, A. L. (1985). Bile in the stomach. *Gastroenterol. Clin. Biol.*, **9**, 72–79

64. Matlin, K. S. and Valentich, J. D. (eds.) (1989). *Functional Epithelial Cells in Culture*. (New York: Alan R. Liss)

65. Terano, A., Mach, T., Stachura, J., Tarnawski, A. and Ivey, K. J. (1984). Effect of 16,16-dimethyl prostaglandin E_2 on aspirin induced damage to rat gastric epithelial cells in tissue culture. *Gut*, **25**, 19–25

66. Hiraishi, H., Terano, A., Ota, S-I., Ivey, K. J. and Sugimoto, T. (1986). Effect of cimetidine on indomethacin-induced damage in cultured rat gastric mucosal cells; comparison with prostaglandin. *J. Lab. Clin. Med.*, **108**, 608–615

67. Terano, A., Ota, S-I., Mach, T., Hiraishi, H., Stachura, J., Tarnawski, A. and Ivey, K. J. (1987). Prostaglandin protects against taurocholate-induced damage to rat gastric mucosal cell culture. *Gastroenterology*, **92**, 669–677

68. Romano, M., Razandi, M., Sekhon, S., Krause, W. J. and Ivey, K. J. (1988). Human cell line for study of damage to gastric epithelial cells in vitro. *J. Lab. Clin. Med.*, **111**, 430–440

69. Terano, A., Mach, T., Stachura, J., Sekhon, S., Tarnawski, A. and Ivey, K. J. (1983). A monolayer culture of human gastric epithelial cells. *Dig. Dis. Sci.*, **28**, 595–603

70. Romano, M., Razandi, M. and Ivey, K. J. (1988). Acetaminophen directly protects human gastric epithelial cell monolayers against damage induced by sodium taurocholate. *Digestion* , **40**, 181–190

71. Romano, M., Razandi, M. and Ivey, K. J. (1989). Effect of cimetidine and ranitidine on drug induced damage to gastric epithelial cell monolayers *in vitro*. *Gut*, **30**, 1313–1322

72. Allen, C. N., Eason, C. T., Bonner, F. W., Simmons, N. L. and Hirst, B. H. (1989). Development of human gastrointestinal cultured cells for predictive toxicology. *Hum. Toxicol.*, **8**, 412

73. Sanders, M. J., Ayalon, A., Roll, M. and Soll, A. H. (1985). The apical surface of canine chief cell monolayers resists H^+ back diffusion. *Nature (London)*, **313**, 52–54

74. Sanders, M. J., Amirian, D. A., Ayalon, A. and Soll, A. H. (1983). Regulation of

pepsinogen release from canine chief cells in primary monolayer culture. *Am. J. Physiol.*, **245**, G641–G646

75. Defize, J. and Hunt, R. H. (1989). Effect of hydrochloric acid and prostaglandins on pepsinogen synthesis and secretion in canine gastric chief cell monolayer cultures. *Gut*, **30**, 774–781

76. Ayalon, A., Sanders, M. J., Thomas, L. P., Amirian, D. A. and Soll, A. H. (1982). Electrical effects of histamine on monolayers in culture from enriched canine gastric chief cells. *Proc. Natl. Acad. Sci. USA*, **79**, 7009–7013

77. Rutten, M. J., Rattner, D. and Silen, W. (1985). Transepithelial transport of guinea pig gastric mucous cell monolayers. *Am. J. Physiol.*, **249**, C503–C513

78. Logsdon, C. D., Bisbee, C. A., Rutten, M. J. and Machen, T. E. (1982). Fetal rabbit gastric epithelial cells cultured on floating collagen gels. *In Vitro, Rockville*, **18**, 233–242

79. Chan, A. B., Allen, C. N., Simmons, N. L., Parsons, M. E. and Hirst, B. H. (1989). Resistance to acid of canine kidney (MDCK) and human colonic (T84) and ileo-caecal (HCT-8) adenocarcinoma epithelial cell monolayers *in vitro*. *Q. J. Exp. Physiol.*, **74**, 553–556

80. McRoberts, J. A. and Barrett, K. E. (1989). Hormone-regulated ion transport in T_{84} colonic cells. In Matlin, K. S. and Valentich, J. D. (eds.) *Functional Epithelial Cells in Culture*, pp. 235–265. (New York: Alan R. Liss)

81. Tompkins, W. A. F., Watrach, A. M., Schmale, J. D., Schultz, R. M. and Harris, J. A. (1974). Cultural and antigenic properties of newly established cell strains from adenocarcinomas of the human colon and rectum. *J. Natl. Cancer Inst.*, **52**, 1101–1110

82. Chan, A. B., Hirst, B. H., Parsons, M. E. and Simmons, N. L. (1989). Resistance of Madin–Darby canine kidney (MDCK) epithelial cell monolayers to acid. *J. Physiol. (London)*, **409**, 58P

83. Chan, A. B., Parsons, M. E., Simmons, N. L. and Hirst, B. H. (1989). Madin–Darby canine kidney (MDCK) cell monolayers as in vitro model for nephrotoxic activity of aspirin. *Proc. Int. Union Physiol. Sci.*, **18**, 497

84. Martin, B. K. (1963). Accumulation of drug anions in gastric mucosal cells. *Nature (London)*, **198**, 896–897

2
Establishment and Characteristics of Human Colorectal Adenocarcinoma Cell Lines

S. C. KIRKLAND

INTRODUCTION

Many biological studies of colorectal carcinoma require the availability of established cell lines. These cell lines provide an unlimited quantity of carcinoma cells without contamination with the other cell types found in profusion in primary tumours. The characteristics of established cell lines differ widely. A bank of these cell lines, reflecting some of the diversity observed in primary human colorectal carcinoma, is a useful tool with which to study the biology of colorectal carcinoma cells. In addition, some of these cell lines retain sufficient of the differentiated features characteristic of the normal epithelium to make them useful model systems with which to study the functions of colorectal epithelium. This is particularly useful in light of the inability to establish differentiating cultures from normal colorectal epithelium.

ESTABLISHMENT OF HUMAN COLORECTAL CARCINOMA CELL LINES

Numerous cell lines have now been established from human colorectal carcinomas using a vast range of dissociation techniques and culture conditions[1-8]. Primay carcinomas, metastases and ascitic fluid have all been used as starting material.

Dissociation of tissue

Cells have been isolated from carcinomas using both mechanical and enzymatic methods. Many colorectal carcinomas have a soft texture and clumps of cells can be readily liberated without enzymatic treatment[3,6]. Successful cultures have been established following mechanical dissociation[6,8], collagenase digestion[5] and trypsin digestion[1,4]. In general, attempts are not made to generate a single cell suspension at this stage as cell–cell contact is thought to be important to cell survival; therefore cell clumps are plated into culture flasks. Even at subculture, it is often preferable to transfer cell clumps to fresh flasks rather than to attempt to generate single-cell suspensions, and, in some cell lines, it is extremely difficult to generate single-cell suspensions without damage to the cells.

Primary culture conditions

In general, the culture conditions used to establish colorectal carcinoma cell lines can be divided into three broad categories:

(1) *Culture on tissue culture plastic*
Cells are dissociated from the tumour and plated directly onto tissue culture plastic. Usually, high plating densities are required for success with this method. Different culture media have been used including 'standard' medium (Dulbecco's Eagles medium, Minimal Eagles Medium, McCoys 5A), with 8–10% fetal calf serum (FCS)[5,6], enriched medium with various additives[3,8] and fully defined medium[7]. The fully defined medium, ACL-4, which was formulated for the growth of lung adenocarcinoma cells, has been successfully used to establish cultures from colorectal carcinomas and was shown to give a better success rate than RPMI medium supplemented with 10% fetal calf serum[7].

(2) *Culture on 'feeder' layers*
Cells are dissociated from the tumour and plated into flasks containing feeder layers of confluent mouse fibroblasts (C3H 10T½)[4]. Cells are weaned away from the feeder layer when they become established *in vitro*. Twenty-one specimens from a total of 27 were successfully established on feeder layers and 20 of these were subcultured at least 10 times and maintained in tissue culture for at least 6 months. Cell lines were established from 3 of these carcinomas. In contrast, cells from only 3 of the same specimens plated onto tissue culture plastic were able to grow and survive passage in culture[4]. These methods offer great potential for the short-term culture of approximately 90% of colorectal carcinomas which permits experimentation on a larger percentage of all tumours and makes study of individual tumours a possibility. It is not clear whether additional cell types which are not found in cultures on tissue culture plastic can be found in association with feeder layers[4]. Later experiments have demonstrated that the presence of feeder layers enhanced the colony formation of carcinoma cells in semi-

solid medium and also increased the plating efficiency of xeno-grafted cells returned to culture[9].

(3)   *Collagen gel cultures*
       Previous success with the culture of various normal and neoplastic epithelial cells on collagen gel prompted the use of this method for the establishment of colorectal carcinomas and adenomas in culture. Explants from 100% of adenomatous polyps and 69% of carcinomas were found to attach to Type 1 collagen gels and produced epithelial outgrowth which could subsequently be subcultured[8]. In this system, cells were grown in Eagle's medium supplemented with 2% fetal calf serum, insulin, transferrin, hydrocortisone, tri-iodothyronine and sodium selenite. 2% FCS was used to reduce the fibroblast contamination associated with higher levels of FCS. Cells could only be subcultured from these cultures with non-enzymatic culture methods. Using such methods, three cell lines were established from 12 adenomas and 15 cell lines were established from 45 carcinomas[8].

Establishment of cell lines

There is general agreement that there should be a delay before initial passage until cells are very tightly packed, and that, following subculture, cells should be replated at high density. Different methods have been successfully used to subculture primary cultures of colorectal carcinoma, these include trypsinization and EDTA dissociation. There is some suggestion that more differentiated tumours and adenomas are more sensitive to trypsin treatment than more malignant tumours[8]. However, it is difficult to compare results on trypsinization obtained from different groups as different preparations of trypsin have been used.

The majority of colorectal carcinomas grow on tissue culture plastic as adherent monolayers with a typical epithelial morphology. However, some carcinoma cells fail to attach to the plastic and proliferate in suspension. This anchorage-independent growth has most frequently been observed in serum-free medium[7] but also occurs in cultures grown in the presence of fetal calf serum[5]. These anchorage-indifferent carcinoma cells appear as aggregates which float in the culture medium. Some can form disorganized multicellular aggregates (i.e. VACO 5[5]) while others form tubular structures containing polarized cells with a central lumen (VACO 6[5], VACO 10MS[5] and NCI-H548[7]) and others organize into spheroidal structures with smooth outlines which contain lumina circumscribed with microvilli and tight junctions (i.e. VACO 3,4 and 5[5], NCI-H508[7] and LIM 1863[10]). In light of these observations, it is important to observe closely the floating material present in primary cultures and to retain this material while any sign of viability is present.

Following the establishment of cell lines, it is essential to characterize fully the cell lines obtained. There have been several reports of the establishment of new colorectal carcinoma cell lines which have subsequently proved to be the result of cross-contamination. The cell line, HTFU, which was originally isolated from the HT29 cell line[11], was subsequently shown by isoenzyme analysis to be a rodent contaminant[5]. Recently, the cell line WiDr[12] has been shown to be a derivative of the HT29 cell line[13]. Consequently, the importance of carefully monitoring the development of cell lines cannot be overemphasized. Such monitoring is particularly important if other cell lines are grown in the laboratory alongside the new cultures or if rodent cells are used to provide a feeder layer in the primary culture stage.

Cell lines should also be regularly screened for mycoplasma contamination using either a Hoechst staining technique[14] or commercially available kits. Mycoplasma contamination can go undetected in cultures for long periods of time. However, in view of the effects that mycoplasma has on various biological properties of the cells[15], cultures should be regularly screened so that experiments are only performed on mycoplasma-free populations.

Establishment of HCA/HRA series of cell lines

The HCA/HRA cell lines were established from primary human colorectal adenocarcinomas as previously described[6]. Briefly, pieces of primary colorectal cancers were obtained at the time of surgery from 46 unselected patients and transported in culture medium on ice for immediate processing. Cells were mechanically removed from the tumour and the resultant cell clumps were plated into 25 cm^2 culture flasks in Dulbecco's Eagles medium (Gibco, Paisley, Scotland) supplemented with 10% fetal calf serum (Gibco) and antibiotics. Seeded flasks were gassed with a 10% CO_2: 90% air mixture and incubated at 37°C. In successful cultures, the cell clumps attached to the culture flask and epithelial outgrowth was observed (Figure 2.1). For the first few medium changes, any floating material was returned to the flask until no suspended viable material could be observed.

These primary cultures were only subcultured when areas of tumour growth became very confluent. Initially, the entire contents of a 25 cm^2 flask were transferred to a fresh flask of the same size. Cells were subcultured using trypsin (three times crystallized and dialysed; Worthington Biochemicals, Lorne laboratories, Bury St. Edmunds, Suffolk, UK) but no attempt was made to produce a single-cell suspension; instead, clumps of tumour cells were transferred to a fresh culture flask. Fibroblasts were removed by scraping or by differential trypsinization of cultures.

Primary cultures from 11 tumours were lost through contamination. However, with careful washing of the tumour pieces, this problem was resolved with later specimens. Cell lines were established from 6 of the remaining 35 tumours (17%). The main problem with the unsuccessful cultures was a failure of the cells to attach to the culture plastic. In this series of tumours, no evidence for the proliferation of tumour cells in suspension was obtained unlike some other studies where anchorage-independent cells were established as cell lines. The histological details of the carcinomas from

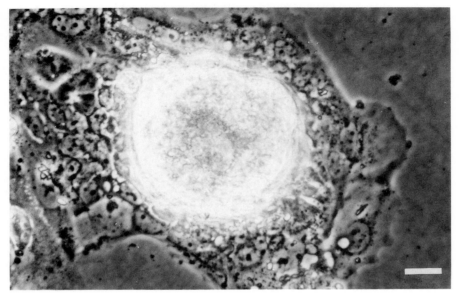

Figure 2.1 Phase contrast micrograph of an 8-day-old primary culture of HCA-46 cells (bar = 21 μm)

which cultures were attempted are shown in Table 2.1. Details of tumours which yielded contaminated cultures were not included in this table. The characteristics of the tumours from which successful cultures were obtained are shown in Table 2.2 No correlation was seen between the histological grade or Duke's stage and the behaviour of the tumour cells *in vitro*. Cell lines were established from both colon and rectum, from well, moderately well and poorly differentiated tumours and from Dukes B and C stage carcinomas.

CHARACTERISTICS OF COLORECTAL CARCINOMA CELL LINES

Although there are now a large number of cell lines derived from colorectal carcinoma[1,8], only a few of these cell lines express differentiated features characteristic of the tissue of origin, the colonic mucosa. This may be due to the characteristics of the primary carcinoma cells or that more poorly differentiated cells have a selective advantage *in vitro* over their more differentiated counterparts.

Characteristics of the HCA/HRA series of cell lines

All of the cell lines from this series had a characteristic morphology. Phase contrast micrographs of the cell lines are shown in Figure 2.2. Some cells have a classical epithelioid appearance with large nuclei (HCA-7: Figure

Table 2.1 Histological details of carcinomas from which cultures were attempted*

Site	Histological grade	No. of tumours	Duke's stage		
			A	B	C
Colon	Well differentiated	5	0	4**	1**
	Moderately differentiated	7	0	7**	0
	Poorly differentiated	1	0	0	1**
Rectum	Well differentiated	5	1	3**	1
	Moderately differentiated	9	0	5**	4
	Poorly differentiated	3	0	1	2
Caecum	Well differentiated	2	0	1	1
	Moderately differentiated	3	0	0	3

*Reproduced from *Br. J. Cancer*, **53**, 779–785 (1986) with the kind permission of Macmillan Press Ltd., Basingstoke, UK
**One cell line was established from each of these groups

Table 2.2 Characteristics of carcinomas which yield cell lines*

Cell line	Patient age (sex)	Site	Histological grade	Duke's stage
HCA-2	83(F)	Sigmoid colon	Well differentiated	C
HCA-7	58(F)	Colon	Moderately differentiated	B
HRA-16	56(M)	Rectum	Moderately differentiated	B
HRA-19	66(M)	Rectum	Well differentiated	B
HCA-24	68(M)	Ascending colon	Well differentiated	B
HCA-46	53(F)	Sigmoid colon	Poorly differentiated	C

*Reproduced from *Br. J. Cancer*, **53**, 779–785 (1986) with the kind permission of Macmillan Press Ltd., Basingstoke, UK

2.2b and HRA-16: Figure 2.2c). Other cell lines form more tightly packed colonies so that their morphology is not easily seen (HCA-24: Figure 2.2e). Most of the cells within a cell line had a uniform morphology but HCA-7 at early passage and HRA-19 (Figure 2.2d) displayed a heterogeneous morphology. The relevance of such heterogeneity will be discussed below. Dome formation is observed in monolayers of HCA-7 cells and occasionally in HRA-19 monolayers. All of the cell lines retain some of the morphological characteristics of colonocytes. All cell lines have microvilli although the number and organization of these microvilli differs widely betwen lines. HCA-7, HRA-16, HRA-19 and HCA-46 cell lines have tight junctions and HCA-7 and HRA-19 cells form morphologically and functionally polarized monolayers. Mucous cell differentiation has not been observed in monolayer cultures of any of the cell lines but is present in xenografts of HRA-19[16] and HCA-46 cells. Endocrine differentiation has been observed both in monolayer cultures (unpublished observations) and xenografts[16,17] of the HRA-19 cell line. A more detailed description of the differentiated properties of the HCA-7 and HRA-19 cell lines is provided in the subsequent relevant sections.

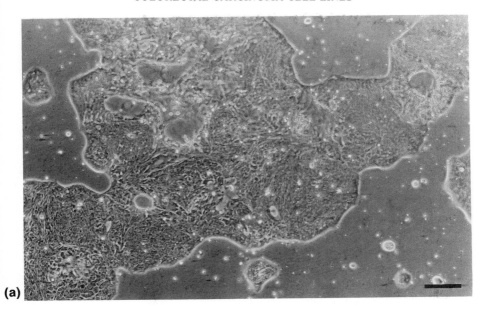

(a)

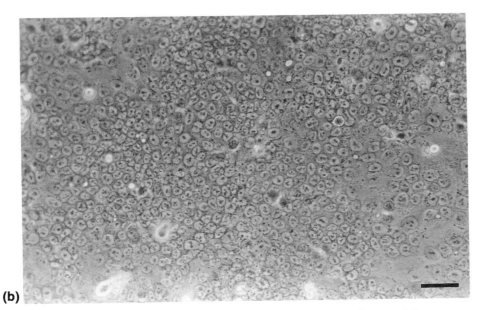

(b)

Figure 2.2　Phase contrast micrographs of human colorectal adenocarcinoma cell lines:

(a)　HCA-2 (passage 16) (bar=84 μm)

(b)　HCA-7 (passage 19) (bar=42 μm)

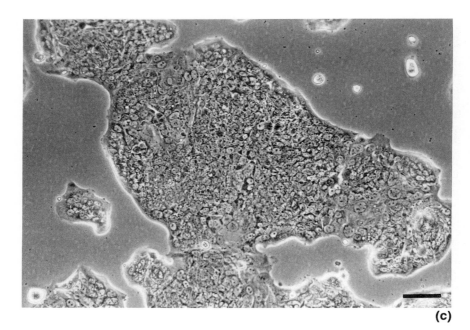

(c)

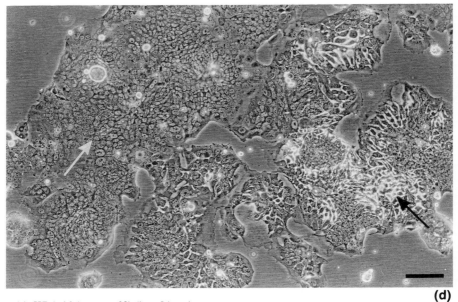

(d)

(c) HRA-16 (passage 32) (bar=84 μm)
(d) HRA⁻ 19 a1.1. cells (passage 25) (bar=84 μm). Monolayers contain tightly packed epithelial cells with indistinct borders (white arrow) and other cells with large intercelular spaces (black arrow) which are refractile under phase contrast microscopy. Increasing numbers of the cells with large intercellular spaces are found with time after trypsinization

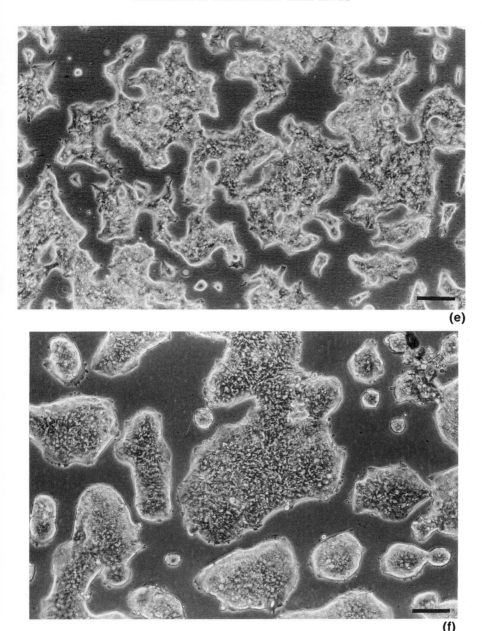

(e)

(f)

(e) HCA-24 (passage 62) (bar=84 μm)

(f) HCA-46 (passage 102) (bar=84 m)

Differentiation in normal colorectal epithelium

Normal mucosa of the large intestine exhibits a regular pattern of differentiation where cells migrate from the crypt base to the surface epithelium. Four basic cell types are present: undifferentiated, columnar, mucous and enteroendocrine[18]. Although this basic pattern of differentiation is retained throughout the large intestine, differences in the ratios of mucous and columnar cells are observed between different sections of the large intestine. All of the differentiated cells in colorectal epithelium are thought to be derived from multipotential stem cells located in the base of the crypt (the Unitarian hypothesis)[19-22].

Differentiated features retained by colorectal carcinoma cell lines

Absorptive cells

Most differentiated colorectal carcinoma cell lines have properties of absorptive or columnar epithelial cells. Some lines form polarized monolayers with a brush border[23-27] while other lines are able to form domes or hemicysts in monolayer culture. Vectorial fluid transport can be responsible for the dome formation observed in these cultures[27-29]. Quantitation of dome formation under different conditions can provide some data on the response of cells to various secretagogues. However, this procedure has limitations in that agents can only be applied to the apical surface of the cell, while many receptors for secretagogues are polarized exclusively to the basolateral surface. To overcome this problem, cells have been grown on millipore filters which allow access to both the apical and the basolateral membranes. Short circuit current measurements of such monolayers have provided a great deal of information about the transport properties of colorectal carcinoma cell lines. Three human colorectal carcinoma cell lines have been used to study epithelial ion transport. Monolayers of T84[25,30], Caco-2[28] and HCA-7[31] have been shown to have small transepithelial potentials and low resistances. These cell lines also respond to the secretagogue vasoactive intestinal polypeptide with an increased short circuit current, which is thought to be due to electrogenic chloride secretion[30,32]. In normal colorectal epithelium, crypts are the site of chloride secretion and the surface epithelium is the site of sodium absorption[33]. As it has not yet been possible to demonstrate sodium absorption in carcinoma cell lines, it suggests that carcinoma cells have characteristics of crypt epithelial cells[32]. A more detailed account of the transepithelial ion transport in colorectal carcinoma cell lines can be found in the next chapter of this volume[34].

In addition to the ion transport properties of colorectal carcinoma cell lines, many of these cell lines express enzymes which are normally associated with small intestinal or fetal large intestinal epithelium. Monoclonal antibodies against sucrase–isomaltase, aminopeptidase and dipeptidylpeptidase, which were unreactive with adult large intestinal epithelium, bound to the brush border of fetal colon, to apical borders of two colorectal carcinoma cells lines grown as xenografts (HT-29 and Caco-2) and to 7/27 primary human colorectal carcinomas[35,36]. These results indicate a fetal pattern of enterocytic differentiation in some colorectal carcinomas, which can be retained by cell lines established from these carcinomas.

Mucous cells

The presence of mucous cells has only rarely been reported in a few colorectal carcinoma cell lines. Mucous cells have been demonstrated in the LIM 1215 cell line using electron microscopy, mucicarmine staining and staining with a monoclonal antibody to colonic mucous[26]. In addition, the LIM 1863 cell line contains morphologically mature polarized columnar cells and mucous cells[10]. LIM 1863 cells grow as organoids and are polarized around a central lumen which is lined by the apical membrane domain of the cells. Mucin production has also been demonstrated in the NCl-H498 cell line[7].

In addition to the cell lines where mucous cells are 'spontaneously' produced, colonies of mucous cells have been reported in monolayers of HT-29 cells following treatment with sodium butyrate[37] or replacement of glucose in the medium with galactose[38].

Endocrine cells

Endocrine cells are a common feature of primary human colorectal carcinomas; however, such cells have only been conclusively demonstrated in one carcinoma cell line (HRA-19). HRA-19 cells were shown to display endocrine differentiation when grown as xenografts in nude mice[16,17] and, to a lesser extent, when grown as monolayers *in vitro* (unpublished observations).

Induction of differentiation in colorectal carcinoma cell lines

Colorectal carcinoma cells can be induced to differentiate by the addition of certain chemicals, such as sodium butyrate, or by changes in nutritional conditions. The fatty acid salt, sodium butyrate, decreased the growth rate[39,40], produced morphological changes[39] and increased enzyme expression[40-42] in colorectal carcinoma cell lines. However, it has been demonstrated that alkaline phosphatase synthesized by HRT-18 cells in response to sodium butyrate is different from that synthesized under control conditions[41]. More recent work has demonstrated that placental-like alkaline phosphatase is synthesized by cells exposed to sodium butyrate[43]. These observations, coupled with the fact that many of the effects of sodium butyrate have been shown to be reversible[41], suggest that caution should be used when comparing differentiation in response to sodium butyrate with the normal differentiation processes in colorectal epithelium. Some permanently differentiated clones have been established from the HT-29 cell line following butyrate treatment[37]. Some of these clones exhibit transepithelial transport with the formation of domes while others exhibit mucous secretion. Whether these clones arise by selection of cells which were more resistant to butyrate or by permanent changes to the cells in response to sodium butyrate remains to be established.

Enterocytic differentiation can also be induced in HT-29 cells by replacing glucose in the culture medium with galactose[44]. HT-29 cells grown in the presence of glucose were undifferentiated while a subpopulation of HT-29, selected for their ability to grow in the absence of sugar (Glc⁻), exhibited an enterocytic differentiation after confluency. This enterocytic differentiation was characterized by polarization of the monolayer with apical brush borders and tight junctions and the presence of sucrase–isomaltase[45]. If these differentiated cells were returned to glucose-containing medium, they gradually lost their differentiated features and returned to the undifferentiated state after 7 passages. Caco-2 cells which remain differentiated in high-glucose medium have a lower glucose consumption and a higher glycogen content than HT-29 cells which suggests that glucose metabolism is involved in the regulation of cell differentiation[45].

Finally, it has recently been shown that treatment with polyethylene glycol results in the appearance of differentiated colonies in HT-29 cultures[46]. The mechanisms underlying this induction of differentiation have yet to be determined.

Study of cell lineage using colorectal carcinoma cell lines

Human colorectal epithelium is composed of columnar, mucous and endocrine cells. All of these cell lineages are thought to arise from a multipotential stem cell at the crypt base (the Unitarian hypothesis)[19-22]; however, this has not yet been demonstrated. Gut endocrine cells have variously been considered to be of neural crest[47] or endodermal[48-50] origin, but conclusive evidence, particularly in humans, is lacking. Study of cell lineage and the control of differentiation in colorectal epithelium has been severely hampered by the lack of differentiated *in vitro* model systems. One approach to this problem has been to grow primary colorectal adenocarcinomas as cell lines and study their differentiation characteristics. Several cell lines have been described as having pluripotential characteristics; one such cell line is HT-29. This cell line expresses enterocytic and mucoussecreting characteristics when grown under glucose-free conditions[38,44] and mucous cell characteristics when treated with sodium butyrate[37]. In addition, a clone of the HT-29 cell line (HT-29-18) isolated by limiting dilution from HT-29 cells also displays both absorptive and mucous cell characteristics[38].

The LIM 1863 cell line which grows as floating organoids also contains both absorptive and mucous cells[10]. Although this may result from the coculture of two different cell types, it is more probable that the organoids contain multipotential precursor cells which give rise to both differentiated cell types. The LIM 1863 cell line is capable of differentiation in serum-free medium which suggests that either the cells are producing their own growth/differentiation factors which act in an autocrine fashion or that they are independent of such factors[10]. LIM 1863 cells could, therefore, provide a model system with which to study the processes of differentiation in colorectal epithelium.

The HRA-19 cell line can now be added to the list of cell lines with multipotential characteristics.

HRA-19 cell line

HRA-19 cells were derived from a primary adenocarcinoma of the rectum[16,17]. HRA-19 cells grow as monolayers on tissue culture plastic and have a heterogeneous morphology even after 150 passages *in vitro*. Clones of this cell line also display morphological heterogeneity (Figure 2.2d). Monolayers contain tightly packed epithelial cells with indistinct borders and other cells with large intercellular spaces which were refractile under phase contrast microscopy (Figure 2.2d)[17]. Increasing numbers of cells with large intercellular spaces are found with time after trypsinization. This phenomenon is thought to be due to vectorial fluid transport which results in local fluid accumulation between the cells. Preliminary experiments show that, like the HCA-7 cells, HRA-19 cells can also maintain a short circuit current (A. W. Cuthbert, personal communication). Dome formation was, however, rarely seen in these monolayers; therefore, cells should not be regarded as non-transporting simply because they do not demonstrate dome formation.

Electron microscopy of monolayers demonstrates that the cells have the morphology of poorly differentiated absorptive cells, neither endocrine nor mucous cells being observed. However, immunocytochemistry performed on monolayers with an antibody to chromogranin (see later) demonstrates the presence of a few endocrine cells (unpublished observations). The failure to demonstrate these cells by electron microscopy is probably a sampling phenomenon as only selected areas of the monolayer can be viewed. On the contrary, immunocytochemistry allows us to scan several million cells for the presence of endocrine cells.

When the HRA-19 cells were grown as xenografts in nude mice, the resulting tumours were composed of columnar and endocrine cells. The columnar cells had a better differentiated phenotype than their *in vitro* counterparts, with a well-organized brush border and associated glycocalyx. Endocrine cells could be demonstrated in xenografts by Grimelius silver staining and electron microscopy[17]. No evidence for mucous cell differentiation was obtained in these xenografts using either Alcian blue staining or electron microscopy. To further study the differentiation pathways in this cell line, the line was single-cell cloned using 'feeder' layers of mouse fibroblasts. Attempts to clone this cell line without the use of 'feeder' cells were unsuccessful. The resulting clones had a similar morphological appearance to the parent cell line. Xenografts of these cloned cells grown in nude mice were composed of columnar, mucous and endocrine cells

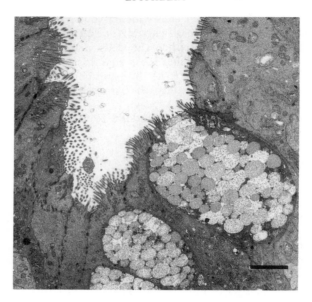

Figure 2.3 Transmission electron micrograph of a xenograft of clone HRA-19a1.1 cells (bar=4.35 μm) showing both mucous cells with apical mucous vacuoles and columnar cells with abundant apical microvilli and tight junctions. (Reproduced from *Cancer*, Vol. 61, p. 1359, with kind permission from Lippincott/Harper & Row, Philadelphia)

(Figures 2.3 and 2.4). Endocrine cells were demonstrated in these xenografts using Grimelius silver staining, electron microscopy and immunocyto-chemistry with antibody, LK2H10, which was raised against human chromogranin and has been shown to be a specific endocrine tissue marker[51]. These results show that a single epithelial cell can give rise to all differentiated cell types present in human colorectal epithelium, i.e. columnar, mucous and endocrine. This demonstrates that colorectal endocrine cells have an endodermal origin, at least in neoplastic epithelium[16].

As HRA-19 cells have a more differentiated appearance *in vivo* compared to their *in vitro* counterparts, the cells must be responsive to factors present in the nude mouse which are absent from the culture system. This makes HRA-19 cells a very useful system with which to elucidate the conditions necessary to induce differentiation in colorectal epithelium.

Heterogeneity

As with many other carcinomas, human colorectal carcinomas have been shown to contain a heterogeneous mixture of cells. Heterogeneity in the membrane antigens of human colorectal carcinoma cells has been demonstrated by immunostaining of sections with monoclonal antibodies[52]. The ability to derive cell lines with different biological characteristics from single carcinomas has confirmed the presence of these subpopulations which

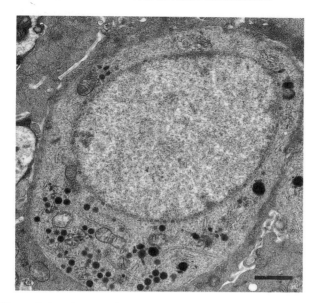

Figure 2.4 Transmission electron micrograph of an endocrine cell with a xenograft of clone HRA-19a1.1 cells (bar= 1.1 μm). (Reproduced from *Cancer*, Vol. 61, p. 1359, with kind permission from Lippincott/Harper & Row, Philadelphia)

retain their unique features over many passages *in vitro*. Two cell colonies which differed in their morphology, karyotype and other biological characteristics were isolated from the DLD-1 colon carcinoma cell line[53]. Two subpopulations have also been isolated from another colonic carcinoma cell line, HCT 116. These subpopulations, designated HCT 116a and HCT 116b, differed in their ability to grow in soft agarose and yielded xenografts in nude mice with distinctive histology[54]. Cellular heterogeneity and possible interactions between the subpopulations is an important aspect of the biology of colorectal carcinoma. However, the mechanisms which generate and maintain this heterogeneity are poorly understood. The availability of cell lines of subpopulations from single carcinomas provides a model system with which to study this phenomenon.

HCA-7

Early passages of the HCA-7 cell line contained many morphological cell types. Two subpopulations were isolated from the cell line, each with a characteristic morphology throughout this period. In addition to their morphological differences, the cells had differing abilities to form xenografts in nude mice (Figure 2.5) and produce tumours of differing histology. The transport properties of these two subpopulations have been extensively investigated[32] and shown to differ markedly, both from each other and from the parent cell line. Therefore, in addition to their obvious use in the

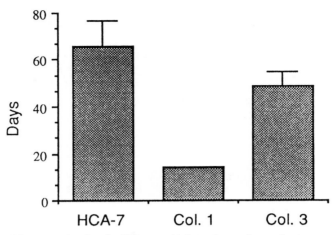

Figure 2.5 Histogram showing the differences in time taken to form subcutaneous xenografts in nude mice following the injection of 10^7 cells from three carcinoma cell populations. These populations are: a human colonic adenocarcinoma cell line (HCA-7) and two subpopulations isolated from the cell line (Col. 1 & Col. 3)

investigation of the basis of heterogeneity in colorectal carcinoma, these subpopulations can provide a means of analysing normal functions of colorectal epithelium, such as chloride transport[32]. Several more subpopulations have now been isolated from the HCA-7 cell lines which are being investigated in the described systems.

SUMMARY

Many colorectal carcinoma cell lines have been established. These are useful for the investigation of the biological properties of colorectal carcinoma cells and the origin and maintenance of heterogeneity within single carcinomas. In addition, increasing numbers of these cell lines retain features which are characteristic of their tissue of origin, i.e. colorectal mucosa. In general, the cell lines which express differentiated features have the characteristics of absorptive cells and form morphologically and functionally polarized monolayers on tissue culture plastic. These cell lines provide a useful system for studying the processes of transport in colorectal epithelium. In addition, multipotential cell lines are now established which differentiate into two or three of the differentiated cell types found in the normal epithelium. Such cell lines are very useful for studying the control of differentiation in colorectal epithelium. In other continually renewing systems, such as the haematopoietic system, a range of proliferation and differentiation factors has been isolated which controls the differentiation of the stem cell into a variety of cell lineages. In contrast, virtually nothing is known about the factors controlling differentiation in the gastrointestinal epithelium although an important role has been demonstrated for epithelial–mesenchymal interactions[55]. The availability of multipotential cell lines will provide the model systems with which to study these differentiation pathways.

Acknowledgements

This work was supported by the Cancer Research Campaign and the Imperial Cancer Research Fund.

References

1. Tompkins, W. A., Watrach, A. M., Schmale, J. D., Schultz, R. M. and Harris, J. A. (1974). Cultural and antigenic properties of newly established cell strains derived from adenocarcinomas of the human colon and rectum. *J. Nat. Cancer Inst.*, **52**, 1101–1110
2. Fogh, J. and Trempe, G. (1975). New human tumour cell lines. In Fogh, J. (ed.) *Human Tumour Cells in Vitro*, pp. 115–141. (New York: Plenum Press)
3. Leibowitz, A., Stinson, J. C., McCombs, W. B., McCoy, C. E., Mazur, K. C. and Mabry, N. D. (1976). Classification of human colorectal adenocarcinoma cell lines. *Cancer Res.* **36**, 4562–4569
4. Brattain, M. G., Brattain, D. E., Fine, W. D., Khaled, F. M., Marks, M. E., Kimball, P. M., Arcolano, A. and Danbury, B. H. (1981). Initiation and characterization of cultures of human colonic carcinoma with different biological characteristics utilizing feeder layers of confluent fibroblasts. *Oncodev. Biol. Med.*, **2**, 355–366
5. McBain, J. A., Weese, J. L., Meisner, L. F., Wolberg, W. H. and Willson, J. K. V. (1984). Establishment and characterisation of human colorectal cancer cell lines. *Cancer Res.*, **44**, 5813–5821
6. Kirkland, S. C. and Bailey, I. G. (1986). Establishment and characterisation of six human colorectal adenocarcinoma cell lines. *Br. J. Cancer*, **53**, 779–785
7. Park, J.-G., Oie, H. K., Sugarbaker, P. H., Henslee, J. G. Chen, T.-R., Johnson, B. E. and Gazdar, A. (1987). Characteristics of cell lines established from human colorectal carcinoma. *Cancer Res.*, **47**, 6710–6718
8. Willson, J. K. V., Bittner, G. N., Oberley, T. D., Meisner, L. F. and Weese, J. L. (1987). Cell culture of human colon adenomas and carcinomas. *Cancer Res.*, **47**, 2704–2713
9. Brattain, M. G., Brattain, D. E., Sarrif, A. M., McRae, L. J., Fine, W. D. and Hawkins, J. G. (1982). Enhancement of growth of human colon tumor cell lines by feeder layers of murine fibroblasts. *J. Nat. Cancer Inst.*, **69**, 767–771
10. Whitehead, R. H., Jones, J. K., Gabriel, A. and Lukies, R. E. (1987). A new colon carcinoma cell line (LIM 1863) that grows as organoids with spontaneous differentiation into crypt-like structures *in vitro*. *Cancer Res.*, **47**, 2683–2689
11. Kimball, P. M. & Brattain, M. G. (1980). Isolation of a cellular subpopulation from a human colonic carcinoma cell line. *Cancer Res.*, **40**, 1574–1579
12. Noguchi, P., Wallace, R., Johnson, J., Earley, E. M., O'Brien, S., Ferrone, S., Pellegrino, M. A., Milstein, J., Needy, C., Browne, W. and Petricciani, J. (1979). Characterisation of WiDr: A human colon carcinoma cell line. *In Vitro*, **15**, 401–408
13. Chen, T. R., Drabkowski, D., Hay, R. J., Macy, M. and Peterson, W. (1987). WiDr is a derivative of another colon adenocarcinoma cell line, HT-29. *Cancer Genet. Cytogenet.*, **27**, 125–134
14. Chen, T. R. (1977). *In situ* detection of mycoplasma contamination in cell cultures by fluorescent Hoechst 33258 stain. *Exp. Cell Res.*, **104**, 255–262
15. McGarrity, G. J. (1982). Detection of mycoplasmal infection of cell cultures. *Adv. Cell Culture*, **2**, 99–131
16. Kirkland, S. C. (1988). Clonal origin of columnar, mucous and endocrine cell lineages in human colorectal epithelium. *Cancer*, **61**, 1359–1363
17. Kirkland, S. C. (1986). Endocrine differentiation by a human rectal adenocarcinoma cell line (HRA-19). *Differentiation*, **33**, 148–155

18. Shamsuddin, A. M., Phelps, P. C. and Trump, B. F. (1982). Human large intestinal epithelium: light microscopy and ultrastructure. *Hum. Pathol.*, **13**, 790–803
19. Chang, W. W. L. and Leblond, C. P. (1971). Renewal of the epithelium in the descending colon of the mouse: 1. Presence of three epithelial cell populations. Vacuolated, columnar and argentaffin. *Am. J. Anat.*, **131**, 73–100
20. Cheng, H. and Leblond, C. P. (1974). Origin, differentiation and renewal of the four main epithelial cell types in the mouse small intestine: V. Unitarian theory of the origin of the four epithelial cell types. *Am. J. Anat.*, **141**, 537–562
21. Cox, W. F. and Pierce, G. (1982). The endodermal origin of the endocrine cells of an adenocarcinoma of the colon of the rat. *Cancer*, **50**, 1530–1538
22. Ponder, B. A. J., Schmidt, G. H., Wilkinson, M. M., Wood, M. J., Monk, M. and Reid, A. (1985). Derivation of mouse intestinal crypts from single progenitor cells. *Nature (London)*, **31**, 689–691
23. Namba, M., Miyamata, K., Hyodoh, F., Iwama, T., Iwama, T., Utsunomiya, J., Fukushima, F. and Kimoto, T. (1983). Establishment and characterisation of a human colon carcinoma cell line (KMS-4) from a patient with hereditary adenomatosis of the colon and rectum. *Int. J. Cancer.*, **32**, 697–702
24. Pinto, M., Robine-Leon, S., Appay, M-D., Kedinger, M., Triadou, N., Dussaulx, E., Lacroix, B., Simon-Assmann, P., Haffen, K. and Zweibaum, A. (1983). Enterocytic-like differentiation and polarization of the human colon carcinoma cell line Caco-2 in culture. *Biol. Cell*, **47**, 323–330
25. Dharmsathaphorn, K., McRoberts, J. A. Mandel, K. G., Tisdale, L. D. and Masui, H. (1984). A human colonic tumor cell line that maintains vectorial electrolyte transport. *Am. J. Physiol.*, **246**, G204–G208
26. Whitehead, R. H., Macrae, F. A., St. John, D. J. B. and Ma, J. (1985). A colon cancer cell line (LIM 1215) derived from a patient with inherited nonpolyposis colorectal cancer. *J. Nat. Cancer Inst.*, **74**, 759–765
27. Kirkland, S. C. (1985). Dome formation by a human colonic adenocarcinoma cell line (HCA-7). *Cancer Res.*, **45**, 3790–3795
28. Grasset, E., Pinto, M., Dussaulx, E., Zweibaum, A. and Desjeux, J-F. (1984). Epithelial properties of human colonic carcinoma cell line Caco-2: electrical parameters. *Am. J. Physiol.*, **247**, C260–267
29. Ramond, M-J., Martinot-Peignoux, M. and Erlinger, S. (1985). Dome formation in the human colon carcinoma cell line Caco-2 in culture. Influence of ouabain and permeable supports. *Biol. Cell*, **54**, 89–92
30. Dharmsathaphorn, K., Mandel, K. G. Masui, H. and McRoberts, J. A. (1985). Vasoactive intestinal-polypeptide induced chloride secretion by a colonic epithelial cell line. *J. Clin. Invest.*, **75**, 462–471
31. Cuthbert, A. W., Kirkland, S. C. and MacVinish, L. J. (1985). Kinin effects on ion transport in monolayers of HCA-7 cells, a line from a human colonic adenocarcinoma. *Br. J. Pharmacol.*, **86**, 3–5
32. Cuthbert, A. W., Egleme, C., Greenwood, H., Hickman, M. E., Kirkland, S. C. and MacVinish, L. J. (1987). Calcium and cyclic AMP-dependent chloride secretion in human colonic epithelia. *Br. J. Pharmacol.*, **91**, 503–515
33. Welsh, M. J., Smith, P. L., Fromm, M. and Frizzell, R. A. (1982). Crypts are the site of intestinal fluid and electrolyte secretion. *Science*, **218**, 1219–1221
34. Cuthbert, A. W. (1988). Transepithelial ion transport in cultured colonic epithelial cell monolayers. In Jones, C. J. (ed.) *Epithelia: Advances in Cell Physiology and Cell Culture*, pp. 49–64. (Lancaster: MTP Press)
35. Zweibaum, A., Triadou, N., Kedinger, M., Augeron, C., Robine-Leon, S., Pinto, M., Rousset, M. and Haffen, K. (1983). Sucrase–isomaltase: A marker of foetal and malignant epithelial cells of the human colon. *Int. J. Cancer*, **32**, 407–412
36. Zweibaum, A., Hauri, H-P., Sterchi, E., Chantret, I., Haffen, K., Bamat, J. and Sordat, B. (1984). Immunohistological evidence, obtained with monoclonal antibodies, of small intestinal brush border hydrolases in human colon cancers and foetal colons. *Int. J. Cancer*, **34**, 591–598
37. Augeron, C. and Laboisse, C. L. (1984). Emergence of permanently differentiated cell clones in a human colonic cancer cell line in culture after treatment with sodium butyrate.

Cancer Res., **44**, 3961–3969

38. Huet, C., Sahuquillo-Merino, C., Coudrier, E. and Louvard, D. (1987). Absorptive and mucus-secreting subclones isolated from a multipotent intestinal cell line (HT-29) provide new models for cell polarity and terminal differentiation *J. Cell Biol.*, **105**, 345–357

39. Tsao, D., Morita, A., Bella, A., Luu, P. and Kim, Y. S. (1982). Differential effects of sodium butyrate, dimethylsulphoxide and retinoic acid on membrane-associated antigen, enzymes and glycoproteins of human rectal adenocarcinoma cells. *Cancer Res.*, **42**, 1052–1056

40. Dexter, D. L., Lev, R., McKendall, G. R. Mitchell, P. and Calabres, P. (1984). Sodium butyrate-induced alteration of growth properties and glycogen levels in cultured human colon carcinoma cells. *Histochem. J.*, **16**, 137–149

41. Morita, A., Tsao, D. and Kim, Y. S. (1982). Effect of sodium butyrate on alkaline phosphatase in HRT-18, a human rectal cancer cell line. *Cancer Res.*, **42**, 4540–4545

42. Chung, Y. S., Song, I. S., Erickson, R. H., Sleisenger, M. H. and Kim, Y. S. (1985). Effect of growth and sodium butyrate on brush border membrane-associated hydrolases in human colorectal cancer cell lines. *Cancer Res.*, **45**, 2976–2982

43. Gum, J. R., Kam, W. K., Byrd, J. C., Hicks, J. W., Sleisenger, M. H. and Kim, Y. S. (1987). Effects of sodium butyrate on human colonic adenocarcinoma cells. Induction of placental-like alkaline phosphatase. *J. Biol. Chem.*, **262**, 1092–1097

44. Pinto, M., Appay, M. D., Simon-Assmann, P., Chevalier, G., Dracopoli, N., Fogh, J. and Zweibaum, A. (1982). Enterocytic differentiation of cultured human colon cancer cells by replacement of glucose by galactose in the medium. *Biol. Cell*, **44**, 193–196

45. Zweibaum, A., Pinto, M., Chevalier, G., Dussaulx, E., Triadou, N., Lacroix, B., Haffen, K., Brun, J-L. and Rousset, M. (1985). Enterocytic differentiation of a subpopulation of the human colon tumor cell line HT-29 selected for growth in sugar-free medium and its inhibition by glucose. *J. Cell. Physiol.*, **122**, 21–29

46. Laboisse, C. L., Maoret, J-J., Triadou, N. and Augeron, C. (1988). Restoration by polyethylene glycol of characteristics of intestinal differentiation in subpopulations of the human colonic adenocarcinoma cell line HT29. *Cancer Res.*, **48**, 2498–2504

47. Pearse, A. G. E. and Polak, J. M. (1971). Neural crest origin of the endocrine polypeptide (APUD) cells of the gastrointestinal tract and pancreas. *Gut*, **12**, 783–788

48. LeDouarin, N. M. and Teillet, M. A. (1973). The migration of neural crest cells to the wall of the digestive tract in avian embryo. *J. Embryol. Exp. Morphol.*, **30**, 31–48

49. Andrew, A. (1974). Further evidence that enterochromaffin cells are not derived from the neural crest. *J. Embryol. Exp. Morphol.*, **31**, 589–598

50. Andrew, A., Kramer, B. and Rawdon, B. B. (1982). The embryonic origin of endocrine cells of the gastrointestinal tract. *Gen. Comp. Endocrinol.*, **47**, 249–265

51. Lloyd, R. V. and Wilson, B. S. (1983). Specific endocrine tissue marker defined by a monoclonal antibody. *Science*, **222**, 628–630

52. Daar, A. S. and Fabre, J. W. (1983). The membrane antigens of human colorectal cancer cells: Demonstration with monoclonal antibodies of heterogeneity within and between tumours and anomalous expression of HLA-DR. *Eur. J. Clin. Cancer Clin. Oncol.*, **19**(2), 209–220

53. Dexter, D. L., Spremulli, E. N., Fligiel, Z., Barbosa, J. A., Vogel, R., VanVoorhees, A. and Calabresi, P. (1981). Heterogeneity of cancer cells from a single human colon carcinoma. *Am. J. Med.*, **71**, 949–956

54. Brattain, M. G., Fine, W. D., Khaled, F. M., Thompson, J. and Brattain, D. E. (1981). Heterogeneity of malignant cells from a human colonic carcinoma. *Cancer*, **41**, 1751–1756

55. Haffen, K., Kedinger, M. and Simon-Assmann, P. (1987). Mesenchyme-dependent differentiation of epithelial progenitor cells in the gut. *J. Ped. Gastroenterol. Nutr.*, **6**, 14–23

3
Transepithelial Ion Transport in Cultured Colonic Epithelial Cell Monolayers

A. W. CUTHBERT

INTRODUCTION

In the last decade, the study of transepithelial ion transport in cultured monolayers grown on pervious supports has become rather commonplace. This is in sharp contrast to the predominance of *in vitro* studies of intact epithelia in the previous thirty years. It is vital to anyone contemplating a study with cultured epithelia to consider both the advantages and disadvantages of such a step and particularly if the questions to be posed are most appropriately answered by such an approach. There are both advantages and disadvantages in using culture systems.

Among the advantages offered by tissue culture are the following. First, cellular heterogeneity typical of most natural epithelia is avoided. For example, in natural colonic epithelia, surface and crypt cells have different functions in relation to ion transport[1]. The structural complexity, especially of intestinal epithelia, is avoided in cultured tissues, which generally form single monolayer structures. Serosal access by chemical probes is often modified by tissues in the lamina propria which not only form a physical barrier but also contain other elements which may alter epithelial function. Of particular importance are nerve cell bodies and nerve endings and cells of the immune system. These may release a plethora of neurotransmitters, neuromodulators or autacoids in response to chemical agents which then act indirectly upon the epithelium. This is avoided with cultured monolayers and so allows more definitive conclusions to be reached about the nature of receptors in the basolateral aspect of the cells. For those who wish to detect single channel currents with the patch clamping technique, then at least the apical surfaces of cultured monolayers are easily accessible. The basolateral surface poses technical problems but no worse than those in intact epithelia. The presence of specific receptors in epithelial membranes can be approached by a combination of functional studies coupled with those for ligand binding. Cultured epithelial cell membranes will be purer than those obtained from tissues where contaminating membranes arise from other cell types. Also, autoradiographic location of bound ligands is less ambiguous in cultured epithelia.

Isolation of mutants by use of selective procedures forms a powerful approach to the study of the molecular mechanisms of transport processes. This does not necessarily require epithelial cells, but somatic cell genetics studies with epithelial cells will allow important questions to be posed about the functional responsibilities of transport proteins in these systems. The interested reader should consult the review by Gargus[2].

Epithelia cultured upon surfaces always grow with the basolateral surface against the substrate end with the apical surface uppermost. In suspension culture, they can form acini with either the apical or basolateral surface facing inwards[3]. In no situation do the cells arrange themselves randomly with respect to polarity. The way epithelial tissues establish this polarity is a vital, and yet, an only partially answered question. Many elegant experiments have been devised to examine how cells address membrane components to the correct domain and cultured epithelia provide a sensible system in which this phenomenon can be investigated. Until now, much of the work was carried out with MDCK cells, but recently generation of cell polarity has been studied in a colonic cell line[4].

Epithelia grown from primary cultures of kidney tubules, mammary glands, sweat glands, the epididymis, etc., have been used to study transport processes but, as yet, there are no studies using primary cultures of intestinal epithelia. Cultured monolayers from the large intestine have been made with cell lines and herein lies a possible major disadvantage with this approach. Such lines, whether derived from naturally occurring tumours or following virus transformation, may have undergone a considerable amount of differentiation and therefore not be representative of naturally occurring cells. Questions addressed to an understanding of cellular mechanisms are therefore likely to be more successful than those which relate to *in vivo* behaviour.

A second disadvantage of cultured systems is that receptor systems normally present may not be expressed in culture or, alternatively, cultured cells may produced receptors not found *in vivo*. Furthermore, expression of receptors may occur only in some circumstances. A classic example of this kind is with the toad kidney cell line, A-6. Grown on a plastic surface, A-6 monolayers do not show amiloride-sensitive sodium uptake or arginine–vasopressin-sensitive adenylate cyclase activity. Both of these properties are found in monolayers grown upon pervious supports where nutrient can bathe both surfaces[5,6].

Culture conditions are often crucial for obtaining satisfactory monolayers for transport studies. The ideal condition is to have a fully defined medium so that the effect of removal of one component at a time on properties may be investigated. More usually, serum (fetal calf mostly) or other mixtures (chick embryo extract) are added to produce satisfactory growth. A difficulty then arises since an unknown component or components may induce properties not normally expressed, making comparison to *in vivo* situations hazardous.

METHODOLOGY

Essentially, the principle of producing cultured epithelial monolayers is a simple one. Cells are persuaded to grow until a confluent monolayer is formed on some sort of pervious support. The monolayer and support are then mounted in an appropriate fashion depending on the protocol to be used.

The earliest studies were pioneered by Misfeldt *et al.*, 1976[7], Cereijido *et al.*, 1978[8] and Rabito *et al.*, 1978[9] with MDCK cells, growing these on Millipore filters. Millipore or Nucleopore filters are still widely used and can be coated with collagen or Matrigel to promote cell attachment. Because of the difficulties referred to earlier with A-6 cells, filter-bottomed cups were devised with small feet so that the culture has access to growth medium from both surfaces[6,10]. This method has the added bonus that the transepithelial potential, in developing monolayers, may be monitored at intervals, using sterile external electrodes dipped into the apical and basolateral bathing solutions. Furthermore, once the cell layer has become 'tight', the cups can be fitted to modified Ussing chambers or used for investigation with microelectrodes.

As the cups are not altogether ideal for all subsequent manipulations, we devised a much simpler system by gluing washers made from Sylgard silicone elastomer (central hole 0.2 cm^2, thickness 2 mm) to precoated Millipore filters, so creating a small well into which a cell suspension could be loaded[11]. Four such units are then floated on the surface of medium in a petri dish. The time to confluence with this method can only be learned from experience as there is no easy way to monitor the transepithelial potential. Nevertheless, with both methods, there is avoidance of edge damage when the monolayers are used in an Ussing chamber configuration as the cells grow up to and sometimes onto the edge of the cup or washer. The washer does, of course, provide an excellent way of sealing the two chamber halves together when slight compression is applied.

There are some variants on these standard methods which are useful in some circumstances. For example, collagen can be spread on the surface of a dish and cells allowed to grow on his. Once the cells are nearly confluent, the collagen layer can be gently eased off the dish and allowed to float to the surface of the medium. Some contraction of the gel matrix takes place, but, importantly, it allows medium to have access to the serosal surface. The resulting epithelium can be lightly clamped between chambers using silicone grease to make a seal[12].

There are other protocols which eventually might prove very useful in transport studies in epithelia. For example, after an epithelial monolayer has been established, what if a second collagen overlay is added on the apical surface? Cells are therefore provided with alternative surfaces upon which to lay down a basement membrane. With MDCK cells, tubules are formed in which the lumen is lined with apical microvilli[13]. Such structures may be useful both for organizational studies as well as for those of transport.

Finally, some epithelial cells, when grown in suspension culture, form free-floating hollow spheres lined with a monolayer of polarized epithelial cells with outward facing microvilli[3]. Some colonic epithelial cell lines can also form cellular microspheres (Kirkland, private communication). Clearly, such cellular spheres could be used in a variety of permeability studies with ions and other substances.

COLONIC EPITHELIAL CELL LINES

Studies with human epithelia are facilitated by using a tissue culture approach. Curiously, all of the colonic epithelial cell lines available for transport studies are derived from human tissues, usually carcinomas. Table 3.1 lists the major types available together with their origins. Eight of the nine listed form confluent monolayers in culture and therefore are potentially useful for transport studies. The three lines derived from HCA-7 (Colonies 1, 3 and 29) were obtained from the parent line using a cloning cylinder and/or sodium butyrate treatment. This suggests that heterogeneity exists in earlier passages of HCA-7[14]. The dome-forming capacity of HCA-7 cells is slowly lost at high passage numbers, suggesting that the proportion of different cells types is changing. On the other hand, Colony 1 cells form domes readily even when subconfluent, while Colony 3 cells rarely form domes.

HRA-19 is of rectal origin and probably arose from malignant progenitor cells as it retains the ability to differentiate even after 120 passages. The pleomorphic nature is evidenced by the ability to differentiate into endocrine cells and cells with absorptive characteristics when implanted into nude mice[15]. Clonal monolayers (i.e. derived from a single epithelial cell) also develop endocrine and mucous cells in monolayer culture[16].

The T84 cell line has been extensively used for transport studies and was derived from lung metastases of a human colon carcinoma grown in nude mice. Cells from mouse tumour tissue had the characteristics of epithelial cells when grown in serum-supplemented medium and again formed identical tumours when returned to the mice[17].

Caco-2, another human cell line, exhibits spontaneous epithelial differentiation *in vitro*. Although there are relatively few transport studies with this line, it has sufficiently different properties to make it of interest.

The original HT 29 cell line showed no tendency to form polarized monolayers, but, after treatment of cultures with sodium butyrate, a number of different morphologies emerged and five clonal cell lines were isolated[18]. Strictly, flat colonies of cells were isolated using cloning cylinders giving stable cell lines but it cannot be absolutely claimed that each line derived from a single cell. All five clonal cell lines produced dome-forming polarized epithelia, with or without mucous secreting cells. The latter may represent a post-replicative population arising from a stem cell pool.

LIM 1863 is included in this section because it is unusual. After explanting small pieces of tumour tissue, free-floating cells accumulate in the medium with no evidence of epithelial cells attached to the plastic surface. The floating cell masses or 'organoids' contain mature columnar

Table 3.1 Colonic epithelial cell lines

Name	Derivation	Formation of polarized epithelium	Reference
HCA-7	Human colonic adenocarcinoma	Yes	14
HCA-7 Colony 1	From HCA-7	Yes	27
HCA-7 Colony 3	From HCA-7	Yes	27
HCA-7 Colony 29	From HCA-7	Yes	Cuthbert and Kirkland (unpublished)
HRA-19	Human rectal adenocarcinoma	Yes	15
T-84 (HC84S)	Lung metastases from a human colon carcinoma	Yes	17
Caco-2	Human colonic carcinoma	Yes	42
HT29	Human colonic adenocarcinoma	Yes	18
LIM 1863	Human colonic carcinoma	No	19

cells and goblet cells, secreting mucus into a central lumen. They represent, in some ways, cultured crypts, although endocrine cells have not been found[19]. As yet, there are no transport studies on these interesting objects, but they do offer a variety of possibilities.

RESPONSES TO SECRETAGOGUES IN CULTURED EPITHELIA

Wherever monolayers of colonic epithelial cells have been studied, the transepithelial potential is small (less than 1 mV) with the apical side negative. In general, basal short circuit currents (SCC) are low and with transepithelial resistances in the range 30–150 Ω cm^2. The apical membrane potential has been measured in Caco 2 cells and has a value of around -60 mV, with a voltage divider ratio of 0.8, indicating that the apical face is the main resistance barrier[20]. Many secretagogues cause an increase in SCC in colonic epithelial monolayers with a corresponding, almost parallel increase in transepithelial voltage. Small increases in electrical conductance have been recorded in response to secretagogues, for example, in Caco 2 cells, there is a 4% increase in conductance with dibutyryl cyclic AMP, corresponding to a conductance change of 3 mS cm^{-2} in the apical membrane[21].

Table 3.2 Secretagogues affecting cultured epithelial monolayers

Name	Secretagogues	Reference
Caco 2 (1.9 μA cm^2, 0.3 mV, 150 Ω cm^2)	Db cAMP, VIP, adrenaline, forskolin, and amphotericin	20, 21
HCA-7 (9-5 μA cm^2 0.3 mV, 50–80 Ω cm^2)	LBK, forskolin, VIP, CCh, A23187, histamine	27, 38, 44
HCA-7 Colony 1 (42 Ω cm^2)	LBK, forskolin, VIP, CCh, A23187	27
HCA-7 Colony 3 (117 Ω cm^2),	LBK, forskolin, VIP CCh, A23187	27
HCA-7 Colony 29	Forskolin, A23187, LBK	Unpublished
HRA-19	Forskolin, A23187, LBK	Unpublished
T84 (Near zero SCC and potential, 100 Ω cm^2)	VIP, ACh, PGE$_1$, A23187, histamine	29, 43

Where available basal values of SCC, transepithelial potential and membrane resistance are given. LBK refers to lysylbradykinin, CCh and ACh to carbachol and acetylcholine respectively and VIP to vasoactive intestinal polypeptide

Table 3.2 lists the secretagogues which have been shown to increase SCC. This increase could be due to electrogenic anion secretion, electrogenic cation absorption or a mixture of the two. Nystatin and amphotericin were shown to increase SCC in a way which is dependent upon mucosal sodium in Caco-2. However, this current was insensitive to amiloride so it cannot be concluded that this line has electrogenic sodium absorbing capacity. This finding is similar to the behaviour of *Necturus maculosus* skin in which the apical face is impermeable to sodium. However, if sodium gains access through the apical face, then it can be expelled by the basolateral sodium pump, thus affecting electrogenic sodium absorption. Amphotericin is able to permeabilize the cell membrane to cations[22]. The SCC responses to all the other secretagogues mentioned in Table 3.2 appear to be due to electrogenic anion secretion. The evidence for this is from either flux studies or by use of inhibitors. Flux studies are rather difficult in cultured monolayers as the areas are generally small and any minute areas of non-confluence provide extra leakage pathways. A typical mammalian colon has a resistance of 200–300 Ω cm^2, somewhat higher than those for cultured monolayers as recorded in Table 3.2. However, electrogenic chloride secretion has been shown to account for a major fraction, at least, of the responses to many secretagogues. For example, VIP and PGE$_1$ cause both an increase in chloride flux in the apical to basolateral direction and vice versa with an overall increase in net chloride secretion. This is found without any significant changes in sodium flux[23,24]. With A23187 and carbachol, only a significant change in the serosal-to-mucosal flux of chloride is found, together with a consequent change in net chloride secretion, again without a significant effect on sodium movement[25,26]. Fluxes have not been measured following treatment of monolayers with lysylbradykinin (LBK), although the effect can be inhibited with piretanide, a loop diuretic[27]. LBK does, however, cause net chloride secretion in many other epithelia, for example the rat colon[28]. In some of the HCA-7 type epithelia, not the whole of the SCC increase to secretagogues is sensitive to loop diuretics, such as piretanide, bumetanide or frusemide. Some of the remaining fraction can be inhibited by acetazolamide suggesting that, at least under some circumstances, part of the secretion is due to bicarbonate.

A convincing demonstration of neutral NaCl absorption or of electrogenic sodium absorption has not been made in cultured colonic monolayers, indicating that they have properties of crypt cells rather than those of surface or villous cells[1]. There are also a number of scattered reports on the lack of effect of aldosterone on these monolayers, which accords with the same view.

The secretagogues listed in Table 3.2 fall into two neat categories: VIP, forskolin and PGE_1 are known to increase cAMP in epithelia and other cells via an action on membrane receptors coupling to adenylate cyclase through G-proteins. On the other hand, LBK, CCh, ACh and histamine are known to increase intracellular calcium concentrations. Indeed, direct measurements of Ca_i with Fura-2 fluoresence has shown this to be so for the action of histamine and carbachol in T84 cells[26,29]. There is another important difference in that the responses of the group which increase cAMP are, in general, sustained while those to the Ca_i-raising group are transient, the peak SCC falling to just above baseline plateau after several minutes. The reasons for these differences are not altogether clear.

Apart from forskolin and A23187, which are very lipophilic, all of the secretagogues show sidedness, with the exception of LBK. In the HCA-7 series of cell lines, LBK acts from both the apical and basolateral faces[44] and the quality of the responses is also different. This difference may point to different mechanisms. In general, however, the other secretagogues act only from the basolateral face.

SECRETORY MECHANISMS IN CULTURED COLONIC EPITHELIA

Electrogenic chloride secretion in mammalian epithelia is thought to require a number of elements. First, chloride must enter the cell through the basolateral side, moving up an electrochemical gradient into the cell. This it does by using the energy stored in the sodium gradient and an $NaKCl_2$ triporter. The sodium can be removed from the cell by the basolateral sodium pump while the potassium ions equilibrate through basolateral K^+ channels. Chloride ions can leave the cell by the apical face through chloride channels, moving down an electrochemical gradient. Thus, one expects the apical surface to be polarized negatively with respect to the basolateral surface, providing a driving force for the movement of a counter-cation in open circuit conditions. Agents might then interact with apical chloride channels, the basolateral K^+ channels or the triporter directly or indirectly.

Some epithelial K^+ channels are Ca-sensitive so that signals increasing Ca_i might be expected to open K^+ channels, resulting in an accelerating voltage for apical chloride exit, as well as increasing the K^+ concentration in the basolateral stationary layer to fuel the triporter. This picture of chloride secretion appears to be accurate for monolayers of colonic cells, mainly through the work of Dharmsathaphorn and his colleagues with T84 cells.

For example, the uptake of Na^+, Rb^+ (as a marker for K^+) and Cl^- was measured in the presence of VIP and in the presence or absence of bumetanide. The bumetanide-sensitive uptake of each ion was found to be interdependent and, from initial velocities, the uptakes approached the ratio of 1:1:2 for $Na:K:Cl$[23]. Further, in T84 monolayers loaded with $^{86}Rb^+$, it was found that PGE_1 increased serosal but not the mucosal efflux of the isotope, an action which was blocked by Ba^{2+} ions indicative of basolateral K^+ channels[24]. A similar result was obtained with carbachol; however, serosal Rb^+ efflux was not blocked[26] by Ba^{2+}. While PGE_1 increased cAMP content of the cells, carbachol did not, although there was a transient rise in Ca_i. A23187 behaves like carbachol with respect to basolateral Rb^+ efflux, with the exception that it is sensitive[30] to Ba^{2+}. The evidence therefore points to two types of K^+ channel in the basolateral membrane, one sensitive to cAMP and one to Ca_i, but the differences between A23187 and CCh in terms of Ba^{2+} sensitivity are difficult to explain.

From measurements of ^{36}Cl uptake into T84 monlayers, it was found that VIP, but not A23187, promoted uptake which was independent of Na^+ or K^+ and therefore distinct from the co-transport uptake, a finding confirmed by the insensitivity to bumetanide. This uptake was selectively at the apical face, where VIP altered the rate of chloride uptake but not the chloride concentration for half maximal saturation[31]. The uptake studies provide excellent evidence for cyclic AMP-dependent chloride channels in the apical face of these cells, now confirmed by patch clamp studies.

Chloride channels with a conductance of 50 pS (at 160 mmol L^{-1} NaCl) have been found in the apical membranes of both T84 and HT29 cells; additionally, a smaller conductance channel of 15 pS was found in HT29 cells. The 50 pS channels in both types of cell show outward rectification with a permeability sequence of $I>Br>Cl>F$. As chloride secretory epithelia transport only Cl^-, and occasionally Br^- to a lesser extent, the anion selectively may reside in the basolateral membrane. The PCl/PNa ratio for the 50 pS channel was around 50. In the cell attached mode, forskolin and PGE_2, as well as cAMP, cause the appearance of channel activity in quiescent cells. In both systems, the open state probabilty was increased by depolarization of the apical membrane. This effect acts to counter the reduced electrochemical conductance gradient occasioned by the increase in apical chloride conductance caused by cAMP. Both types of chloride channel in HT29 cells were blocked by NPPB (5-nitro-2-(3-phenylpropylamino)-benzoate)[32,33]. In a very recent study, it appears that NPPB affects the rate constant for channel opening or closing rather than blocking the channels directly[34]. At present, there do not seem to be any really potent chloride-channel blockers, and some commonly used agents (e.g. anthracene-9-carboxylate) affect the generation of cAMP in some epithelial cell lines.

SYNERGISM BETWEEN MESSENGER SYSTEMS

Chloride secretion requires that the ion be moved through two barriers in series in order to complete the transport process. Agents like VIP increase both the chloride conductance of the apical barrier and the potassium conductance at the basolateral side (see above). The latter then both assists chloride exit and may prime the NaKCl$_2$ exchanger. Agents such as VIP produced sustained responses in T84 and HCA-7 monolayers. On the other hand, agents like carbachol only produce rather transient responses and affect only basolateral K$^+$-conductance. Interactions between these two sorts of agents have been studied in T84 cells[25] and in HCA-7 monolayers (Cuthbert, unpublished). In general, the findings are that pairs of agents, such as VIP and A23187, carbachol and VIP, carbachol and PGE$_1$, produce responses which are more than additive. The major question which arises is whether or not this is because the apical membrane remains a rate-limiting barrier in the case of an agonist acting through Ca^{2+} or if cAMP-producing agonists are promoted by an extra increase in basolateral K$^+$ conductance, or again if there is a more subtle interplay between the two separate signalling systems. Some obvious interactions have been eliminated, for example A23187 does not affect the generation of cAMP by VIP and basolateral K$^+$ efflux for these two is simply additive. The question remains to some extent unanswered, for example responses to carbachol and histamine are transient, yet that for A23187 is sustained. Agonists acting through Ca^{2+} may act through intracellular Ca^{2+} pools and not be dependent solely upon external calcium, as is A23187. It has been reported for some systems that cAMP can increase Ca$_i$, or, alternatively cAMP may promote the release of Ca^{2+} from internal stores when acted upon IP$_3$, a messenger generated by carbachol, histamine, etc. Most impressive is the finding that the potentiation is seen when maximally effective concentrations of different Ca$_i$ and cAMP-generating agonists are used.

Figure 3.1 illustrates some results from unpublished work from my laboratory in which some features of the potentiation phenomena are explored.

Lysylbradykinin, which in HCA-7 monolayers acts through Ca$_i$ and not the production of eicosanoids[27], is an unusual agonist in that it has receptors on both the apical and basolateral faces of the cell[44]. The illustration shows that basolaterally applied LBK is significantly potentiated by forskolin. On the other hand, thapsigargin, an agent which increases Ca$_i$ but not by an ionophoric mechanism[35], inhibits the action of LBK, on this occasion applied apically. The effect of thapsigargin is maintained, although there is some decline from the original peak response, yet this agent is not known to affect cAMP metabolism. More puzzling is why the terpene is virtually ineffective in some lines (e.g. Colony 1 and Colony 3) and the target for this agent needs identifying. Important questions also remain about how thapsigargin (and A23187) give sustained increases in chloride secretion. While hyperpolarization of the apical membrane, caused by opening of Ca^{2+}-sensitive basolateral K$^+$ channels, will increase the gradient for chloride exit, the open state probability of the chloride channels will be reduced.

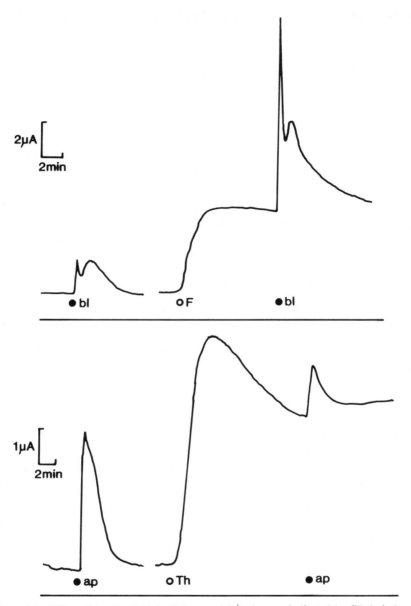

Figure 3.1 Effect of lysylbradykinin (0.1 μmmol L⁻¹) given as indicated by filled circles on either the basolateral (bl) or apical (ap) side of the monolayer. Upper record shows the potentiating effect of forskolin (F) (10 μmmol L⁻¹) on LBK responses in a 7-day HCA-7 monolayer. Lower record shows the inhibitory effect of thapsigargin (170 nmol L⁻¹, apical) on LBK responses in a 21-day HCA-7 monolayer. Each monolayer was 0.2 cm². Horizontal lines indicate zero SCC (Brayden and Cuthbert, unpublished)

Earlier suggestions of Ca^{2+}-sensitive chloride channels are worthy of further investigation.

Interesting information can be gathered, in relation to the role of Ca^{2+} and cAMP as second messengers, by examining the behaviour of mutant cell lines. The results given in Figure 3.2 confirm and extend the data given by Cuthbert et al.[27]. Three mutants derived from HCA-7 cells, namely Colonies 1, 3 and 29 are examined alongside the parent cell line and together with the human rectal carcinoma line, HRA-19. The responses are given as charge transfer in nEq in 8 min which takes into account both the SCC increase and how well the increase is maintained. Only HCA-7 and HRA-19 monolayers give substantial responses to both secretagogues. Colony 1 monolayers give large responses to forskolin but very minor responses to A23187. For Colonies 3 and 29, the reverse is true. Clearly, there must be apical chloride channels in these preparations for chloride secretion to take place at all, but the mechanisms by which they are regulated are still obscure. Whether effective chloride secretion depends entirely upon an increased gradient for apical exit, modified by the potential dependence of channel open time, has already been discussed. A major question is why are the apical chloride channels not equally well operated by cAMP in different lines. Figure 3.2 also shows that this is not because the cells fail to accumulate cAMP. Interestingly, those lines which respond well to forskolin only accumulate modest amounts of cAMP, while the unresponsive lines accumulate 500–600% more. The failure of chloride channels to respond to cAMP at a point downstream of nucleotide formation is a characteristic of transporting tissues in cystic fibrosis[36,37]. It was found by Cartwright et al.[25] that there was a reasonable parallelism between cAMP generation and the chloride secretory response in T84 monolayers, which accumulate cAMP to the same levels as those found in HCA-7, Colony 1 and HRA-19 tissues. Clearly, there is no parallelism between the cAMP accumulation and the chloride secretory response in different epithelial systems.

One of the major advantages of using cultured monolayers for transport studies is, as stated earlier, the ability to perform experiments with a single known cellular type. However, there have already been attempts to reconstruct epithelial monolayers into more complex structures. An example is given in Figure 3.3 in which a 'sandwich' was constructed from rat peritoneal mast cells and an HCA-7 monolayer. In this instance, the rat had been previously sensitized to ovalbumin. Challenge of the 'sandwich' with ovalbumin resulted in a SCC response, whereas this did not happen with mast cells from unsensitized animals. In experiments such as these, it was shown tht the SCC response was dependent on the number of mast cells present and that, when using a fixed number of mast cells, a concentration-dependent reduction of the response could be achieved with the H_1 antagonist, mepyramine[38]. Models such as the one described may be useful as microbioassays for investigating the mediators in some disease states where the availability of cells is limited. For example, airway epithelia might be sandwiched with a variety of cells from human lung lavage. Use of various allergens combined with cell sorting and use of specific inhibitors

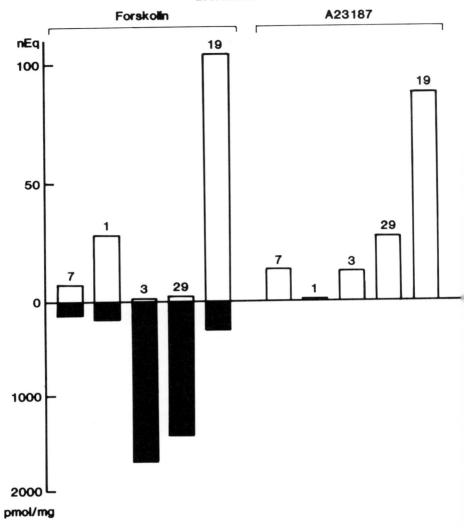

Figure 3.2 SCC responses in monolayers (0.2 cm^2) of HCA-7, Colony 1, Colony 3, Colony 29 and HRA-19 (designated 7, 1, 3, 29, 19) to forskolin (10 μmol L^{-1}) or A23187 (1 μmol L^{-1}) (both applied apically) expressed as nEq of charge transfer in 8 min (open columns). The closed columns indicate the tissue accumulation of cAMP during 15 min after exposure to forskolin (10 μmol L^{-1}) in pmol/mg protein. Basal values were around 10 pmol/mg protein. Each column represents the mean values for 4 to 17 observations

my shed light onto the nature of the mediators affecting epithelial function in allergic asthma. It should also be possible to culture defined types of enteric neurones and overlay these with intestinal epithelia to complement

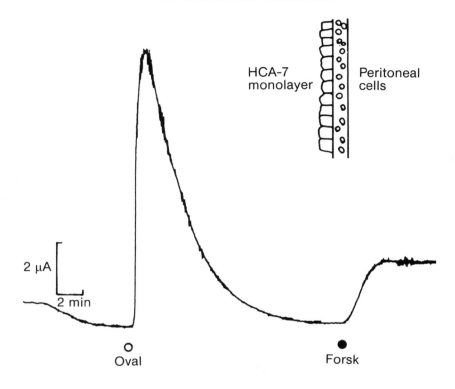

Figure 3.3 SCC responses in an HCA-7 monolayer (0.2 cm^2) sandwich with peritoneal mast cells taken from rat sensitized to ovalbumin. Ovalbumin (oval 20 µg/ml) was applied on the basolateral side. Forskolin (10 µmol L^{-1}) was added after the hypersensitivity reaction had subsided.

other types of study on the role of enteric nervous system in epithelial function.

OF OTHER THINGS

This review has concentrated upon the transporting functions of cultured colonic epithelial monolayers. However, it should be remembered that the major advantage of cellular homogeneity makes them useful for other types of problem, for example the nature of epithelial receptors and the biochemistry of second messenger systems. Three examples will suffice to demonstrate the diversity of these possibilities. Intracellular calcium has been implicated in having a role in apolipoprotein B secretion in Caco 2 cells[39]; the turnover and metabolism of α_2-adrenoceptors has been

61

investigated in HT-29 cells[40]; and nuclear VIP receptors have been shown in HT-29 cells[41].

CONCLUSIONS AND THE FUTURE

The picture of chloride secretion which emerges from the investigation of colonic epithelial cells in culture is much as would have been predicted from our knowledge of mechanisms of chloride secretion gained from *in vitro* studies of intact animal epithelia. The isolation of somatic cell mutants, with unique properties, together with the appliation of cellular and molecular genetic approaches offers considerable scope for the isolation and molecular characterization of molecules which are crucial to the transporting process as a whole.

Acknowledgements

Work by the author and reported here was supported by a grant, NIH HL17705.

References

1. Welsh, M. J., Smith, P. L., Fromm, N. and Frizzell, R. A. (1982). Crypts are the site of intestinal fluid and electrolyte secretion. *Science*, **218**, 1219–1221
2. Gargus, J. J. (1987). Mutant isolation and gene transfer as tools in the study of transport proteins. *Am. J. Physiol.*, **252**, C457–C467
3. Wohlwend, A., Montesano, R., Vassalli, J-D. and Orci, L. (1985). LLC-PK₁ cysts: A model for the study of epithelial cell polarity. *J. Cell Physiol.*, **125**, 533–539
4. Le Bivic, A., Hirn, M. and Reggis, H. (1988). HT-29 cells are an *in vitro* model for the generation of cell polarity in epithelia during embryonic differention. *Proc. Natl. Acad. Sci. USA*, **85**, 136–140
5. Lang, M. A., Muller, J., Preston, A. S. and Handler, J. S. (1986). Complete responses to vasopressin requires epithelial organisation in A6 cells in culture. *Am. J. Physiol.*, **250**, C138–C145
6. Sariban-Sohraby, S., Burg, M. B. and Turner, R. J. (1983). Apical sodium uptake in toad kidney epithelial cell line A6. *Am. J. Physiol.*, **245**, C167–C171
7. Misfeldt, D. S., Hamamoto, S. T. and Pitelka, D. R. (1976). Transepithelial transport in cell culture. *Proc. Natl. Acad. Sci. USA*, **73**, 1212–1216
8. Cereijido, M., Robbins, E. S., Dolan, W. J., Rotunno, C. A. and Sabatini, D. D. (1978). Polarized monolayers formed by epithelial cells on a permeable and translucent support. *J. Cell Biol.* **77**, 853–880
9. Rabito, C. A., Tchao, R., Valentich, J. and Leighton, J. (1978). Distribution and characteristics of the occluding junctions in a monolayer of a cell line (MDCK) derived from canine kidney. *J. Memb. Biol.*, **43**, 351–365
10. Perkins, F. M. and Handler, J. S. (1981). Transport properties of toad kidney epithelia in culture. *Am. J. Physiol.*, **241**, C154–C159
11. Cuthbert, A. W., George A. M. and MacVinish, L. (1985) Kinin effects on electrogenic ion transport in primary cultures of pig renal papillary collecting tubule cells. *Am. J. Physiol.*, **249**, F439–F447
12. Bisbee, C. A., Machen, T. E. and Bern, H. A. (1979). Mouse mammary epithelial cells on floating collagen gels: Transepithelial ion transport and effects of prolactin. *Proc. Natl. Acad. Sci. USA*, **76**, 536–540

13. Hall, H. G., Farson, D. A. and Bissell, M. J. (1982). Lumen formation by epithelial cell cultures in response to collagen overlay. A morphogenic model in culture. *Proc. Natl. Acad. Sci. USA*, **79**, 4672–4676

14. Kirkland, S. C. (1985). Dome formation by a human colonic adenocarcinoma cell line (HCA-7). *Cancer Res.*, **45**, 3790–3795

15. Kirkland, S. C. (1986). Endocrine differentiation by a human rectal adenocarcinoma cell line (HRA-19) *Differentiation*, **33**, 148–155

16. Kirkland, S. C. (1988). Clonal origin of columnar, mucous and endocrine cell lineages in human colorectal epithelia. *Cancer*, **61**, 1359–1363

17. Murakami, H. and Masui (1980). Hormonal control of human colon carcinoma cell growth in serum-free medium. *Proc. Natl. Acad. Sci. USA*, **77**, 3454–3458

18. Augeron, C. and Laboisse, C. L. (1984). Emergence of permanently differentiated cell clones in a human colonic cancer cell line in culture after treatment with sodium butyrate. *Cancer Res.*, **44**, 3961–3969

19. Whitehead, R. H., Jones, J. K., Gabriel, A. and Lukies, R. E. (1987). A new colon carcinoma cell line (LIM 1863) that grows as organoids with spontaneous differentiation into crypt-like structures *in vitro*. *Cancer Res.*, **47**, 2683–2689

20. Grasset, E., Pinto, M., Dessaulx, E., Zweibaum, A. and Desjeux, J-F. (1984). Epithelial properties of human colonic carcinoma cell line Caco-2: electrical parameters. *Am. J. Physiol.*, **247**, C260–C267

21. Grasset, E., Bernabeu, J. and Pinto, M. (1985) Epithelial properties of human colon carcinoma cell line Caco-2: effect of secretagogues. *Am. J. Physiol.*, **248**, C410–C418

22. Bentley, P. J. and Yorio, T. (1977). The permeability of the skin of neotenous urodele amphibian, the mudpuppy Necturus maculosus. *J. Physiol.*, **265**, 537–548

23. Dharmsathaphorn, K., Mandel, K. G., Masui, H. and McRoberts, J. A. (1985) Vasoactive intestinal polypeptide-induced chloride secretion by a colonic epithelial cell line. *J. Clin. Invest.*, **75**, 467–471

24. Weymer, A., Huott, P., Liu, W., McRoberts, J. A. and Dharmsathaphorn, K. (1985). Chloride secretory mechanism induced by prostaglandin E$_1$ in a colonic epithelial cell line. *J. Clin. Invest.*, **76**, 1828–1836

25. Cartwright, C. A., McRoberts, J. A., Mandel, K. G. and Dharmsathaphorn, K. (1985). Synergistic action of cyclic adenosine monophosphate and calcium mediated chloride secretion in a colonic epithelial cell line. *J. Clin. Invest.*, **76**, 1837–1842

26. Dharmsathaphorn, K. and Pandol, S. J. (1986). Mechanism of chloride secretion induced by carbachol in a colonic epithelial cell line. *J. Clin. Invest.*, **77**, 348–354.

27. Cuthbert, A. W., Egléme, C., Greenwood, H., Hickman, M. E., Kirkland, S. C. and MacVinish, L. J. (1987). Calcium and cyclic AMP dependent chloride secretion in human colonic epithelia. *Br. J. Pharmacol.*, **91** 503–515

28. Cuthbert, A. W. and Margolius, H. S. (1982). Kinins stimulate net chloride secretion by the rat colon. *Br. J. Pharmacol.*, **75**, 587–598

29. Wasserman, S. I., Barrett, K., Huott, P. A., Beuerlein, G., Kagnoff, M. F. and Dharmsathaphorn, K. (1988). Immune-related intestinal Cl$^-$ secretion. I. Effect of histamine on the T84 cell line. *Am. J. Physiol.*, **254**, C53–C62

30. Mandel, K. G., McRoberts, J. A., Beuerlein, G., Foster, E. S. and Dharmsathaphorn, K. (1986). Ba^{2+} inhibition of VIP and A23187-stimulated Cl$^-$ secretion by T84 cell monolayers. *Am. J. Physiol.*, **250**, C486–C494

31. Mandel, K. G., Dharmsathaphorn, K. and McRoberts, J. A. (1986). Characterisation of a cyclic AMP-activated Cl$^-$ transport pathway in the apical membrane of a human colonic epithelial cell line. *J. Biol. Chem.*, **261**, 704–712

32. Hayslett, J. P., Gogelein, H., Kunzelmann, K. and Greger, R. (1987). Characteristics of apical chloride channels in human colon cells (HT29). *Pflügers Arch.*, **410**, 487–494

33. Halm, D. R., Rechkemmer, G. R., Schoumacher, R. A. and Frizzell, R. A. (1988). Apical membrane chloride channels in a colonic cell line activated by secretory agonists. *Am. J. Physiol.*, **254**, C505–C511

34. Dreinhöfer. J., Gögelein, H. and Greger, R. (1988). Blocking kinetics of Cl$^-$ channels in colonic carcinoma cells (HT29) as revealed by 5-nitro-2- (3-phenylpropylamino)-benzoic acid (NPPB). *Biochim. Biophys. Acta*, **956**, 135–142

35. Hanley, M. R., Jackson, T. R. Vallego, T. J., Patterson, S. I., Thastrup, O., Lightman, S.,

Rogers, J., Henderson, G. and Pini, A. (1988). Neural function: metabolism and actions of inositol metabolites in mammalian brain. *Phil. Trans. R. Soc. London*, B **320**, 237–238

36. Frizzell, R. A., Rechkemmer, G. and Shoemaker, R. L. (1986). Altered regulation of airway epithelial cell chloride channels in cystic fibrosis. *Science*, **253**, 558–560

37. Welsh, M. J. and Liedtke, C. M. (1986). Chloride and potassium channels in cystic fibrosis airway epithelia. *Nature (London)*, **322**, 467–470

38. Baird, A. W., Cuthbert, A. W. and MacVinish, L. J. (1987). Type 1 hypersensitivity reactions in reconstructed tissues using syngeneic cell types. *Br. J. Pharmacol.*, **91**, 857–869

39. Hughes, T. E., Ordovas, J. M. and Schaefer, E. J. (1988). Regulation of intestinal apolipoprotein B synthsis and secretion by Caco-2 cells. *J. Biol. Chem*, **263**, 3425–3431

40. Paris, H., Taouis, M. and Galitzky, J. (1987). In vitro study of α_2-adrenoceptor turnover and metabolism using the adenocarcinoma cell line HT29. *Mol. Pharm.*, **32**, 646–654

41. Omary, M. B. and Kagnoff, M. F. (1987). Identification of nuclear receptors for VIP on a colonic adenocarcinoma cell line. *Science*, **238**, 1587–1581

42. Pinto, M. S., Robine-Leon, S., Appay, M. D., Keddinger, M., Triadou, N., Dussaulx, E., Lacroix, B., Simon-Assmann, P., Haffen, K., Fogh, J. and Zweibaum, A. (1983). Enterocyte-like differentiation and polarisation of the human colon-carcinoma cell line Caco-2 in culture. *Biol. Cell*, **47**, 323–330

43. Dharmsathaphorn, K., McRoberts, J. A., Mandel, K. G., Tisdale, L. D. and Masui, H. (1984). A human colonic tumour cell line that maintains vectoral electrolyte transport. *Am. J. Physiol.*, **246**, G204–G208

44. Cuthbert, A. W., Kirkland, S. C. and MacVinish, L. J. (1985). Kinin effects on ion transport in monolayers of HCA-7 cells, a line from a human adenocarcinoma. *Br. J. Pharmacol.*, **86**, 3–5

Cambridge, January 1989

Section II
PANCREAS

Introduction

The pancreatic section of the book contains two chapters, the first by Barry Argent and Michael Gray and the second by Ann Harris.

In the first chapter, a technique for isolating ducts from the copper-deficient rat pancreas is described. Copper deficiency causes a non-inflammatory atrophy of pancreatic acinar cells (80% by volume of the gland) but leaves the ducts (2% by volume) structurally and functionally intact. As a starting point for duct isolation, this preparation has two advantages over the copper-replete rat pancreas. First, the proportion of duct cells which can be isolated is markedly increased and secondly, the content of potentially harmful digestive enzymes is significantly reduced. The isolated ducts are well preserved morphologically and exhibit biochemical characteristics that are typical of ducts within the intact copper-replete rat pancreas. Furthermore, the isolated ducts can be maintained in tissue culture where their ends seal up, allowing micropuncture techniques to be used to study fluid secretion. In addition to this work, conventional microelectrode and patch-clamp techniques have been applied to the isolated ducts in order to elucidate the cellular mechanism of bicarbonate secretion. The authors present the latest model.

Dr Harris' chapter deals with the culture and characterization of epithelial cells derived from human fetal pancreas. Two main epithelial cell types grow out from the primary explants: small tightly-packed cells and larger streaming cells. The smaller cell type appears first and is considered to be a precursor population that differentiates into the larger cell type after a variable period in culture. Immunocytochemical studies performed with antibodies to mucins and to a range of cytokeratins have indicated that the cells in culture are most probably of ductal origin. Full-length messenger RNA has been isolated from these cells, an achievement which would have been virtually impossible with intact fetal pancreas where degradative enzymes released from the acini postmortem and upon homogenization would have cleaved the RNA. The isolated mRNA has been used to construct cDNA libraries. Attempts have been made to establish immortalized cell lines from the primary cultures and these are described.

4
Pancreatic Ducts: Isolation, Culture and Bicarbonate Transport

B. E. ARGENT AND M. A. GRAY

INTRODUCTION

The pancreatic ductal tree is a network of branching tubules whose primary fuction is to conduct digestive enzymes into the duodenum. Strictly speaking, it begins with the centroacinar cells which line the middle of each enzyme-secreting acinus, and which connect to the smallest element of the true ductal system, the intercalated ducts. In most species, these intercalated ducts open into intralobular ducts, which run within the pancreatic lobules, and then eventually into larger interlobular ducts[1]. However, in the rat, intercalated ducts open directly into interlobular ducts, and morphologically distinct intralobular ducts are absent[2] (Figure 4.1A and B). The final division of the ductal tree is usually the main pancreatic duct; however, in the rat, mouse and guinea-pig, a variable number of large interlobular ducts open into the bile duct forming a common bile–pancreatic duct[1].

In addition to acting as a simple conduit, the ductal tree secretes mucins[3], and is also an ion-transporting epithelium[4]. Little is known about pancreatic mucin biosynthesis and secretion[3]; however, ductal ion transport is relatively well understood, at least at the whole gland level[4]. In response to stimulation with the hormone, secretin, the ducts produce a bicarbonate-rich isotonic fluid which flushes enzymes towards the gut, and is also partly responsible for neutralizing acid chyme which enters the duodenum from the stomach. However, not all secreted bicarbonate reaches the duodenum; some is exchanged for blood chloride in a flow-rate-dependent manner. Thus, at high flow rates, when there is less time for exchange, the juice bicarbonate concentration is high, whereas the reverse is true at low flow rates[5] (Figure 4.2). Although a description of this anion exchange process features in all basic textbooks of physiology, its function remains a mystery. We think it may play a role in protecting the pancreas against autodigestion. One can easily imagine that during inter-digestive periods (when the pancreas is unstimulated), the ductal tree will hold a static column of fluid containing a potentially harmful mixture of digestive pro-enzymes.

69

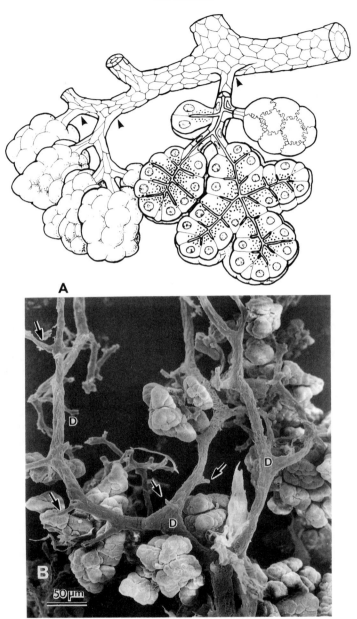

Figure 4.1 **A** Three dimensional representation of the rat exocrine pancreas. Small intercalated ducts (arrowheads) branch from a larger interlobular duct. In the rat, there are no distinctive intralobular ducts. Enzyme secreting acini are located at the end of the intercalated ducts. **B** Scanning electron-microscope overview of rat pancreas after removal of most acini by ultrasonic vibration. Interlobular ducts (D) branch to form intercalated ducts (arrowheads), which eventually terminate in acini. (From reference 2, with permission)

Obviously, if these pro-enzymes became active, the consequences for the ductal epithelium would be serious. Ductal anion exchange might protect against this eventuality by lowering luminal bicarbonate, which, in turn, would move liminal pH in the acid direction, i.e. away from the pH optimum of the enzymes. At high flow rates, this mechanism would not be required, since activated enzymes would be rapidly flushed from the gland.

Why should we be interested in studying the ductal epithelium? Well, apart from its ability to secrete large amounts of bicarbonate, defects in ductal function may underlie the pancreatic pathology that occurs in cystic fibrosis[6-8], and perhaps acute pancreatitis[9]. Furthermore, 90% of pancreatic carcinomas, which now account for one quarter of all cancer deaths in the USA[10], are ductal in origin[11]. Unfortunately, studying duct cell function *in situ* is not easy. This is because the duct cells comprise only a small proportion of the pancreas: about 14% by volume in man, 4% in guinea-pig, and only 2% in the rat[12]. Thus, metabolic and biochemical observations on whole glands must largely reflect the properties of acinar cells, which form 74–85% of the tissue[12]. Furthermore, the smaller ducts, which are probably the major sites of bicarbonate secretion[4], are not easily accessible, making biophysical studies of the type required to elucidate ion transporting mechanisms virtually impossible, Obviously, the solution to all these problems is to isolate either ductal epithelial cells[13-16] or intact pancreatic ducts[16-31]. For most transport studies, the latter approach is preferred since the structural and functional characteristics of the ductal epithelium are retained. At the time of writing, reconstitution of a functional pancreatic ductal epithelium in monolayer culture has not been reported, although studies directed at this goal are in progress[12,32] (see Chapter 5 by A. Harris).

Several comprehensive reviews have recently appeared dealing with the cell biology[12], monolayer culture[12,32] (see Chapter 5 by A. Harris), permeability properties of the main duct[9] and ion transport functions of the ductal epithelium[1,4,33-36] Here, we focus on the techniques available for the isolation and culture of small pancreatic ducts and show how these new preparations are beginning to advance our understanding of the mechanisms by which the pancreas transports bicarbonate.

ISOLATION OF SMALL PANCREATIC DUCTS

Table 4.1 summarizes the techniques available for the isolation of small interlobular and intralobular pancreatic ducts which are probably the major sites of bicarbonate secretion[4]. Essentially, the methods fall into two categories:

(1) Dissociation of the gland with enzymes and mechanical shearing, followed by isolation of ducts by either manual selection, centrifugation or microdissection.
(2) Microdissection without prior tissue dissociation.

Unfortunately, the yield of ducts obtained by microdissection alone is low,

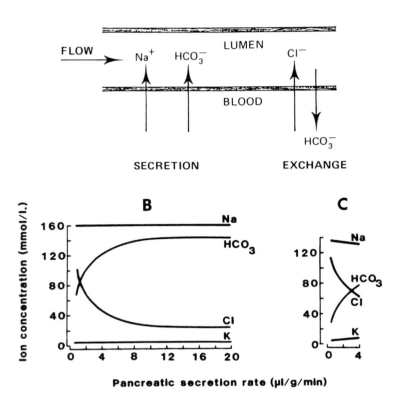

Figure 4.2 **A** Ion transport functions of pancreatic ducts. The major site of bicarbonate secretion (sometimes called the primary secretion) is probably the smaller interlobulr ducts, whereas flow-rate-dependent Cl^-/HCO_3^- exchange probably occurs throughout the ductal tree. **B and C** Electrolyte composition of pancreatic juice collected from the anaesthetized cat (B) and rat (C). The glands were stimulated to secrete at different rates by continuous infusion of different doses of secretin. See text for explanation of these plots. The bicarbonate concentration in pancreatic juice varies markedly with species. Maximum values are 140–150 mmol L^{-1} in cat, dog, pig, guinea-pig, Syrian golden hamsters and humans; 120 mmol L^{-1} in the rabbit; and 80 mmol L^{-1} in the rat[4]. Whether this difference reflects a species variation in the composition of the primary secretion, or a difference in the rate of Cl^-/HCO_3^- exchange, is unknown. In all species, the concentrations of sodium and potassium in pancreatic juice are about equal to those in plasma and do not vary much with flow rate[4]. For all species studies so far, pancreatic juice $[Na^+] + [K^+] = [Cl^-] + [HCO_3^-]$ to within a few mmol/L. (**B** and **C** from reference 5, with permission)

and may be a limiting factor for biochemical studies. Prior dissociation of the gland with enzymes gives a much higher yield, although it is our experience and the experience of others[19], that this approach produces ducts in which the epithelial cells are morphologically poorly preserved. We suspect the damage results largely from the action of digestive enzymes, particularly lipases, which are difficult to inhibit pharmacologically. These are released from acinar cells during the prolonged dissociation periods employed by most workers. Furthermore, ducts isolated in this way always contain adherent acinar tissue.[19-22].

In our laboratory, we have successfully combined a high yield of ducts with excellent morphological preservation of the ductal epithelium, by starting with glands taken from copper–deficient rats[23,24]. Feeding young rats (approx. 125 g) a copper-deficient diet has an intriguing, and as yet unexplained, effect on the pancreas, causing a non-inflammatory atrophy of the acinar cells but leaving the ducts structurally and functionally intact[37-39]. As a starting point for duct isolation, this preparation has two advantages. First, the proportion of duct cells in the gland is markedly increased, and secondly, the content of potentially harmful digestive enzymes is markedly reduced. The possibility that copper deficiency affects the ductal epithelium has been ruled out by extensive morphological, biochemical and secretory studies (see below), all of which show that these isolated ducts possess functional characteristics similar to those of ducts *in situ* within the pancreas of copper-replete animals.

A collection of ten freshly isolated interlobular ducts is shown in Figure 4.3A. Their external diameters vary between approximately 50–100 μm and their length between about 300–900 μm. Twenty to fifty ducts of this size can be isolated from each gland, and occasionally ducts up to 2 mm in length are obtained. Note the complete absence of adherent acinar tissue.

Figure 4.3B shows a high-magnification view of an interlobular duct. The epithelium rests on a thick layer of connective tissue which can be dissected away to expose the basolateral surface of the epithelial cells (Figure 4.3C, D and E). In Figure 4.3E, the typical 'cobblestone' outline of individual epithelial cells at their apical surfaces is clearly visible.

From an ultrastructural point of view, isolated interlobular ducts are well preserved[24]. Their epithelium consists of a single layer of either columnar or cuboidal epithelial cells which vary from 8–12 μm in height and 1.5–4.5 μm in width, and which have a characteristic lobed nucleus positioned towards their basal pole[24]. In addition to the epithelial cells, endocrine and caveolated cells are occasionally observed within the epithelium of the interlobular ducts[24].

Microdissection of the smaller intercalated ducts (there are no distinctive intralobular ducts in the rat pancreas[2]; see Figure 4.1A) is technically difficult, but, conveniently, they are often obtained as branches of the interlobular ducts (Figure 4.3D, 5A and B)[24]. Epithelial cells within intercalated ducts are 5–7 μm in height and appear pear shaped in transverse section with the lobed nucleus occupying the major portion of the cytoplasm[24].

In addition to morphology, simple biochemical criteria, such as O_2 consumption[24], adenine nucleotide levels[24], and incorporation of amino acids into proteins[18] have been employed to show that isolated duct preparations are viable (Figure 4.4). One important observation is that isolated ducts increase their cyclic AMP content when stimulated by the hormones, secretin (Figure 4.4) and VIP, whereas pancreozymin, glucagon,

Table 4.1 Summary of techniques employed to isolate intct pancreatic ducts.

Duct isolation technique	Species	Cultured [reference]	Type of studies performed [reference]
Gland dissociation with collagenase and chymotrypsin, then fractionation on albumin gradients or sieves, followed by manual selection of ducts	Rat	Yes [20,21,25] No [19,20,31]	Morphology [19–21,31] Marker enzymes [19,21,26,25] Polypeptide composition [25]
	Hamster	Yes [21,22]	Morphology [21,22] Marker enzymes [21]
Gland dissociation with collagenase and mechanical shearing, then isolation of ducts by manual selection	Rat	No [30]	Intracellular pH [30]
	Guinea-pig	Yes [27]	Morphology [27]
Gland dissociation with collagenase, hyaluronidase and mechanical shearing, then isolation of duct fragments by centrifugation	Copper-deficient rat	No [18]	Effect of secretin, VIP, pancreozymin, glucagon, BBP, and CCh on cyclic AMP levels [18]
Gland dissociation with collagenase and hyaluronidase, then microdissection of ducts	Copper-deficient rat	No [24] Yes [16,23]	Morphological [23,24] O$_2$ consumption [24] ATP, ADP and AMP levels [24] Effect of secretin on cyclic AMP levels [23,24] Secretory (micropuncture) [23] Electrophysiological (patch clamp) [16]
Microdissection	Cat	No [17]	Marker enzymes [17]
	Rat	No [28,29]	Electrophysiological (microelectrode) [28,29]

Only studies involving small interlobular and intralobular ducts, which are probably the major sites of bicarbonate secretion, have been included. For studies dealing with the isolation and explant culture of the main pancreatic duct see reference 12.

Abbreviations: BPP, bovine pancreatic polypeptide; CCh, carbamylcholine; AMP, adenosine monophosphate; ADP, adenosine diphosphate; ATP, adenosine triphosphate; cyclic AMP, cyclic 3′, 5′-adenosine monophosphate

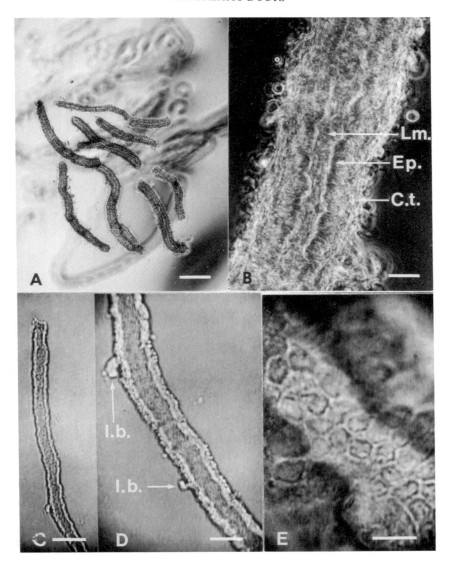

Figure 4.3 Small interlobular ducts isolated from the rat pancreas. **A** Collection of ten ducts. Bar, 250 μm. **B** Higher magnification of an interlobular duct. Lumen (Lm.), epithelium (Ep.) and connective tissue (C.t.) are visible. Bar, 25 μm. **C** Interlobular duct after removal of the connective tissue. This low-magnification view shows that the duct was about 1 mm in length. Bar, 100 μm. **D** Small projections (I.b.) on the left side of the duct are intercalated branches that have fractured close to their site of origin as the connective tissue was stripped off. Bar, 50 μm. **E** Cobblestone appearance of the apical region of the epithelial cells. Bar, 10 μm. (From reference 24, with permission)

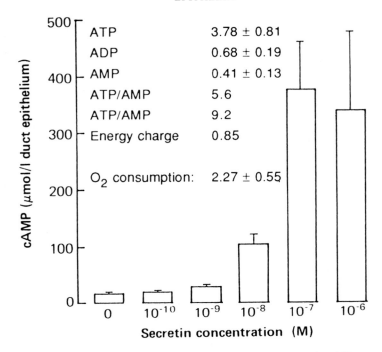

Figure 4.4 Dose–response curve for the effect of secretin on the cyclic AMP content of rat interlobular ducts. Each column is the mean ± SE of 3 or 4 observations made on different preparations. A statistically significant ($p < 0.05$) rise in cyclic AMP content occurred with 1 nmol L[-1] secretin, while the hormone dose required for a half-maximal effect was about 20 nmol L[-1]. Inset shows adenine nucleotide levels (mmol (L duct epithelium)[-1]), and O_2 consumption (mlO_2. min[-1]. (100 g duct epithelium)[-1]) in isolated rat ducts. Energy charge was calculated as $[ATP] + 0.5 [ADP]/[ATP] + [ADP] + [AMP]$, and a value of 0.85 is indicative of a viable healthy cell population. (From reference 24, with permission)

bovine polypeptide and carbamycholine have no effect[18,24]. These important findings that the isolated ducts retain secretin and VIP receptors, and that these receptors are functionally coupled to adenylate cyclase. Secretin is probably the most important physiological regulator of pancreatic bicarbonate secretion[1,4], and there is very good evidence from intact gland studies that this peptide uses cyclic AMP as an intracellular messenger[1,4].

CULTURE OF ISOLATED DUCTS

Githens and his collaborators first described techniques for the culture of ducts isolated from the rat and hamster pancreas in agarose or collagen gels for up to 20 weeks[12,20–22,25]. In our laboratory, we find it more convenient to culture ducts for short periods (2–3 days) on polycarbonate filter rafts which float on the growth medium[23]. This has the advantage that individual ducts can easily be removed from the filters using micropipettes. Maintaining the ducts in culture results in a marked dilatation of their lumens, a flattening of the epithelium against the surrounding connective tissue layer, and an overall swelling of the ducts[12,20–23,25] (Figure 4.5). These morphological changes are caused by the ends of the ducts sealing during the early stages of culture, followed by fluid secretion into the closed luminal space. The speed with which this occurs probably reflects the morphological and biochemical preservation of the freshly isolated ducts, and takes 2–4 days for ducts isolated from the normal rat and hamster pancreas[20–22,25], but occurs within 8 h for ducts isolated from the copper-deficient rat gland[23]. If swollen ducts are punctured, their lumen collapses and the epithelium returns to its normal dimensions (Figure 4.5). Whether the cut ends of the ducts simply stick together or whether sealing is associated with cell growth is unknown. However, epithelial cells in cultured ducts certainly retain a proliferative capability as shown by the incorporation of labelled thymidine into DNA[12,25,31].

Apart from a reduction in their height, a fall in the number of intracellular fat droplets and a widening of intercellular spaces, epithelial cells within the cultured ducts retain all of the ultrastructural features of those in freshly isolated preparations[23]. Cultured interlobular ducts from rat and hamster glands also retain carbonic anhydrase activity[25], which is a biochemical marker for the ductal epithelium[40], and also accumulate mucins in their lumens[25]. Furthermore, these cultured ducts increase their cyclic AMP content when stimulated with secretin[23], which must indicate that they retain receptors for this hormone.

BIOCHEMICAL STUDIES ON ISOLATED DUCTS

Excluding those aimed at assessing viability (see above), biochemical studies on isolated pancreatic ducts have hardly begun. Most of the work published so far reflects attempts to identify 'marker' enzymes, or duct-specific peptides, which could be used for the biochemical identification of ductal tissue[12,17,19–21,25]. It is probably fair to say that little new information has been gained from these studies that was not previously known from work on the intact gland. However, it is worth emphasizing that freshly isolated and cultured ducts retain carbonic anhydrase, Na^+/K^+-ATPase, and hormone-stimulated adenylate cyclase, three enzymes which play a central role in pancreatic bicarbonate secretion[4].

SECRETORY STUDIES ON ISOLATED DUCTS

The availability of isolated pancreatic ducts offers a unique opportunity to sample ductal secretions which are uncontaminated by acinar fluids. Previously, this has only been possible by applying renal micropuncture

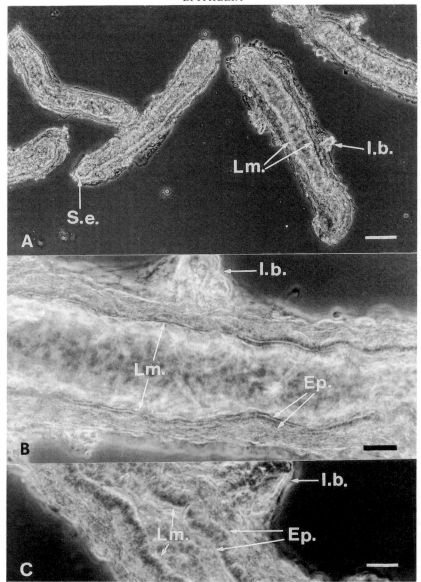

Figure 4.5 Light micrographs of interlobular ducts that had been maintained in culture for 24 h. Phase contrast. **A** Low-magnification view of a group of five ducts. Note the sealed ends of the ducts (S.e.) and the dilated lumens (Lm.). On one duct, the remnant of an intercalated branch (I.b.) is visible. Bar, 100 μm. **B** Higher magnification view of the duct in **A** with the intercalated branch (I.b.). The dilated lumen (Lm.) and the flattened epithelial layer (Ep.) are clearly visible. Bar, 25 μm. **C** The same duct immediately after being punctured. Photographed in the region of the intercalated branch (I.b.) and at the same magnification as **B**. Note the decrease in lumen width (Lm.) and increased epithelium height (Ep.). Bar, 25 μm. (From reference 23, with permission)

technology to the intact gland[1,4] which is technically very difficult due to the scarcity and inaccessibility of the smaller ducts. The ease with which isolated ducts can be micropunctured offers the possibility of establishing un-equivocally which hormones and neurotransmitters stimulate the duct cells, how these agents interact, and the true composition of ductal fluid.

Figure 4.6 shows the sequence of events during micropuncture of a cultured interlobular duct.[23]. The basal rate of fluid secretion from ducts that had been maintained in culture for 16–50 h was 0.16 ± 0.03 nl h^{-1} nl^{-1} duct epithelium ($n=12$), and this was increased fourteen-fold by 10^{-8} mol L^{-1} secretin. The full dose-response curve for secretin is illustrated in Figure 4.7, and shows that fluid transport was significantly increased by 10^{-11} mol L^{-1} secretin, and that the dose required for half-maximal secretion was about 2.0×10^{-11} mol L^{-1}. For comparison, the dose–response curve for secretin-stimulated fluid secretion from the perfused rat pancreas is also plotted. Clearly, the response of isolated ducts is very similar to the response of duct cells within the intact gland. As predicted from studies on the intact gland[4], secretin-stimulated fluid secretion is abolished if bicarbonate ions are removed from the fluid bathing isolated ducts (Figure 4.7). Figure 4.7 also shows that dibutyryl cyclic AMP increased the basal rate of fluid secretion about 6-fold; however, the decapeptide, caerulein, which stimulates enzyme and fluid secretion from the acinar cells[4], has no effect. Finally, it should be noted that the secretin dose–response curve for fluid secretion lies far to the left of that for cyclic AMP accumulation (compare Figures 4.4 and 4.7). This may indicate that only very small changes in intracellular cyclic AMP are required to maximally activate ion transport, or that another intracellular messenger system is involved (see below).

Overall, these results show that, in terms of secretory responses, isolated pancreatic ducts behave exactly as predicted from studies on intact glands. However, we were surprised to find that the chloride concentration in fluid collected from maximally stimulated ducts was about 130 mmol L^{-1}, and that the bicarbonate concentration[23] was no greater than 25 mmol L^{-1}. This was not what we had expected; pancreatic juice collected from a maximally stimulated rat pancreas contains about 60 mmol L^{-1} chloride and about 80 mmol L^{-1} bicarbonate[4]. To understand the reason for this apparent discrepancy, it must be recalled that there are two anion transport processes operating in pancreatic ducts: (1) the secretion of a bicarbonate-rich fluid, and (2) the exchange of secreted bicarbonte for blood chloride in a flow-rate-dependent manner (Figure 4.2). When interlobular ducts are isolated, it is not unreasonable to suppose that their inherent secretory rate is much lower than the flow rate through them in the intact gland. This must occur because *in situ* they will act as conduits for secretions originating from the more distal regions of the ductal tree. Thus, there will be adequate time for chloride/bicarbonate exchange to occur in isolated ducts, which explains the high chloride concentration, and low bicarbonate concentration, in ductular fluid.

Secretory studies on isolated ducts are still in their infancy; however, this area is likely to be a rewarding one for future research in terms of understanding the mechanism and control of ductal secretions. In this

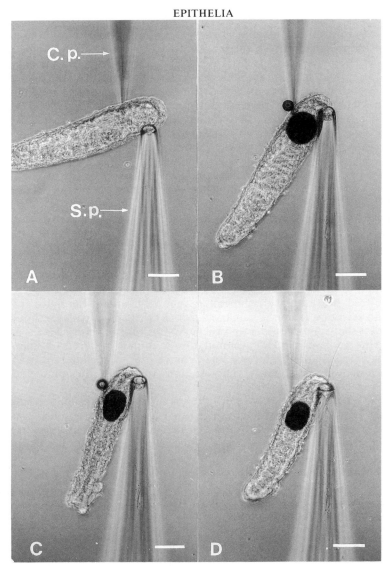

Figure 4.6 Measurement of fluid secretion from a cultured pancreatic duct using micropuncture techniques. **A** The duct is first immobilized by applying a suction pipette (S.p.) to its outer connective tissue layer, then micropunctured using a bevelled oil-filled collection pipette (C.p.). **B** Success is confirmed by the injection of a small volume of coloured oil into the lumen. **C** The duct is then deflated by aspirating the luminal fluid into the collection pipette. This pipette is then withdrawn from the lumen, the fluid ejected to waste in the tissue bath, and the duct immediately repunctured along the same entry track. Usually, the collection period is then started by application of a subatmospheric pressure to the pipette. **D** If suction is not applied to the collection pipette, secreted fluid accumulates within the closed lumen of the duct causing it to dilate. This photograph was taken 40 min after **C**, during which time the duct was perifused with Krebs–Ringer bicarbonate buffer (pH 7.4) at 37°C. Phase contrast. Bars, 200 μm. (From reference 23, with permission)

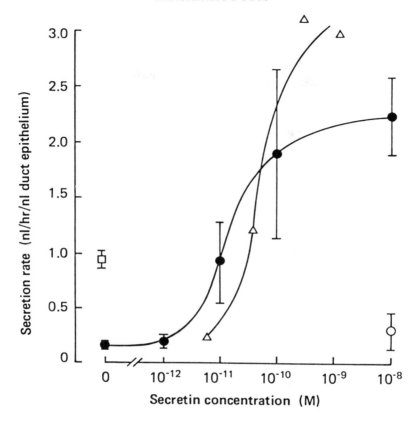

Figure 4.7 The effects of secretin, bicarbonate ions and dibutyryl cyclic AMP on fluid secretion from cultured interlobular ducts. Individual ducts were micropunctured as shown in Figure 4.6. At the end of the collection period, the dimensions of the epithelium were measured and the epithelial volume calculated. Secretion rates, plotted as mean ± SEM, are expressed as nl h^{-1} (nl duct epithelium)$^{-1}$. Dose–response curve for the effect of secretin on fluid secretion from isolated ducts (●–●). For comparison, the dose–response curve for secretin-stimulated fluid secretion from the perfused rat pancreas is also shown (△—△). Effect of replacing bicarbonate ions in the perifusion buffer with HEPES on secretin-stimulated fluid secretion (○). Effect of 2×10^{-4} mol L^{-1} dibutyryl cyclic AMP on fluid secretion (□) To obtain the data in this figure, only one fluid collection was made from each duct. However, ducts could be returned to the tissue culture incubator after an initial measurement, allowed to reflate (which usually takes 2–2.5 h) and then repunctured without any loss of secretin responsiveness. (Drawn from data in reference 23)

respect, we have recently found that bombesin is as effective as secretin in stimulating fluid secretion from isolated ducts, and that substance P has an inhibitory effect (N. Ashton, B. E. Argent and R. Green, unpublished observations). Since bombesin stimulates enzyme secretion from pancreatic acinar cells[41] by mobilizing intracellular Ca^{2+}, this finding may indicate the presence of a previously unsuspected Ca^{2+}-activated pathway for ductal anion transport.

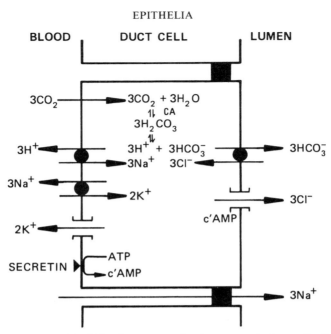

Figure 4.8 Cellular model of bicarbonate secretion by pancreatic duct cells. See text for details. (From reference 16, with permission)

BIOPHYSICAL STUDIES ON ISOLATED PANCREATIC DUCTS

Over the last 20 years, several models have been proposed to explain the cellular mechanisms by which pancreatic duct cells actively transport bicarbonate[1,4]. Unfortunately, these schemes were largely based on whole-gland studies and were impossible to evaluate without biophysical data on the duct cell. Recently, this goal has been achieved by applying conventional microelectrode and microperfusion, patch clamp, and fluorescent probe techniques to isolated pancreatic ducts[16,28-30,42]. So far, this approach has:

(1) Shown that bicarbonate secretion is electrogenic, i.e. causes an alteration in duct cell membrane potential;

(2) Localized potassium and chloride conductances to the basolateral and apical membranes of the duct cell, respectively, and identified the ion channels responsible;

(3) Identified at least one site, the apical chloride channel, at which cyclic AMP acts to regulate bicarbonate secretion; and

(4) Identified basolateral Na^+/H^+ exchangers, and apical Cl^-/HCO_3^- exchangers on the duct cell. Previously, their presence had been inferred from inhibitor studies on intact glands[4].

The current cellular model for duct cell bicarbonate secretion is shown in Figure 4.8. The experimental evidence for each of the transport elements is summarized in Table 4.2, and Table 4.3 details the biophysical properties of this epithelium. At present, the model is largely based on the spatial

Table 4.2 Evidence for the spatial distribution of transport elements on pancreatic duct cells

Transport element		Evidence	Reference
Basolateral membrane			
Na$^+$/K$^+$ ATPase (electrogenic)	(1)	Cytochemical/immunochemical localization	See ref. 4
	(2)	Basolateral ouabain depolarizes V_{bl}	28
K$^+$ channel	(1)	Increased extracellular [K$^+$] depolarizes V_{bl},	28
	(2)	Ba^{2+}, and to a lesser extent TEA, depolarize V_{bl}	28
	(3)	K$^+$ channel characterized using patch clamp technique	44
Na$^+$/H$^+$ exchanger	(1)	Basolateral amiloride depolarizes V_{bl} (intracellular acidification decreases basolateral K$^+$ conductance)	28
	(2)	Reducing extracellular Na$^+$ causes intracellular acidification (blocked by amiloride)	30*
Apical membrane			
Secretin- and cyclic AMP-regulated Cl$^-$ conductance and Cl$^-$ channel	(1)	Luminal application of Cl$^-$ channel blockers (NPPB, DCl-DPC) hyperpolarizes V_{bl} and increases fractional resistance of the apical membrane	29
	(2)	Secretin and dibutyryl cyclic AMP depolarize V_{bl} (predicted result for an increase in apical Cl$^-$ conductance), reduce input resistance, and reduce fractional resistance of the apical membrane	29,42
	(3)	Cl$^-$ channel characterized using patch clamp technique	16,47
Cl$^-$/HCO$_3$ exchanger	(1)	Luminal application of SITS hyperpolarizes V_{bl} (lowers intracellular Cl$^-$ and decreases Cl$^-$ efflux through the channel)	29
	(2)	Reducing extracellular Cl$^-$ causes intracellular alkalinization (blocked by SITS)	30*

*Study did not distinguish between apical and basolateral location.
Abbreviations: V_{bl}, basolateral membrane potential; NPPB, 5-nitro-2-(3-phenylpropylamino)-benzoic acid; DCl-DPC, 3′, 5-dichlorodiphenylamine-2-carboxylic acid; SITS, 4-acetamido-4′-isothiocyanatostilbene-2,2′-disulphonic acid; TEA, tetraethylammonuim

distribution of the transport elements and the way in which these are modulated by stimulants. Its confirmation will require the electrochemical gradients for chloride and bicarbonate to be measured, and additional information about the permeability properties of the paracellular pathway in smaller ducts.

The initial step in bicarbonate secretion is viewed as diffusion of CO_2 into the duct cell and its hydration by carbonic anhydrase (CA) to carbonic acid.

Table 4.3 Electrophysiological properties of small pancreatic ducts

Property	Value	Reference
Basolateral membrane potential of duct cells	−33 to −63 mV	28,42
Transepithelial potential	−2.6 mV	28
Transepithelial resistance	88 $\Omega \cdot cm^2$	28

This dissociates to form a proton and a bicarbonate ion, and the proton is translocated back across the basolateral membrane on the Na^+/H^+ exchanger. Effectively, this is the active transport step for bicarbonate, the energy being derived from the sodium gradient established by the Na^+/K^+ ATPase. Bicarbonate ions are then thought to exit across the apical membrane on the Cl^-/HCO_3^- exchanger. The rate at which this exchanger cycles will depend on the availability of luminal chloride, which in turn depends on the opening of the apical Cl^- channel. This channel is activated following secretin stimulation of the duct cell, causing a dose-dependent depolarization of the membrane potential (Figure 4.9), and is probably the main control point in the secretory mechanism. Since bicarbonate exit at the apical membrane is rheogenic, i.e. it generates a current, there must be equal current flow across the basolateral membrane during secretion. Two thirds of this current is accounted for by potassium efflux through the voltage-dependent K^+ channel, and one third by cycling of the electrogenic sodium pump. Finally, the negative transepithelial potential, generated by activation of the apical chloride conductance, draws sodium and a small amount of potassium into the lumen via a cation-selective paracellular pathway.

In our laboratory, we have been using the patch clamp technique to study the ion channels whose activity underlies the potassium and chloride conductances on the basolateral and apical membranes of the duct cell (Figure 4.8)[28,29], Patch clamping involves sealing a glass micropipette onto the plasma membrane, and then measuring the very small currents that flow when single ion channels open[43] (see chapters by Malcolm Hunter and Jeffrey J. Smith for further details). It has the advantage that patches can be obtained selectively on either the apical or basolateral membranes of the duct cell, and thus the spatial distribution of channels can be unequivocally determined. Although we have focused mainly on the regulated K^+ and Cl^- channels which play an important role in bicarbonate secretion, a number of other channels, whose functions remain unclear, have also been identified on the duct cell (Figure 4.10).

Basolateral potassium conductance

The role of this conductance is to recycle potassium accumulated by the Na^+/K^+ ATPase, and to provide a pathway for current flow across the basolateral membrane (Figure 4.8). It also offsets the depolarization that occurs when the apical membrane chloride conductance increases following stimulation, and thus helps maintain the driving force for electrogenic anion secretion (Figure 4.8 and 4.9). The best candidate for the channel responsible is a voltage-dependent Ca^{2+}-activated maxi-K^+ channel[44], which has many properties in common with those previously described in other glandular epithelia[45]. There are about twenty of these large-conductance K^+ channels on each duct cell, giving a channel density of about 1 per 19 μm^2 of plasma membrane.

What is the evidence that activity of this channel underlies the basolateral potassium conductance of the intact cell? From the pharmacological point of view, it is not conclusive. Both the potassium conductance[28] and the maxi-K^+ channel are blocked by barium; however, tetraethylammonium (TEA), which also blocks the channel, has only a small depolarizing effect on the duct cell membrane potential[28]. Moreover, we have recently found that whole-cell potassium currents can be completely blocked by barium, whereas, in the presence of TEA, a small residual current persists. Taken together with the fact that neither barium nor TEA has any effect on secretin-stimulated fluid secretion from the perfused rat pancreas[46], these data may indicate that other, as yet unidentified, K^+ channels are present in the basolateral membrane.

However, strong evidence for a role of the maxi-K^+ channel in bicarbonate transport comes from the effect of secretory stimulants on channel activity. One can predict from the model shown in Figure 4.8 that channel activity should increase following stimulation in order to balance the increased secretory current flowing across the apical membrane. This is, in fact, the case: the proportion of total time that maxi-K^+ channels on intact duct cells are open (their open-state probability) is markedly elevated by secretin, dibutyryl cyclic AMP and forskolin. As activity of this channel is voltage dependent[44], such an effect might result from the depolarization that occurs following exposure of the duct cell to these stimulants[29,42] (Figure 4.9). However, this is not the only explanation since whole cell potassium currents can be markedly activated by intracellular cyclic AMP, and, in excised patches, the maxi-K^+ channel can be activated by exposing the cytoplasmic face of the membrane to a solution containing ATP and the catalytic subunit of cyclic AMP-dependent protein kinase (M. A. Gray, J. R. Greenwell, A. J. Garton and B. E. Argent, unpublished observations). Under these experimental conditions, the membrane potential is clamped, so both results provide evidence for an additional activation pathway involving cyclic AMP-dependent protein phosphorylation.

Apical chloride conductance

Three lines of evidence suggest that a small (5 pS) Cl^- channel is responsible for the apical membrane chloride conductance.

(1) The channel is located only on the apical membrane and is active in cell-attached patches[16].

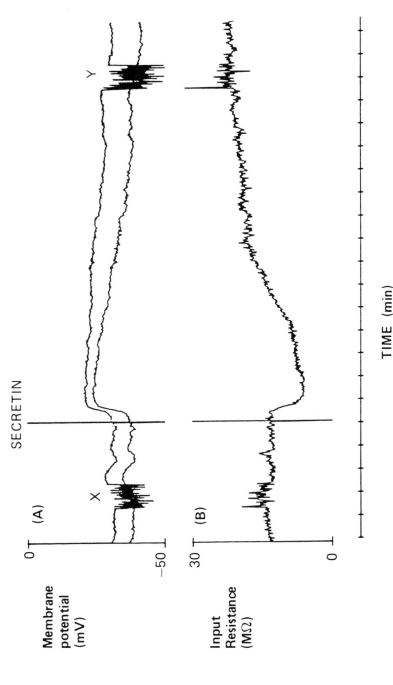

Figure 4.9 Effects of secretin on basolateral membrane potential, and input resistance, of a rat pancreatic duct cell. A cultured interlobular duct was held in a tissue bath by drawing each end into a glass micropipette. Since the lumen was not cannulated and drained in these experiments, accumulation of secreted fluid often caused an obvious dilatation of the ducts. The bath (volume 1 ml) was perfused with a Krebs–Ringer bicarbonate buffer (3 ml/min) at 37°C. A potassium acetate-filled microelectrode (resistance = 100 MΩ) was then used to impale a duct cell across its basolateral membrane. **A** The lower of the two traces shows the duct cell membrane potential. Every two seconds, a depolarizing square wave current pulse (0.5 nA, 100 ms) was applied to the microelectrode. The upper trace shows the maximum change in membrane potential associated with these current pulses. For the periods marked 'X' and 'Y', the magnitude of the injected current was varied in order to construct a current/voltage plot. At the vertical line marked 'secretin', a single dose of the hormone was injected into the bath in order to give an instantaneous concentration of 10^{-8} mol L^{-1}. **B** Input resistance calculated from the deflections in membrane potential caused by injected current

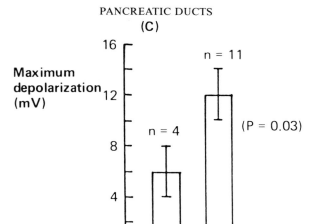

(C)

Maximum
depolarization
(mV)

SECRETIN (M)

Figure 4.9 C Dose dependency of the effect of secretin on basolateral membrane potential. Each column indicates the mean ± SEM of the maximal depolarization obtained with 10^{-9} and 10^{-8} mol L^{-1} secretin applied to the duct cell as described in section A. (From reference 42, with permission)

(2) Both channel activity and a chloride conductance can be detected on unstimulated cells[16,29].

(3) Secretin, cyclic AMP and forskolin all increase channel activity[16].

This last effect causes the stimulant-induced depolarization of the duct cell[29,42] (Figure 4.9), reduces the input resistance[42] (Figure 4.9) and also decreases the fractional resistance of the apical plasma membrane[29]. We originally identified this channel on rat duct cells[16], but it is also present in the human pancreas[47]. It provides luminal chloride for the Cl^-/HCO_3^- exchanger, and is probably the main control point in the secretory mechanism.

Under the most favourable experimental conditions, only 23% of all patches on the apical plasma membrane contain this small conductance channel. However, when present, it typically occurs in clusters which usually contain two or three, but occasionally up to seven, active channels (Figures 4.11A and 4.12A). The open-state probability is not markedly voltage dependent (Figures 4.11A and 4.12A), indicating that the channel will be active at potentials (−50 to −60 mV) measured across the apical membrane in microperfused ducts[29].

Proof that this channel selects for chloride is shown in Figure 4.11. When equal chloride concentrations are present on both sides of excised patches, the current/voltage (I/V) relationship is linear and the currents always reverse at a membrane potential of 0 mV, that is when there is no chemical or electrical driving force for chloride diffusion across the patch (Figure 4.11B). However, if a three-fold chloride concentration gradient is created

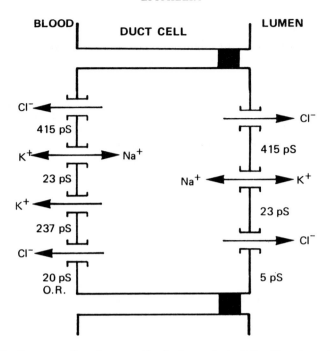

Figure 4.10 Summary of conductance, distribution and selectivity of ion channels on rat and human pancreatic duct cells as determined using the patch clamp technique. Conductance data are taken from experiments on excised patches. The channels are: 23 pS non-selective cation[44]; 237 pS maxi-K[44]; 415 pS voltage-dependent anion[50]; 20 pS Cl⁻, O.R. = outward rectifier, the slope conductance at a membrane potential of 0 mV is given[47]; 5 pS Cl⁻[16,47]. The 5 pS Cl⁻ channel on the apical membrane, and the 237 pS maxi-K⁺ channel on the basolateral membrane are involved in electrogenic bicarbonate secretion

across the membrane, there is a marked leftward shift on the I/V plot, and the channel currents now reverse at a membrane potential of −26 mV (Figure 4.11B). This is exactly the predicted shift (calculated from the Nernst equation) for a channel that selects for chloride over sodium and potassium. Using a similar approach, it is possible to estimate that the channel is virtually impermeable to bicarbonate ions, ruling out the possibility that significant amounts of bicarbonate enter the lumen directly via the channel.

As illustrated in Figure 4.12, short-term exposure of duct cells to secretin causes a marked increase in total current flow across cell-attached patches. This effect is accompanied by an increase in the number of channels that are open simultaneously, as can be seen from the current records themselves (Figure 4.12A), and from the associated current amplitude histograms (Figure 4.12B). We believe that this is caused by an increase in the open-state probability of individual ion channels (Figure 4.12C), rather than an increase in the total number of channels in the patch. This change in open-state probability is achieved by a large reduction in the time that the channel spends in the closed state (Figure 4.12E), coupled with a slight increase in the open time (Figure 4.12D). Said another way, each channel opens more frequently and stays open for slightly longer. All these effects of secretin were fully reversed upon withdrawal of the hormone, but this usually took between eight and ten minutes (Figure 4.12A–E). Similar results were obtained with forskolin (a drug which activates adenylate cyclase) and also with dibutyryl cyclic AMP. Since there is ample evidence that secretin uses cyclic AMP as an intracellular messenger[1,4] (see Figure 4.4), physiological regulation of the channel is probably achieved by protein phosphorylation mediated by cyclic AMP-dependent protein kinase.

Recently, the characteristics of Ca^{2+}- and cyclic AMP-activated chloride channels on the apical membranes of a number of chloride secreting epithelial cells have been described[48,49]. Most of these channels have much larger conductances than the secretin-regulated channel we have identified on pancreatic duct cells. Two other, larger conductance, chloride channels are present on the duct cell[47,50] (Figure 4.10), but neither appears to be hormonally regulated, and, at the moment, their functions are unknown. A small-conductance (1–2 pS) Ca^{2+}-activated anion channel has been identified on the chloride secreting lacrimal acinar cell by noise analysis of whole-cell currents[51]. However, this channel has not been further characterized or localized to the apical membrane using single-channel recording techniques.

PERSPECTIVES FOR FUTURE RESEARCH

Now that pure pancreatic ductal tissue is available in the form of intact ducts or monolayer cultures of epithelial cells (see Chapter 5 by A. Harris), it will be possible to study the cellular physiology/biochemistry of this epithelium directly. In this respect, it has been known for a long time that pancreatic bicarbonate secretion can be stimulated, or modulated, by a number of hormones and neurotransmitters other than secretin[4]. However, we have little information about receptor types present on the duct cell, so data from receptor binding assays, and micro-assays for hormone-stimulated adenylate cyclase[52], would help clarify this point. In terms of understanding the control of bicarbonate secretion, we also need data on intracellular calcium concentration and whether this is affected by secretory stimulants. While it is impossible to obtain this information using the intact gland, fluorescent probe technology can easily be applied to isolated pancreatic ducts[30]. We also know virtually nothing about the metabolism of pancreatic duct cells. Again, the microtechniques required for this type of investigation have already been developed and applied to single renal tubules[53].

Biophysical studies on the cellular mechanism of duct cell bicarbonate secretion are now advancing rapidly. Perhaps the most exciting aspect of this work is the realization that chloride ions play a central role in bicarbonate transport and that a cyclic AMP-regulated chloride channel is present on the apical membrane of both rat[16] and human[47] pancreatic duct

A

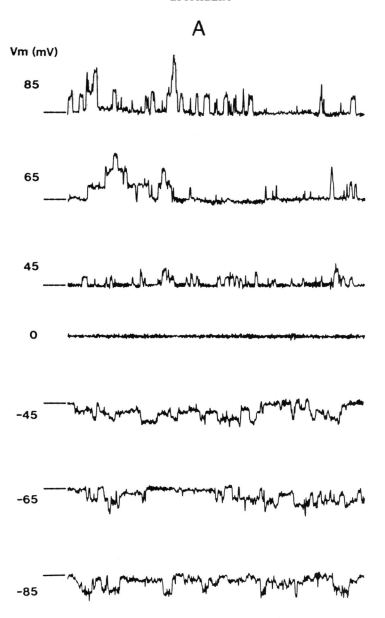

Figure 4.11 Small conductance Cl⁻ channels on the apical membrane of rat pancreatic duct cells. **A** Single-channel currents recorded from an excised inside-out patch. Both extracellular (pipette), and intracellular (bath), faces of the membrane were bathed in a solution containing 150 mmol L⁻¹ chloride. the membrane potential (Vm) is indicated adjacent to the records. An upward deflection from the closed state (horizontal line adjacent to traces) represents outward current

B

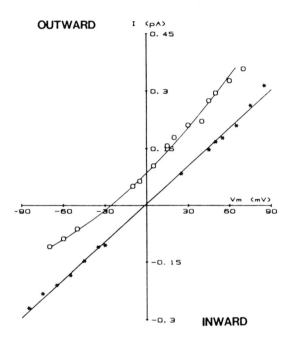

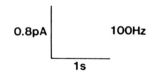

Figure 4.11 B Single-channel current/voltage plots. *Data from the experiment shown in **A**; ○, same experiment after a threefold chloride concentration gradient had been created across the patch by replacing bath chloride with sulphate. (From reference 16, with permission)

91

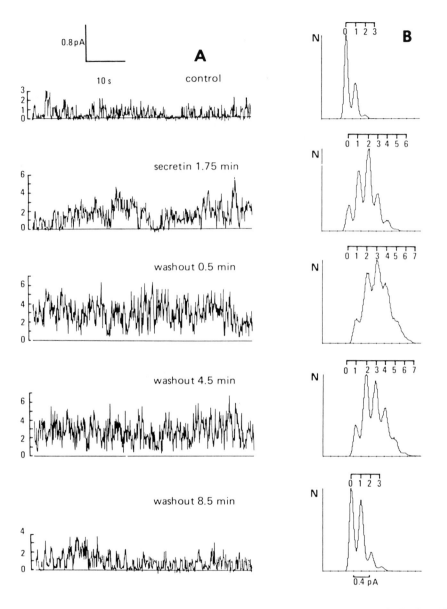

Figure 4.12 Regulation of small conductance Cl channels by secretin. **A** Single-channel currents in a cell-attached patch on an isolated rat pancreatic duct cell. The times indicate the period between either addition, or washout, of secretin and the start of each trace. Vertical scale indicates the number of channels open simultaneously. **B** Current amplitude histograms derived by analysis of the corresponding tracings. Horizontal scale indicates number of channels open simultaneously.

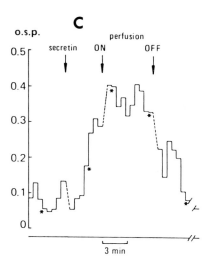

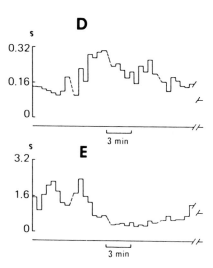

Figure 4.12 C, D and E Effects of 10 nmol L^{-1} secretin on the open-state probability (**C**), mean open time (**D**) and mean closed time (**E**) of the Cl^{-} channel. Same experiment as **A** and **B**. Arrow 'secretin' indicates when the hormone was added to the bath (volume 1.5 ml) and arrows 'perfusion on' and 'perfusion off' when the perfusion flow (5 ml/min) was switched. Dashed lines indicate access to screened cage, and stars the mid-point of recordings shown in **A**. For illustrative purposes, data collected over 4.5 min towards end of experiment has been omitted. (From reference 16, with permission)

93

cells. As in airways epithelia[54,55] abnormal regulation of this channel might explain the reduced pancreatic bicarbonate secretion that occurs in the inherited disease, cystic fibrosis[6-8]. Now that techniques have been developed for the growth and maintenance of human pancreatic duct cells in culture[56,57] (see also chapter by A. Harris), it will be possible to test this hypothesis directly. Since the main function of ductal secretions is to flush digestive enzymes out of the gland, stasis, followed by activation of these enzymes within the ductal tree, may well initiate the pathological changes that occur in the pancreas of cystic fibrosis patients.

Acknowledgements

Work in our laboratory is supported by the Medical Research Council (UK), the Cystic Fibrosis Research Trust (UK) and the Cystic Fibrosis Foundation (USA).

References

1. Case, R. M. and Argent, B. E. (1986). Bicarbonate secretion by pancreatic duct cells: mechanisms and control. In Go, V. L. W., Gardner, J. D., Brooks, F. P., Lebenthal, E., Di Magno, E. P. and Scheele, G. A. (eds.) *The Exocrine Pancreas: Biology, Pathology and Diseases*, pp. 213-243. (New York: Raven Press)
2. Takahashi, H. (1984). Scanning electron microscopy of the rat exocrine pancreas. *Arch. Histol. Jpn.*, **47**, 387-404
3. Forstner, G. and Forstner J. (1986). Mucus: biosynthesis and secretion. In Go, V. L. W., Gardner, J. D., Brooks, F. P., Lebenthal, E., Di Magno, E. P. and Scheele, G. A. (eds.) *The Exocrine Pancreas: Biology, Pathology and Diseases*, pp. 283-286. (New York: Raven Press)
4. Case, R. M. and Argent B. E. (1989). Pancreatic secretion of electrolytes and water. In Schultz, S. G., Forte, J. G. and Rauner, B. B. (ed.) *Handbook of Physiology. The Gastrointestinal System*, Vol. III, Section 6, pp. 383-417. (New York: Oxford University Press)
5. Case, R. M. (1978). Pancreatic secretion: cellular apsects. In Duthie, H. L. and Wormsley, K. G. (eds.) *Scientific Basis of Gastroenterology*, pp. 163-198. (London: Churchill Livingstone)
6. Gaskin, K. J., Durie, P. R., Corey, M., Wei, P. and Forstner, G. G. (1982). Evidence for a primary defect of pancreatic HCO_3^- secretion in cystic fibrosis. *Pediatr. Res.*, **16**, 554-557
7. Kopelman, H., Durie, P., Gaskin K., Weizman, Z. and Forstner, G. (1985). Pancreatic fluid secretion and protein hyperconcentration in cystic fibrosis. *N. Engl. J. Med.*, **312**, 329-334
8. Kopelman, H., Corey, M., Gaskin, K., Durie, P., Weizman, Z. and Forstner, G. (1988). Impaired chloride secretion, as well as bicarbonate secretion, underlies the fluid and secretory defect in the cystic fibrosis pancreas. *Gastroenterology*, **95**, 349-355
9. Reber, H. A., Adler, G. and Wedgwood, K. R. (1986). Studies in the perfused pancreatic duct in the cat. In Go, V. L. W., Gardner, J. D., Brooks, F. P., Lebenthal, E., Di Magno, E. P. and Scheele, G. A. (eds.) *The Exocrine Pancreas: Biology, Pathology and Diseases*, pp. 255-273. (New York: Raven Press)
10. Gordis, L. and Gold, E. B. (1986). Epidemiology and etiology of pancreatic cancer. In Go, V. L. W., Gardner, J. D., Brooks, F. P., Lebenthal, E., Di Magno, E. P. and Scheele, G. A. (eds.) *The Exocrine Pancreas: Biology, Pathology and Diseases*, pp. 621-636. (New York: Raven Press)
11. Klöppel, G. and Fitzgerald, P. J. (1986). Pathology of nonendocrine pancreatic tumors. In Go, V. L. W., Gardner, J. D., Brooks, F. P., Lebenthal, E., Di Magno, E. P. and Scheele,

G. A. (eds.) *The Exocrine Pancreas: Biology, Pathology and Diseases*, pp. 649–674. (New York: Raven Press)

12. Githens, S. (1988). The pancreatic duct cell: proliferative capabilities, specific characteristics, metaplasia, isolation, and culture. *J. Pediatr. Gastroenterol. Nutr.*, **7**, 486–506

13. Schulz, I., Heil, K., Miltinović, S., Haase, W., Terreros, D. and Rumrich, G. (1979). Preparation of duct cells from the pancreas. In Reid, E. (ed.) *Cell Populations, Methodological Surveys, (B)*, Vol. 9, pp. 127–135. (Chichester: Ellis Harwood)

14. Schulz, I., Heil, K., Kribben, A., Sachs, G. and Haase, W. (1980). Isolation and functional characteristics of cells from the exocrine pancreas. In Ribet, A., Pradayrol, L. and Susini C. (eds.) *Biology of Normal and Cancerous Exocrine Pancreatic Cells*, pp. 3–18. (Amsterdam: Elsevier/North Holland Biomedical Press)

15. Tsao, M-S. and Duguid, W. P. (1987). Establishment of propagable epithelial cell lines from normal adult rat pancreas. *Exp. Cell Res.*, **168**, 365–375

16. Gray, M. A., Greenwell, J. R. and Argent, B. E. (1988). Secretin-regulated chloride channel on the apical plasma membrane of pancreatic duct cells. *J. Membr. Biol.*, **105**, 131–142

17. Wizemann, V., Christian, A-L., Wiechmann, J. and Schulz, I. (1974). The distribution of membrane bound enzymes in the acini and ducts of the cat pancreas. *Pfleugers Arch.*, **347**, 39–47

18. Fölsch, U. R., Fischer, H., Söling, H-D. and Creutzfeldt, W. (1980). Effects of gastrointestinal hormones and carbamylcholine on cAMP accumulation in isolated pancreatic duct fragments from the rat. *Digestion*, **20**, 277–292

19. Githens, S., Holmquist, D. R. G., Whelan, J. F. and Ruby, J. R. (1980). Characterization of ducts isolated from the pancrease of the rat. *J. Cell Biol.*, **85**, 122–135

20. Githens, S., Holmquist, D. R. G., Whelan, J. F. and Ruby, J. R. (1980). Ducts of the rat pancreas in agarose matrix culture. *In Vitro*, **16**, 797–808

21. Githens, S., Holmquist, D. R. G., Whelan, J. F. and Ruby, J. R. (1981). Morphologic and biochemical characteristics of isolated and cultured pancreatic ducts. *Cancer*, **47**, 1505–1512

22. Githens, S. and Whelan, J. F. (1983). Isolation and culture of hamster pancreatic ducts. *J. Tiss. Culture Meth.*, **8**, 97–102

23. Argent, B. E., Arkle, S., Cullen, M. J. and Green, R. (1986). Morphological, biochemical and secretory studies on rat pancreatic ducts maintained in tissue culture. *Q. J. Exp. Physiol.*, **71**, 633–648

24. Arkle, S., Lee, C. M., Cullen, M. J. and Argent, B. E. (1986). Isolation of ducts from the pancreas of copper-deficient rats. *Q. J. Exp. Physiol.*, **71**, 249–265

25. Githens, S., Finley, J. J., Patke, C. L., Schexnayder, J. A., Fallon, K. B. and Ruby, J. R. (1987). Biochemical and histochemical characterization of cultured rat and hamster pancreatic ducts. *Pancreas*, **2**, 427–438

26. Madden, M. E. and Sarras, M. P. (1987). Distribution of Na⁺, K⁺-ATPase in rat exocrine pancreas as monitored by K⁺-NPPase cytochemistry and [³H]-ouabain binding: a plasma membrane protein found primarily to be ductal cell associated. *J. Histochem. Cytochem.*, **35**, 1365–1374

27. Hootman, S. R. and Logsdon, C. D. (1988). Isolation and monlayer culture of guinea pig pancreatic duct epithelial cells. *In Vitro*, **24**, 566–574

28. Novak, I. and Greger, R. (1988). Electrophysiological study of transport systems in isolated perfused pancreatic ducts: properties of the basolateral membrane. *Pfleugers Arch.*, **411**, 58–68

29. Novak, I. and Greger, R. (1988). Properties of the luminal membrane of isolated perfused rat pancreatic ducts. Effect of cyclic AMP and blockers of chloride transport. *Pfleugers Arch.*, **411**, 546–553

30. Stuenkel, E. L., Machen, T. E. and Williams, J. A. (1988). pH regulatory mechanisms in rat pancreatic ductal cells. *Am. J. Physiol.*, **254**, G925–G930

31. Githens, S., Schexnayder, J. A., Desai, K. and Patke, C. L. (1989). Rat pancreatic interlobular duct epithelium: isolation and culture in collagen gel. *In Vitro*, **25**, 679–688

32. Jones, R. T., Trump, B. F. and Stoner, G. D. (1980). Culture of human pancreatic ducts. *Meth. Cell Biol.*, **21B**, 429–439

33. Kuijpers, C. A. J. and De Pont, J. J. H. H. M. (1987). Role of proton and bicarbonate transport in pancreatic cell function. *Ann. Rev. Physiol.*, **49**, 87–103

34. Schulz, I. (1987). Electrolyte and fluid secretion in the exocrine pancreas. In Johnson, L. R. (ed.) *Physiology of the Gastrointestinal Tract*, 2nd Edn., pp. 1147–1171. (New York: Raven Press)

35. Novak, I. (1988). Pancreatic bicarbonate secretion. In Häussinger, D. (ed.) *pH Homeostasis: Mechanisms and Control*, pp. 447–470. (New York: Academic Press)

36. Gray, M. A., Greenwell, J. R. and Argent, B. E. (1989). The role of ion channels in the mechanism of pancreatic bicarbonate secretion. In Young, J. A. and Wong, P. Y. D. (eds.) *Epithelial Secretion of Electrolytes and Water*, pp. 253–265. (Heidelberg: Springer-Verlag)

37. Fölsch, U. R. and Creutzfeldt, W. (1977). Pancreatic duct cells in rats: secretory studies in response to secretin, cholecystokinin-pancreozymin, and gastrin *in vivo*. *Gastroenterology*, **73**, 1053–1059

38. Fell, B. F., King, T. P. and Davies, N. T. (1982). Pancreatic atrophy in copper-deficient rats: histochemical and ultrastructural evidence of a selective effect on acinar cells. *Histochem. J.*, **14**, 665–680

39. Smith, P. A., Sunter, J. P. and Case, R. M. (1982). Progressive atrophy of pancreatic acinar tissue in rats fed a copper-deficient diet supplemented with D-penicillamine or triethylene tetramine: morphological and physiological studies. *Digestion*, **23**, 16–32

40. Kumpulainen, T. and Jalovaara, P. (1981). Immunohistochemical localisation of carbonic anhydrase isoenzymes in the human pancreas. *Gastroenterology*, **80**, 796–799

41. Gardner, J. D. and Jensen, R. T. (1986). Receptors mediating the actions of secretagogues on pancreatic acinar cells. In Go, V. L. W., Gardner, J. D., Brooks, F. P., Lebenthal, E., Di Magno, E. P. and Scheele, G. A. (eds.) *The Exocrine Pancreas: Biology, Pathology and Diseases*, pp. 109–122. (New York: Raven Press)

42. Gray, M. A., Greenwell, J. R. and Argent, B. E. (1988). Ion channels in pancreatic duct cells: characterization and role in bicarbonate secretion. In Mastella, G. and Quinton, P. M. (eds.) *Cellular and Molecular Basis of Cystic Fibrosis*, pp. 205–216. (San Francisco: San Francisco Press)

43. Hamill, O. P., Marty, A., Neher, R., Sakmann, B. and Sigworth, F. J. (1981). Improved patch-clamp techniques for high-resolution current recording from cells and cell-free membrane patches. *Pfleugers Arch.*, **391**, 85–100

44. Argent, B. E., Arkle, S., Gray, M. A. and Greenwell, J. R. (1987). Two types of calcium-sensitive cation channels in isolated rat pancreatic duct cells. *J. Physiol.*, **386**, 82P

45. Petersen, O. H. (1986). Calcium-activated potassium channels and fluid secretion by exocrine glands. *Am. J. Physiol.*, **251**, G1–G13

46. Evans, L. A. R., Pirani, D., Cook, D. I. and Young, J. A. (1986). Intraepithelial current flow in rat pancreatic secretory epithelia. *Pfleugers Arch.*, **407** (Suppl. 2), S107–S111

47. Gray, M. A., Harris, A., Coleman, L., Greenwell, J. R. and Argent, B. E. (1989). Two types of chloride channel on duct cells cultured from human fetal pancreas. *Am. J. Physiol.*, **257**, C240–C251

48. Frizzell, R. A. (1987). Cystic fibrosis: A disease of ion channels. *Trends Neurosci.*, **10**, 190–193

49. Gögelein, H. (1988). Chloride channels in epithelia. *Biochim. Biophys. Acta*, **947**, 521–547

50. Argent, B. E., Gray, M. A. and Greenwell, J. R. (1987). Characteristics of a large conductance anion channel in membrane patches excised from rat pancreatic duct cells *in vitro*. J. Physiol., **394**, 146P

51. Marty, A., Tan, Y. P. and Trautmann, A. (1984). Three types of calcium-dependent channel in rat lacrimal glands. *J. Physiol.*, **357**, 293–325

52. Griffiths, N. M., Chabardès, D., Imbert-Teboul, M., Siaume-Perez, S., Morel, F. and Simmons, N. L. (1988). Distribution of vasoactive intestinal peptide-sensitive adenylate cyclase along the rabbit nephron. *Pfleugers Arch.*, **412**, 363–368

53. Hus-Citharel, A. and Morel, F. (1986). Coupling of metabolic CO_2 production to ion transport in isolated rat thick ascending limbs and collecting tubules. *Pfleugers Arch.*, **407**, 421–427

54. Schoumacher, R. A., Shoemaker, R. L., Halm, D. R., Tallant, E. A., Wallace, R. W. and Frizzell, R. A. (1987). Phosphorylation fails to activate chloride channels from cystic fibrosis airway cells. *Nature (London)*, **330**, 752–754

55. Li, M., McCann, D., Liedtke, C. M., Nairn, A. C., Greengard, P. and Welsh, M. J. (1988). Cyclic AMP-dependent protein kinase opens chloride channels in normal but not cystic fibrosis airway epithelium. *Nature (London)*, **331**, 358–360
56. Harris, A. and Coleman, L. (1987). Establishment of a tissue culture system for epithelial cells derived from human pancreas. *J. Cell Sci.*, **87**, 695–703
57. Harris, A. and Coleman, L. (1988). Cultured epithelial cells derived from human foetal pancreas as a model for the study of cystic fibrosis: further analysis on the origins and nature of cell types. *J. Cell Sci.*, **90**, 73–77

5
Cultured Epithelial Cells Derived from Human Fetal Pancreatic Duct

A. HARRIS

INTRODUCTION

The human pancreatic duct is a complex system that consists of a branching network of channels between the acini of the pancreas and the duodenum. Adjacent to acini are centroacinar cells which form the termini of intralobular ducts. The smallest branches of these join to form larger intralobular ducts which in turn pass out of the pancreatic lobules to form interlobular ducts. Interlobular ducts join to form ducts of increasing diameter which eventually lead into the main pancreatic duct. The pancreatic duct not only provides a channel for the passage of digestive enzymes from the acini to the duodenum, but also actively secretes bicarbonate-rich fluid and mucins[1,2].

The human pancreas is one of the first organs to autolyse postmortem and, as a result, the cell biology and development of this organ has received less attention than many others. In particular, studies on the development of the ductal portion of the pancreas are few. However, the pancreatic duct is intimately involved in several important human diseases. One of the main features of the autosomal recessive disease, cystic fibrosis, is pancreatic insufficiency caused by the replacement of acinar tissue with cystic spaces. This occurs due to blockage of pancreatic ducts, preventing normal passage of digestive enzymes from the acini to the duodenum. As a result, the enzymes gradually autodigest the organ. The pancreatic duct is also the site of the vast majority of tumours of the exocrine pancreas[3,4].

Against this background of the importance of the pancreatic duct in medicine, we set out to establish a culture system for human pancreatic duct cells to enable *in vitro* experimentation on components of the ductal tree.

Previous attempts at culturing ductal epithelial from bovine, hamster, mouse and guinea-pig pancreas have been successful[5-9]. The cultured cells have characteristic surface microvilli and secrete mucins as they do *in vivo*. There is also one report in the literature of a short-term *in vitro* human pancreatic duct cell culture system[9].

For two main reasons, we decided to try to culture pancreatic duct cells from human mid-trimester terminations, both normals and with various known abnormalities. First, at the beginning of the second trimester of pregnancy, the pancreas is made up of a simple ductal tree surrounded by loose interstital tissue containing largely undifferentiated cell types. Between 12 and 14 weeks, cells from the ducts migrate into the interstitial tissue and start to form lobular structures. Though primitive acini may already be seen at this stage[10], the first mature acinar cells do not appear until around 20 weeks of gestation and the acini do not appear to have a lumen before 24 weeks[11]. It thus seemed likely that the mid-trimester pancreas would be less affected by postmortem autolysis than postnatal pancreas. Second, our research interests are centred around the basic defect in cystic fibrosis. It appears that the cystic fibrosis gene is already functional at 18 weeks of gestation since fetuses diagnosed as having CF (on the basis of amniotic fluid microvillar enzyme assays)[12,13] often show abnormal pancreatic histology[13,14]. Studies on the CF pancreas later in life are further complicated by the disease process.

The pancreatic duct system is complex and it is not clear which of the epithelial cell types contained within it might be functionally abnormal in CF. However, though this system is not fully differentiated at 18 weeks, the population of cells expressing the CF gene defect is apparently functional, since abnormal pancreatic histology may already be seen[13,14]. We have established a reliable tissue culture system for epithelial cells derived from mid-trimester human fetal pancreas[15,16].

EXPERIMENTAL DETAILS

The source of pancreases is critical to the success of establishing mid-trimester human pancreatic epithelial cell cultures. Pancreases obtained within 48 h of mid-trimester prostaglandin-induced terminations or spontaneous abortions (either normal or with known abnormalities) have yielded viable cell cultures. Cultures have been established from both normal fetuses and those with cystic fibrosis (CF having been diagnosed on the basis of amniotic fluid microvillar enzyme and alkaline phosphatase activities)[12,13]. However, it is essential that the termination procedure does not involve urea in any form, administered by any route, since this renders all pancreatic cells non-viable. Pancreases obtained from fetuses terminated following a diagnosis of CF by DNA-based tests in the first trimester have not yielded ductal epithelial cells. We believe that, prior to about 12 weeks gestation, these cells have not yet developed within the pancreas and so cell cultures are predominantly fibroblastic.

The pancreas, still attached to the duodenum, is removed from the fetus and washed in tissue culture medium containing antibiotics. Mesentery around the pancreas is trimmed away and then 1–2 ml of collagenase (Sigma type IA at 0.5 mg/ml) injected into the organ. When visible, the main duct is microdissected from the inflated pancreas. In early mid-trimester pancreases, where the main duct cannot be clearly distinguished, the whole pancreas is set up in culture, following physical and enzymatic disruption. The cell types that grow from the main duct alone or from the whole pancreas at this age are indistinguishable. In other words, the majority of viable cells within the pancreas are clearly of ductal epithelial origin in early mid-trimester organs.

Microdissected tissue is washed in tissue culture medium, cut into pieces of about 1–2 mm diameter and plated out in CMRL 1066 medium containing 20% fetal calf serum (FCS) (Gibco UK); penicillin (100 iu/ml); streptomycin (100 µg/ml); L-glutamine (4 mmol L^{-1}); insulin (0.2 iu/ml); cholera toxin (10^{-10} mol L^{-1}) and hydrocortisone (1 µg/ml), all from Sigma. Cultures are routinely maintained at 37°C in a saturated 5% CO_2 incubator. Various cell types are seen migrating from the primary explants after 3 to 10 days.

Cell types

Two main epithelial cell types grow out of primary pancreatic explants and are maintained in subsequent passages as illustrated in Figure 5.1: small tightly packed cells and larger streaming cells. The cultures usually form characteristic structures *in vitro* (Figure 5.1A) with areas of the smaller cells separated by the large streaming cells which, though they often look fibroblastic, in fact express epithelial cell cytokeratin markers, such as those detected by the monoclonal antibody LE61. (These markers will be discussed further below.) The smaller cell type appears first in culture and seems to be a precursor population that differentiates into the larger cell type after a variable period of time in culture. There is no apparent morphological difference between ductal epithelial cells from CF and normal pancreases (Figures 5.1B and C).

Culture media and substrates

The growth characteristics of the pancreatic epithelial cell colonies have been monitored in a variety of tissue culture media (see Table 5.1) and on several different culture substrates (see Table 5.2). Tissue culture plastics used were as follows: multiwells from Nunclon, Gibco; flasks from Sterilin and Falcon, Becton Dickinson; and Primaria flasks from Falcon, Becton Dickinson.

As is shown in Table 5.1, CMRL 1066 with 20% FCS, insulin (0.2 iu/ml, cholera toxin (10^{-10} mol L^{-1}) and hydrocortisone (1 µg/ml) was the best culture medium. Primaria tissue culture flasks or collagen-coated glass or plastic were the most efficient substrates (Table 5.2).

Fibroblast contamination of epithelial cell cultures frequently occurs in both primary cultures and in subsequent passages. The most successful method of eliminating fibroblasts in our hands is physical removal with a rubber policeman. It is essential to remove fibroblasts as soon as they appear in culture (usually at the periphery of epithelial cell colonies), since,

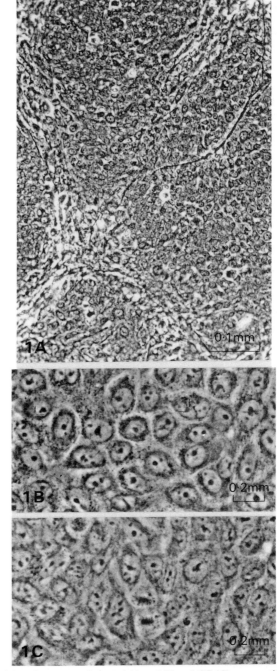

Figure 5.1 Light microscopy of normal (A and B) and CF (C) pancreatic epithelial cell types in culture. Panel A shows the 2 main cell types and their spatial arrangement (× 100). Panels B and C show the small tightly packed cells only (× 425)

Table 5.1 Tissue culture medium

Medium	Epithelial Cell Growth	Fibroblast Growth
CMRL 1066 (Gibco) + 20% FCS with supplements*	++++	++
CMRL 1066 + 20% FCS	++	++
Ham's F10 (Gibco) + 20% FCS	+	+++
Dulbecco's MEM (Gibco) + 20% FCS	+	+++
RPM1 1640 (Flow) + 20% FCS	+	+++
MCDB† – no FCS; no supplements	–	++

*Insulin   (0.2 iu/ml),   cholera   toxin   (10^{10} mol L^1), hydrocortisone (1 µg/ml)
†Hammond et al., 1984, PNAS **81**, 5435–5439

Table 5.2 Culture substrates

	Primary Explants	Passaged colonies
Plastic multiwells or plastic flasks	+	–
Primaria	+++	+++
Collagen-coated* plastic multiwells	++	+++
Collagen-coated* glass coverslips	++	+++
Glass coverslips	–	+
Feeders-layers† – (pancreas-derived fibroblasts or NIH 3T3 cells)	–	+/–

* Collagen (Sigma type IV – human placenta) at 1 mg/ml in 1:1000 glacial acetic acid: water, placed on culture substrate and dried at 37°C for 16 h
†Treated with mitomycin C (4 µg/ml) for 2 h

once they have become integrated into epithelial colonies, they rapidly overgrow the latter and are impossible to remove.

Passaging of epithelial cell colonies has been achieved by treating the latter with dispase, a neutral protease (Boehringer, Mannheim) at 2 iu/ml^1 in CMRL 1066 medium for 15–40 min at 37°C. Passaging has not been successful with routine methods utilising trypsin and EDTA. Cell colonies survive passaging if they remain partially intact (for example by physical division into smaller colonies) but not if they are disrupted into single cells or clumps of very small numbers of cells. Cells survive up to five successive passages before reaching terminal crisis and they show no alterations in morphology during this period, which may be up to about 20 weeks in vitro.

Morphological characteristics of cells

The pancreas-derived epithelial cells when analysed by electron microscopy (Figure 5.2) show characteristic features of simple epithelial cells. They have

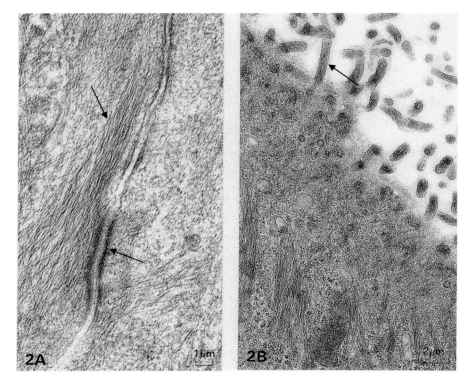

Figure 5.2 Electron microscopy of pancreatic epithelial cells. **A** Cytokeratin intermediate filaments and desmosomal plaques (× 48 000). **B** Surface microvilli on epithelial cells (× 16 500)

extensive cytokeratin intermediate filaments (Figure 5.2A) with associated desmosomal plaques and tight junctions (Figure 5.2C). Further, they have surface microvilli (Figure 5.2B), numerous large mitochondria (Figure 2C and D) and abundant mucin-filled secretory vacuoles (Figure 5.2E). Alcian Blue–periodic acid–Schiff staining of the epithelial cells suggests that the bulk of these vacuoles contain neutral mucins.

Biochemical characteristics of the cells

An extensive characterization of the epithelial cells has been carried out by immunocytochemical procedures. Furthermore, sections from intact 17 to 19-week-old pancreases have been analysed simultaneously, with the same markers, to confirm the origin of the cultured cells.

Cytokeratins

Expression of cytokeratins, 8, 18 and 19 has been investigated using anticytokeratin 8, LE61[17] and anticytokeratin 19 respectively. In adult

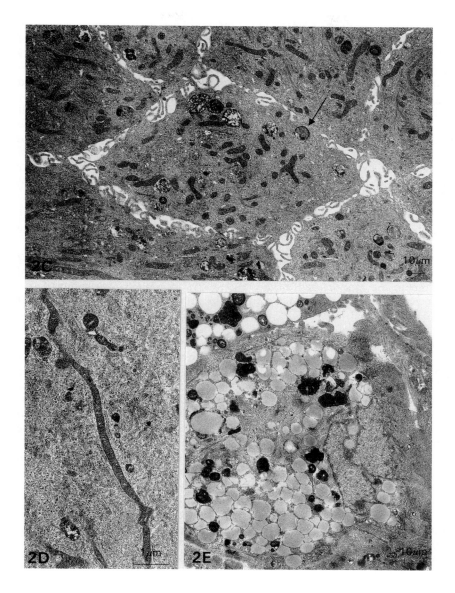

Figure 5.2C Interdigitated cells showing tight junctions and abundant mitochondria (× 4800).
D Elongated mitochondria (× 7650). **E** Secretory vacuoles (× 4800)

pancreas, acini express predominantly cytokeratins 8 and 18, while the ductal tree also contains cytokeratins 7 and 19. Both the small tight-packed cells and the larger streaming cells in our cultures express cytokeratins 18, 8 and 19 (Figure 5.3A,B and C respectively). Cytokeratin production remains stable from 2.5 weeks after the cultures are established to at least 10 weeks (the earliest and latest dates at which cytokeratin production has been analysed). In sections through 19-week fetal pancreas, expression of cytokeratin 18 (Figure 5.4A), cytokeratin 8 (Figure 5.4B) and cytokeratin 19 (Figure 5.4C) is seen to be mainly localized in ductal epithelium. Particularly strong expression of cytokeratin 18 (as detected by LE61) and cytokeratin 19 in the epithelium of the main pancreatic duct and nearby smaller ducts gives support to a ductal origin for the cells we are culturing. It is unclear why cytokeratin 8 is expressed in a much higher level in the cultured cells than in intact fetal pancreas. The expression of cytokeratins *in vitro* is known to be affected by culture conditions. However, the stability of cytokeratin expression in our cells during at least 10 weeks in culture suggests that *in vitro* modulation of phenotype is not playing an important role in our results.

Mucins

It has already been mentioned that the ductal epithelial cells we are culturing are rich in neutral mucins. Studies of mucin production have been monitored further using monoclonal antibodies specific for particular mucins. Three different mucins that are detected by the monoclonal antibodies, Ca2 (or HMFG2)[18,19], DU-PAN-2[20,21]* and 19-9[22], have been analysed.

The smaller epithelial cells in the cultures express low levels of the antigen detected by Ca2 (Figure 5.5A) and HMFG2 (Figure 5.5B). However, differentiation of these cells into the streaming cell types that separate the smaller epithelial cell colonies is accompanied by a dramatic increase in production of this mucous glycoprotein. Since the antigen detected by Ca2 and HMFG2 is known to be expressed on certain ductal epithelia[23], this observation lends further support for the suggestion that the tightly packed cells are indeed precursors of fully differentiated ductal epithelial cells. The streaming cell types may correspond more closely to the latter.

In sections through 19-week fetal pancreas, only cells lining the main pancreatic duct are seen to produce the mucous glycoprotein detected by Ca2 and HMFG2 (Figure 5.5D). These data provide additional evidence that the cultured epithelial cells are ductal in origin.

The pancreatic ducts are known to become blocked with inspissated mucous secretions in CF. Since the mucous glycoprotein detected by Ca2 and HMFG2 is secreted from the epithelial cells expressing it, we compared the levels of this antigen produced by CF and normal duct epithelial cells. The CF gene is known to be expressed by 18 weeks gestation[14], so this comparison could be informative. We saw considerable variation in the amount of this mucous glycoprotein present in different cultures (as

---

*There is now evidence that the Ca2 and DU-PAN-2 antibodies recognize the same mucin.

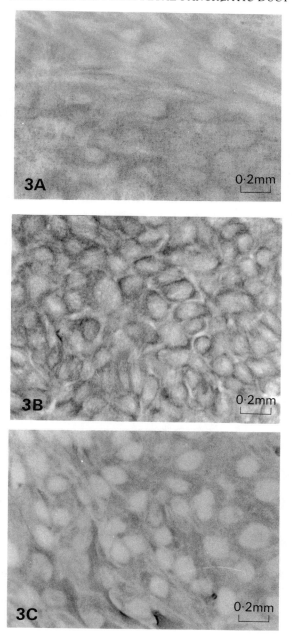

Figure 5.3 Immunocytochemistry of pancreatic epithelial cells. **A** Immunoperoxidase-conjugated LE61. × 390 **B** Immunoperoxidase-conjugated anti-cytokeratin 8. × 390. **C** Immunoperoxidase-conjugated anti-cytokeratin 19. × 390

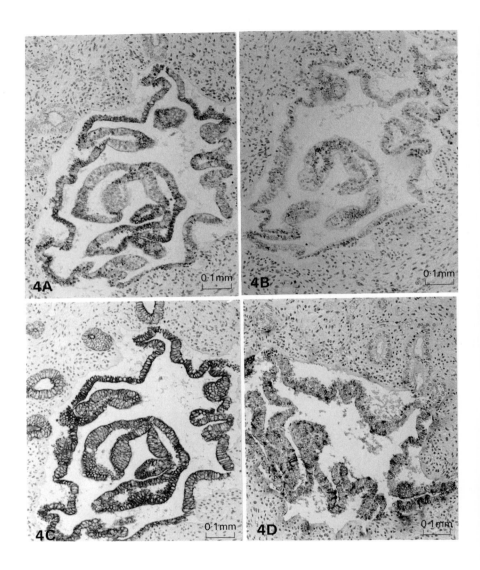

Figure 5.4 Immunocytochemistry of sections through 19-week fetal pancreas. **A** Immuno-peroxidase-conjugated LE61. × 80. **B** Immunoperoxidase-conjugated cytokeratin 8. × 80. **C** Immunoperoxidase-conjugated cytokeratin 19. × 80. **D** Immunoperoxidase-conjugated DU-PAN-2. × 80

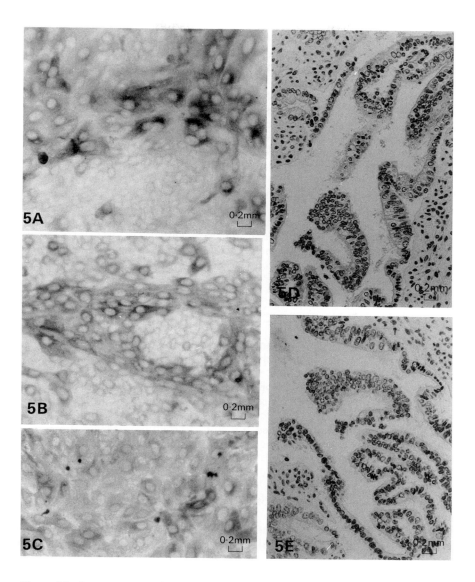

Figure 5.5 Immunocytochemistry of pancreatic epithelial cells (A, B and C) and sections through 19-week-old fetal pancreas (D and E). **A** Immunoperoxidase-conjugated Ca2. × 160. **B** Immunoperoxidase-conjugated HMFG2. × 160. **C** Immunoperoxidase conjugated DU-PAN-2. × 160. **D** Immunoperoxidase-conjugated HMFG2. × 160. **E** Background staining with second antibody only. × 160

109

measured on a gross level by immunocytochemistry). This variation was dependent on both the length of time in culture and on the particular pancreas from which a culture was derived; however, there were no consistent differences between CF and normal cells. It is possible that this lack of discrimination between CF and normal cells is due to the molecule that is detected by Ca2 and HMFG2 being only a minor species of mucin in the cells we are culturing; hence, it may not reflect alterations in levels of more abundant mucins produced by CF and normal duct epithelial cells. However, it is more probable that mucus inspissation may only be a problem encountered within the restricted micro-environment of the pancreatic duct *in vivo*. This could perhaps be due to abnormal ion movements across the duct epithelium or be a secondary reaction to deposits of other secretions in the duct lumen.

Two other mucin antigens, DU-PAN-2* that is associated with human pancreatic adenocarcinoma and 19-9 that is associated with human gastro-intestinal and pancreatic cancer, have also been studied. DU-PAN-2 is expressed at about the same level by both epithelial cell types in culture (Figure 5.5C), though there is wide variation in the amounts of this mucin in individual cells. 19-9 shows a similar pattern of expression (data not shown). In sections through 19-week fetal pancreas, high levels of DU-PAN-2 expression are seen in the epithelial cells lining the main pancreatic duct (Figure 5.3D), while no other cells are seen to express signficant levels of this antigen.

IMMORTALIZATION OF PANCREATIC DUCT EPITHELIAL CELLS

One major drawback to our pancreatic duct epithelial cell culture system has been the relatively small numbers of cells that can be cultured from an individual fetal organ (about 5×10^5 to 1×10^6 cells). This has, to some extent, limited the types of experiment that can be carried out on them. Clearly this problem might be circumvented if long-term 'immortalized' cell lines could be established from the primary cultures, so long as they maintained the fully differentiated functions of the latter. Establishment of pancreatic duct epithelial cell lines would also relax the need for a constant source of fetal tissue.

A considerable amount of time and effort has gone into trying to achieve this aim; however, to date we do not have useful long-term pancreatic duct epithelial cell lines. An outline of the 'immortalization' protocols we have employed is shown in Table 5.3.

An initial problem encountered in the immortalization experiments was lack of success in inserting foreign DNA into the ductal epithelial cells. Protocols employed depended on either infection by intact viruses or transfection of foreign DNA in the form of a calcium phospate precipitate, which may be ingested by the cells[24]. This problem is likely to have been caused by the epithelial cell monolayer secreting and being covered by a thick layer of mucins. Infectious viral particles and foreign DNA entrapped

*See footnote on p. 106.

110

Table 5.3 Immortalization experiments

Immortalizing vector/virus	Method of introduction into cells	Presence of vector, viral DNA or viral proteins in cells	Outcome
SV40	Infection	NO	Fibroblast overgrowth
SV40	Infection	N.D.	—
pX-8*	Transfection**	YES	Altered morphology/ growth crisis
E₁A†	Transfection	N.D.	Fibroblast overgrowth
pX-8+E₁A	Transfection	N.D.	Normal lifespan
ts SV40	Retrovirus Infection	—	In progress

*pX-8: origin defective mutant of SV40[25]
**Transfection protocol[24]
†E₁A (adenovirus early antigen 1A) 0–4.5 map units of adenovirus 2 in Pst site of pBR322
N.D. = not done
ts = temperature sensitive

in calcium phosphate precipitates were unable to cross this layer of mucins. This problem was overcome by growing the cells in a relatively simple cell culture medium (McCoy's), containing few nutrients, for 48 h prior to transfection or infection. The ductal epithelial cells normally grow in a much richer culture medium (CMRL 1066) containing many supplements and do not survive long in simple media; hence it is likely that they shut down productive mucous secretion in McCoy's and a thorough washing of the cells prior to transfection stripped off sufficient amounts of the surface mucus to enable the foreign DNA or virus particules to come into contact with at least parts of the cell membrane (and hence enter the cell). In at least one experimental series (see Table 5.3), SV40 transfection resulted in transformed epithelial cells. These cells persistently formed mounds above the monolayer and eventually budded off into the culture media as balls of cells. We have evidence from dot blot analysis of DNA extracted from transformed cells, followed by probing with SV40 DNA, that the transfected vector, pX-8 (SV40 without an origin of replication)[25], did enter the ductal epithelial cells and hence was probably responsible for their changed morphology and growth characteristics. However, this cell line grew in a manner not suitable for long-term culture or useful for analysis (see Figure 5.6) and, further did not pass through crisis. We would speculate that the lack of success in immortalizing our pancreatic duct epithelial cells may be accounted for by two main problems. First, the numbers of cells being transfected is too low, given that the efficiency of transforming vectors is often low. Second, our epithelial cultures sometimes suffer from fibroblast contamination after prolonged periods in culture, such as those necessary for immortalization experiments. As a result, potentially immortalized epithelial cells are then competing for survival in culture with much more rapidly growing fibroblasts, a competition they are unlikely to win.

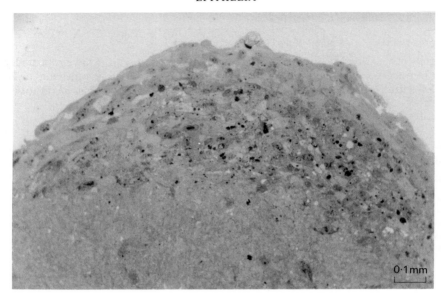

Figure 5.6 Section through pX-8-transformed pancreatic epithelial cells. × 80.

ELECTROPHYSIOLOGICAL ANALYSIS OF HUMAN FETAL PANCREATIC DUCT EPITHELIAL CELLS

The cells in this culture system are clearly an ideal model in which to study aspects of the electrophysiology of the human pancreatic duct epithelium. However, this is beyond the scope of this chapter and data are covered well by Gray et al.[26].

MOLECULAR ANALYSIS OF CELL FUNCTION

One of the major applications of a culture system for pancreatic duct epithelial cells is the ability to undertake a detailed analysis of specialized cell function at a molecular level. It enables us to ask specific questions about what genes are switched on and contribute to the particular properties of these specialized epithelial cells (for example genes involved in ion transport systems and mucus secretion). Comparatively little is known about the molecular biology of ductal epithelia *in vivo*. Culture systems such as the one we have described should make a substantial contribution to their understanding.

Isolation of good-quality intact (full-length) messenger RNA directly from whole fetal pancreas has been relatively unsuccessful. This is likely to be due to degradation processes that commence immediately postmortem following release of degradative enzymes. Furthermore, RNA made from whole pancreas clearly contains a large number of messages produced by cells other than those in the ductal epithelium. However, we have repeatedly obtained good-quality messenger RNA from cultured ductal epithelial cells. Essentially, the methods used are based on those of Chirgwin et al.[27] though procedures have been scaled down to optimize the yield of full-length mRNA from small quantities of material.

The epithelial cell monolayer is removed from the substrate with dispase (as for passaging), washed in PBS and lysed immediately by resuspension in 2.0 ml 4 mol L^{-1} guanidinium thiocyanate. A homogeneous mixture is made by pipetting through a wide-bore glass pipette, and this is then layered on top of a 1.6 ml caesium chloride cushion (56% in 0.1 mol L^{-1} EDTA) in Beckmann SW56 polyallomer tubes. Centrifugation at 30 000 rpm, 22°C for 18 h results in a clean RNA pellet at the bottom of the tube.

This RNA has been used to construct cDNA libraries from normal and CF pancreatic duct epithelial cells. These libraries are being screened by a variety of approaches to isolate messages which code for proteins that are important in pancreatic duct epithelial cell function.

FUTURE DEVELOPMENTS AND APPLICATIONS

It is likely that the studies outlined above will throw substantial light on the functioning of the pancreatic duct epithelium, both at electrophysiological and molecular levels. It will also be an important key in understanding the physiological effects of the CF gene defect.

With the recent cloning of the CF gene[28,29,30], a new chapter in CF research has begun. We now have the tool with which to start asking important biological questions as to how a change in amino acid sequence in the CF gene product, the cystic fibrosis transmembrane conductance regulator (CFTR) protein, causes epithelial cell dysfunction in CF. Not only can the CF defect be measured phenotypically as an electrophysiological defect, but now can also be measured in terms of gene activity. Already, levels of CF gene message can be monitored, and, as soon as a monoclonal antibody is generated against the CFTR protein, this too can be monitored.

The pancreatic duct epithelial cell culture system will provide one of the basic resources in which to elucidate the role of the CF gene product. Analysis of the ductal epithelia from the pancreas, sweat gland and respiratory system will provide important information on regulation of the CF gene. In the longer term, these specialized epithelial cell culture systems will enable attempts at gene therapy; in other words, correction of the CF gene defect. We already know how to get foreign DNA into cultured pancreatic duct cells (see above). It remains to be seen whether insertion of a normal CF gene counterpart (in a vector where it is accompanied by the appropriate regulatory signals) could correct a mutant CF gene in vitro in the same cell. It is also uncertain whether it will ever be practical to attempt to apply this to the in vivo situation.

Acknowlegements

This work is supported by the Cystic Fibrosis Research Trust, UK and the Spastics Society. The project has been approved by Guy's Hospital Ethical Committee. The author thanks the Company of Biologists for permission to reproduce Figures 5.1–5.5 inclusive and Tables 5.1 and 5.2; Lindsay Coleman for excellent technical assistance; also Drs D. Lane and S. Chang, P. Berg and K. Quade for SV40, pX-8 and pA_1-A respectively.

References

1. Case, R. M. and Argent, B.E. (1986). Bicarbonate secretion by pancreatic duct cells: mechanism and control. In Go, V. L. W., Gardner, J. D., Brooks, F. P., Lebenthal, E., Di Magno, E. P., Scheele, G. A. (eds.) *The Exocrine Pancreas: Biology, Pathobiology and Diseases*, pp. 213–244 (New York: Raven Press)
2. Forstner, G. and Forstner, J. (1986). Mucus: biosynthesis and secretion. In Go, V. L. W., Gardner, J. D., Brooks, F. P., Lebenthal, E., Di Magno, E. P., Scheele, G. A. (eds.) *The Exocrine Pancrease: Biology, Pathobiology and Diseases*, pp. 283–286. (New York: Raven Press)
3. Cubilla, A. L. and Fitzgerald, P. J. (1975). Morphological patterns of primary non-endocrine human pancreas carcinoma. *Cancer Res.*, **35** 2234–2248
4. Cubilla, A. L. and Fitzgerald, P. J. (1978). Pancreas cancer. 1. Duct adenocarcinoma. *Pathol. Annu.*, **13**, 241–289
5. Stoner, G. D., Harris, C. C., Bostwick, D. G., Jones, R. T., Trump, B. F., Kingsbury, E. W., Fineman, E. and Newkirk, C. (1978). Isolation and characterization of epithelial cells from bovine pancreatic duct. *In Vitro*, **14**, 581–590
6. Hirata, K., Oku, T. and Freeman, A. E. (1982). Duct, exocrine, and endocrine components of cultured fetal mouse pancreas. *In Vitro*, **18**, 789–799
7. Resau, J. H., Hudson, E. A. and Jones, R. T. (1983). Organ explant culture of adult Syrian golden hamster pancreas. *In Vitro*, **19**, 315–325
8. Sato, T., Mamoru, S., Hudson, E. A. and Jones, T. (1983). Characterization of bovine pancreatic ductal cells isolated by a perfusion-digestion technique. *In Vitro*, **19**, 651–660
9. Jones, R. T., Hudson, E. A. and Resau, J. H. (1981). A review of in vitro and in vivo culture research for the study of pancreatic carcinogenesis. *Cancer*, **47**, 1490–1496
10. Laitio, M., Lev, R. and Orlic, D. (1974). The developing human fetal pancreas: an ultrastructural and histochemical study with special reference to exocrine cells. *J. Anat.*, **117**, 619–634
11. Adda, G., Hannoun, L. and Loygue, J. (1984). Development of the human pancreas: variations and pathology. A tentative classification. *Anat. Clin.*, **5**, 275–283
12. Brock, D. J. H., Bedgood, D. and Hayward, C. (1984). Prenatal diagnosis of cystic fibrosis by assay of amniotic fluid microvillar enzymes. *Hum. Genet.*, **65**, 248–251
13. Boué, A., Muller, F., Nezelof, C., Oury, J. F., Duchatel, F., Dumez, Y., Aubry, M. C. and Boué, J. (1986). Prenatal diagnosis in 200 pregnancies with a 1-in-4 risk of cystic fibrosis. *Hum. Genet.*, **74**, 288–297
14. Ornoy, A., Arnon, J., Katznelson, D., Granat, M., Caspi, B. and Chemke, J. (1987). Pathological confirmation of cystic fibrosis in the fetus following prenatal diagnosis. *Am. J. Med. Genet.*, **28**, 935–947
15. Harris, A. and Coleman, L. (1987). Establishment of a tissue culture system for epithelial cells derived from human pancreas: a model for the study of cystic fibrosis. *J. Cell Sci.*, **87**, 695–703
16. Harris, A. and Coleman, L. (1988). Cultured epithelial cells derived from human fetal pancreas as a model for the study of cystic fibrosis: further analysis on the origins and nature of the cell types. *J. Cell Sci.*, **90**, 73–77
17. Lane, E. (1982). Monoclonal antibodies provide specific intramolecular markers for the study of epithelial tonofilament organization. *J. Cell Biol.*, **92**, 665–673

18. Bramwell, M. E., Bhavanandan, V. P., Wiseman, G. and Harris, H. (1983). Structure and function of the Ca antigen. *Br. J. Cancer*, **48**, 177
19. Burchell, J., Durbin, H. and Taylor-Papadimitriou, J. (1983). Complexity of expression of antigenic determinants recognised by monoclonal antibodies HMFG1 and HMFG2, in normal and malignant human mammary epithelial cells. *J. Immunol.*, **131**, 508–513
20. Borowitz, M. J., Tuck, F. L., Sindelar, W. F., Fernsten, P. D. and Metzgar, R. S. (1984). Monoclonal antibodies against human pancreatic adenocarcinoma: distribution of DU-PAN-2 antigen on glandular epithelia and adenocarcinomas. *J. Natl. Cancer Inst.*, **72**, 999–1003
21. Lan, M. S., Finn, O. J., Fernsten, P. D. and Metzgar, R. S. (1985). Isolation and properties of a human pancreatic adenocarcinoma-associated antigen, DU-PAN-2. *Cancer Res.*, **45**, 305–310
22. Magnani, J. L., Nilsson, B., Brockhaus, M., Zopf, D., Steplewski, Z., Koprowski, H. and Ginsburg, V. (1982). A monoclonal antibody-defined antigen associated with gastrointestinal cancer is a ganglioside containing sialated lacto-*N*-fucopentaose II. *J. Biol. Chem.*, **257**, 14635–14639
23. Harris, H. (1987). The Ca antigen: structure, function and clinical application. In Daar, A. S. (ed.) *Tumour Markers in Clinical Practice*, pp. 115–128 (Oxford: Blackwell Scientific Publications).
24. Graham, F. L. and Van der Eb, A. J. (1973). A new technique for the assay of infectivity of human adenovirus 5 DNA. *Virology*, **52**, 456–467
25. Fromm, M. and Berg, P. (1982). Deletion mapping of DNA regions required for SV40 early region promoter function in vivo. *J. Mol. Appl. Genet.*, **1**, 437–481
26. Gray, M. A., Harris, A., Coleman, L., Greenwell, J. R. and Argent, B. E. (1989). Two types of chloride channel on duct cells cultured from human fetal pancreas. *Am. J. Physiol.*, **257**, C240–251
27. Chirgwin, J. M., Przbyla, A. E., MacDonald, R. J. and Rutter, W. J. (1979). Isolation of biologically active ribonucleic acid from sources enriched in ribonuclease. *Biochemistry*, **18**, 5294–5299
28. Rommens, J. M. *et al.* (1989). Identification of the cystic fibrosis gene: chromosome walking and jumping. *Science*, **245**, 1059–1065
29. Riorden, J. R. *et al.* (1989). Identification of the cystic fibrosis gene: cloning and characterization of complementary DNA. *Science*, **245**, 1066–1073
30. Kerem, B. *et al.* (1989). Identification of the cystic fibrosis gene: genetic analysis. *Science*, **245**, 1073–1080

Section III
KIDNEY

Introduction

The renal section of the book contains three chapters, the first of which by Malcolm Hunter deals with the major transport and electrophysiological properties of the three nephron segments, proximal, diluting and distal, and reviews the literature on ion channels in these epithelia. After describing the principles of the patch-clamp technique, Dr Hunter presents a cell model for each transporting segment and details the principal ion channels in the apical and basolateral membranes. Of particular interest is the multibarrelled potassium channel found in the apical membrane of the amphibian diluting segment. This consists of four parallel equiconductive channels which, in addition to being independently gated, are also controlled by a main gate, exposing or occluding all of the channels in unison. Some of the latest observations on this and other ion channels are presented.

The remaining two chapters by Marshall Montrose and Nicholas Simmons examine the contribution of established renal (and intestinal) epithelial cell lines to our understanding of cell function in the native epithelia.

Dr Montrose considers the approach which should be applied to justify the use of a cultured cell line as a model for a specific epithelial function and he assesses the contribution of established renal and intestinal epithelial cell lines to the study of transport physiology. The chapter contains a compilation of the principal epithelial cell lines, and sub-clones where appropriate, their tissue of origin, expressed membrane enzymes and transport functions. The application of some of the cell lines to the study of cellular ion homoeostasis, hormonal regulation of transport, the isolation of transport proteins and the understanding of certain disease mechanisms are described. There is an extensive bibliography.

Dr Simmons addresses the hormonal responsiveness of Strain 1 and Strain 2 MDCK cells and their clonal lines. The action of VIP on transepithelial ion transport and, particularly anion secretion via a cyclic AMP-mediated mechanism, is discussed. The results of comparative studies with plasma membranes isolated from canine renal cortex and with isolated segments of renal tubules are consistent with the notion of a VIP receptor capable of interacting either directly or indirectly with adenylate cyclase. Many questions remain concerning the physiological role of VIP in different segments of the renal tubular epithelium but studies with MDCK cell lines have prompted useful avenues of enquiry.

As part of his chapter in the gastrointestinal section of the book, Dr Hirst compares the proton permeability of renal cell membranes with those of parietal, duodenal and jejunal epithelial cells and considers the barrier properties of MDCK cells in relation to two human intestinal epithelial cell lines, HCT-8 and T84.

6
Electrophysiology of the Nephron: New Insights gained from the Patch-clamp Technique

M. HUNTER

INTRODUCTION

The prime function of renal epithelia is the transport of solutes either into or out of the tubular fluid; in this way, the major ions present in the filtered load are selectively reabsorbed (as in the case of sodium and bicarbonate) or secreted (e.g. potassium) depending upon the needs of the body[1]. As far as the transport of ions is concerned, it is thought that the majority of the transepithelial flux passes through (transcellular) rather than between (paracellular) the tubule cells. It is obvious, however, for those ions which undergo transcellular transport, that they must traverse both the apical and basolateral membranes.

Ions may cross the cell membranes by a number of mechanisms, including cotransport with other ions or organic molecules, countertransport with other ions, simple or carrier-mediated diffusion or active transport[2]. The ions which are present in solution are not naked but are surrounded by water ions held in place by electrostatic attraction, forming a hydrated ion which is much larger than one would predict from the atomic mass. This makes simple ionic diffusion through the membrane highly improbable since the energy cost in stripping the water molecules from the ion, so that it would be able to pass through the hydrophobic interior of the membrane, is enormous. Instead, ions pass through the membrane at very little energy cost by diffusing through water-filled pores or channels which span the membrane[2,3]. Channels offer a number of advantages over simple diffusion and other transport mechanisms:

(1) Channels are usually highly selective for a particular ionic species; indeed, channels are generally classified by the ion which normally permeates them, e.g. potassium or sodium channels, thus allowing

selective transport of an individual ion species across a membrane or epithelium.

(2) They are not always open but can be closed by one or more 'gates' which are thought to be integral parts of the protein structure forming the channel and act either to stop or allow ions to flow through the channel. As a general rule, the probability of the gates being open or closed is controlled by some factor, e.g. membrane potential or intracellular calcium activity. Thus, channels can be regulated, allowing for the controlled flow of ions across the membrane or epithelium.

(3) Very large fluxes of ions may be accomplished in short periods of time. The measured and predicted flux per unit time through a channel is several orders of magnitude greater than that mediated by a carrier transport protein.

However, it must be stressed that ion flow through a channel can only proceed *down* an electrochemical gradient which must have been established by another active or secondary-active transport process.

Prior to the introduction of the patch clamp technique in the first part of this decade[4], the main evidence that channels could be responsible for at least some of the transmembrane movement of ions in the kidney was that chemicals that inhibited Na^+ (amiloride) and K^+ (barium) channels in tight epithelia caused predictable changes in the trans-tubular fluxes or intracellular activities of ions in the kidney. Unfortunately, the tubular structures of the nephron do not lend themselves readily to noise analysis techniques, since it is very difficult to obtain adequate voltage- or current-clamps in these tissues, and hence these types of data are not available for renal tubules. In 1984, the results of the first renal patch clamp experiments were reported, in which single tubules were dissected free from kidneys and apical membranes exposed by tearing the tubule, and channels were identified in the apical membranes of the proximal and distal tubules[5-7]. Since that time, a number of other experimental approaches have been used. Investigation of channels in the basolateral membrane has been carried out following removal of the basement membrane, either by manual dissection or enzymatic digestion[5,8,9]. In addition, single-cell preparations have proved useful[10] – the main drawback of this technique being that potentially harmful enzymes are commonly used in the isolation procedure.

The patch clamp technique allows one to examine single channel and cellular conductances in a number of different ways. Figure 6.1 shows the configurations that have been used in the study of renal epithelia. In the first instance, a high-resistance ($G\Omega$) seal is formed between the pipette and a small patch of cell membrane. Single-channel activity can now be observed in the *cell-attached* mode. If the pipette is withdrawn from the cell, and the cell is firmly fixed to the bottom of the chamber, the patch of membrane contained within the tip of the pipette becomes detached from the cell, forming an *inside-out* patch. The inside-out patch has its cytosolic (inside)

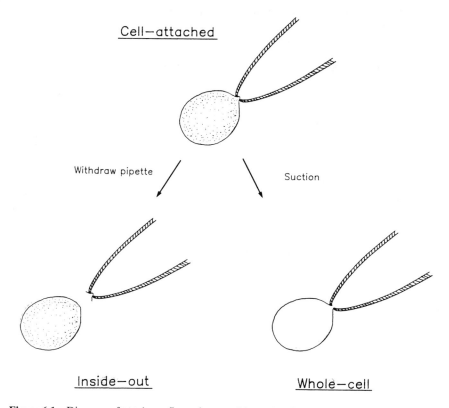

Figure 6.1 Diagram of patch-configurations used in study of conductances of renal epithelia (adapted from references 3 and 4). See text for details

surface facing the bath solution (outwith the pipette) and possible cytosolic regulators of channel function can be assessed by addition to the bath solution. If, instead of withdrawing the pipette, suction is applied to the back of the pipette, the patch of membrane ruptures and the *whole-cell* configuration is obtained. In this case, the pipette comes into direct contact with the cell interior and dialysed with the pipette solution; the cell can be either voltage- or current-clamped. This dialysis of the cell with the pipette solution offers both dis- and advantages. The major disadvantage is that diffusible cytosolic regulators of channel activity may be lost. This is offset by the major advantages that the ionic composition of the internal cytosolic solution can be determined and even large macromolecules can be introduced into the cell through the wide ($>$1 μm) pipette tip.

In this chapter, I will describe the major transport and electrophysiological properties of the three nephron segments, proximal, diluting and distal, and review the literature concerning channels in these epithelia.

PROXIMAL TUBULE

Cell model

The role of the proximal tubule is to selectively reabsorb the bulk of the solutes delivered to it via the glomerulus[11]. The driving force for this reabsorption is provided for by the ubiquitous Na^+, K^+-ATPase which is situated on the basolateral membrane of the cells (Figure 6.2). This pump acts to maintain a low intracellular sodium activity that provides the electrochemical gradient (or chemical in the case of electroneutral transport mechanisms) necessary for the uptake of sodium into the cell. This sodium entry may be via a channel, or via co-transport (e.g. together with amino acids or sugars) or countertransport (e.g. Na^+/H^+ exchange). Sodium ions are then pumped out of the cell by the Na^+, K^+-ATPase in exchange for potassium. Potassium is maintained above electrochemical equilibrium within the cell and may exit either by the apical or basolateral potassium conductance. The basolateral potassium conductance is much larger than that of the apical membrane and thus most of the potassium exits across the basolateral membrane, only to be transported back into the cell by the Na^+, K^+-ATPase. Thus potassium is said to 'recycle' across the basolateral membrane. It is this basolateral potassium conductance that is largely responsible for the negative potential of the proximal tubule cells. In addition to the basolateral potassium conductance, there is also an electrogenic Na^+/HCO_3^- co-transporter (which carries net charge, having a stoichiometry of 1 NA^+ and 3 HCO_3^- ions per cycle) which may account for a considerable amount of the current carried across the basolateral membrane, although the relative magnitude of these two conductive pathways varies along the length of the proximal tubule in both the mammal[12] and the amphibian[13]. This axial heterogeneity applies not only to the basolateral membrane but also to the transporters of the apical membrane. However, when we talk about heterogeneity within the proximal tubule, it is supposed that it is the relative proportion of the various conductances and transporters that are expressed in the cell membrane that vary with distance along the tubule, rather than there being different cell types present.

Apical channels

In the mammal, both potassium[5] and sodium[14] selective channels have been discovered, whereas in the amphibian only K^+ channels have been described[8].

The potassium channel of the apical membrane of *Necturus* (amphibian) proximal tubule has a large conductance (60 pS), is highly selective for K^+ over Na^+, is inhibited by barium when applied to the cytoplasmic face of excised patches and is voltage-sensitive, such that depolarizing potentials cause an increase in open probability[8] (the proportion of time that the channel spends in the open state; Figure 6.3). These properties are typical of calcium-activated potassium channels which are acutely sensitive to changes

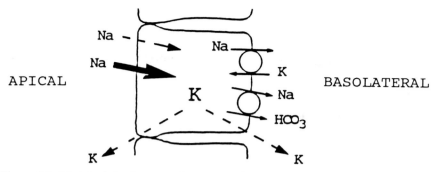

Figure 6.2 Model of dominant conductances and transport processes in proximal cells. Broken lines indicate conductance. The sodium movement from the lumen into the cell indicated by the large arrow encompasses all sodium movements other than through the conductance. See text for details

in the intracellular calcium activity[15]. In our studies, the probability of finding these channels was very low (<0.01) and we were unable to test whether the channels were sensitive to changes in the intracellular calcium activity, although it should be appreciated that there was a very large increase in open probability following excision of the channels from the cell into the bathing solution which contained 1 mmol L^{-1} calcium. Given the above facts, we fully expect the channel to be sensitive to changes in the cytoplasmic calcium concentration but this point is not yet proven.

In the rabbit, the apical potassium channel has a smaller conductance of 33 pS and can also be blocked by barium[5]. However, the kinetic appearance of the channel is dissimilar to that of the amphibian and so is presumably a different channel type (Figure 6.3). Additionally, the channel shows a slight inward rectification, i.e. inward current is conducted more easily than outward current, with the slope conductance decreasing as the reversal potential of the channel is approached[16].

The physiological role of the potassium channels is still uncertain. The proximal tubule is a site of net K^+ absorption, not secretion; thus, we must look to other roles for the apical K^+ channels. They may be involved in the K^+-efflux necessary for the volume-regulatory decrease response seen in proximal tubules or they may be important in the repolarization of the apical membrane following the sustained stimulation of transport by luminally applied substrates.

Sodium channels were found on the apical membrane of late proximal tubules of the rabbit[14]. The channels have a small conductance of 12 pS and a selectivity coefficient of 19:1 (Na^+:K). Like the sodium channel of tight epithelia, it was reversibly inhibited by the diuretic amiloride. The apparent K_d of the channel is in the range of 10^{-4} mol L^{-1}, which is similar to that for the Na^+/H^+ exchanger isolated from the apical membrane of mammalian proximal tubule cells[17], and markedly different from the K_d of around 10^{-7} mol L^{-1} for sodium channels from tight epithelia[19]. As with most channel blockers, amiloride was seen to inhibit the channel by reducing the

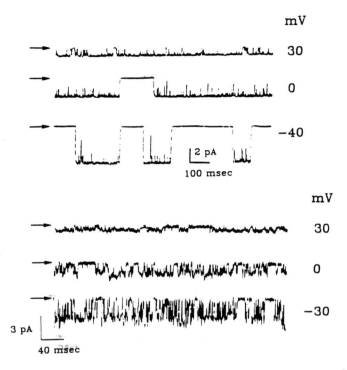

Figure 6.3 Single-channel currents of apical potassium channels from *Necturus*[8] (top) and rabbit[16] (bottom) proximal tubule. Potentials (mV) at which the excised patches were voltage clamped are shown at right of each trace. Arrows indicate the closed channel currents, openings are downwards in all of these traces. (Reproduced with kind permission of the American Physiological Society and Cambridge University Press)

amount of time that the channel remained in the open state, presumably by binding to the open channel. The channel is insensitive to changes in the cytoplasmic calcium concentration[16]. These channels are thought to make only a minor contribution to the total sodium absorption in this segment[14].

Basolateral channels

All of the channels so far described have been selective for potassium or cations, as was expected from the conductive properties of the epithelium.

In the amphibian, two distinct K[+] channels have been described, one of which has a short open time (<1 ms) and whose open probability is increased by stretching the cell membrane[19-21] and the other of which has a longer mean open time of 58 ms at the resting membrane potential[8].

The K[+] channel with the longer open[8] time has a selectivity coefficient of around 10:1 (K:Na[+]). The conductance of the channel was found to be dependent on the potassium concentration of the fluid bathing the channel,

Figure 6.4 Activation of stretch-activated channels in frog proximal tubule cells by the application of suction for the duration of the bar[21]. Figures at right indicate the number of open channels. (Reproduced with kind permission of the Journal of Physiology)

showing saturation as the potassium concentration was increased. With 100 mmol L^{-1} K^+ in the pipette, the conductance was 31 pS. This indicates that there is a binding site for K^+ within the channel and that the rate of binding or unbinding to this site can become rate limiting to the translocation of ions across the membrane. The channel is sensitive to the membrane potential, with hyperpolarizing potentials causing an increase in open probability. Kinetic analysis of the channel events showed that this increase in open probability resulted from a decrease in the time that the channel spent in the closed state, with no change on open time; i.e. the rate constant for leaving the closed state was increased upon hyperpolarization. The channel was also markedly inhibited by the application of barium to the extracellular face of the channel; this blockade was found to be both concentration- and voltage-dependent such that increasing the driving force favouring Ba^{2+} entry into the channel caused a greater degree of block. Barium reduced the mean open time of the channel without affecting the single-channel conductance, showing that barium blocks the open channel.

The stretch-activated channel is similarly sensitive to voltage, with hyperpolarization causing an increase in open probability[19]. The channel has a conductance of 47 pS in cell-attached patches and a selectivity ratio of 12:1 (K:Na$^+$). It may be activated by the application of negative pressure to the rear of the patch pipette or by osmotic swelling of the cells[21] (H. Sackin, personal communication). This phenomenon can be seen in Figure 6.4. The channel open probability in the absence of an applied stretch is minimal but increases with both suction and osmotic swelling.

It is still a matter of debate as to which of these channels is responsible for the K^+ conductance of the basolateral membrane of the amphibian proximal tubule since one can account for the conductance with either of them alone. At the moment, it would be prudent to assume that they both contribute to the potassium conductance of the basolateral membrane, but that the contribution of each might be expected to vary at least during changes in cell volume. Both of the channels show the voltage dependence exhibited by the membrane in the intact tubule, where depolarization causes a decrease in the membrane conductance[22]. However, this voltage dependence cannot account for the delayed repolarization seen following the readmission of phenylalanine to the luminal fluid of perfused tubules[22].

Two types of channel have been described in the basolateral membrane of the rabbit: a K-channel[16,23] and a non-selective cation (CANS) channel which is selective for cations over anions but which does not discriminate between sodium and potassium[24]. The K^+ channel has a conductance of around 50 pS in the presence of saturating concentrations of K, but this conductance is reduced to around 10 pS under physiological conditions (i.e. at the resting membrane potential and with low K^+ concentrations). Both inward and outward currents have been recorded from this channel; in cell-attached patches with a high potassium concentration in the pipette solution, the channel demonstrates inward rectification, i.e. current flows more easily into than out of the cell. However, with physiological concentrations of K^+ bathing the extracellular face of the channel, the conductance is nearly ohmic (linear) since the inward rectification offsets the smaller inward currents expected from Goldman rectification*. The channel is sensitive to potential, but, in this case, the open probability *increases* with depolarization, exactly the opposite to those of the amphibian. However, in the isolated perfused rabbit tubule the voltage dependence of the basolateral potassium conductance is the same as that in the amphibian, i.e. during depolarization the potassium conductance decreases (Dr J. S. Beck, personal communication). The channels are highly selective for potassium over sodium with a selectivity ratio in the order of 16:1 $(K:Na^+)^{23}$. The channel events occur in bursts with rapid flickering closed events during the open periods. Kinetic analysis of the open and closed time distributions show that there are one open state of the channel and two closed states[16,24].

Many previous experimenters had postulated that calcium may modulate the basolateral potassium conductance. Using excised, inside-out patches, varying the bath (cytosolic) calcium concentration over a range larger than that expected to be encountered physiologically had no effect on channel open probability[16]. Furthermore, addition of the calcium ionophore A23187 to the bath solution did not affect the open probability of channels in the cell-attached mode[16].

The CANS (non-selective cation) channel of the rabbit basolateral membrane has a conductance of some 25 pS in the cell-attached mode. In the cell-attached condition with no potential applied to the pipette, the channel was largely closed but was activated by depolarizing the patch potential; this voltage dependence was also evident in excised patches. As mentioned earlier, it does not discriminate between Na^+ and K, but is twice as permeable to each of these ions as it is to chloride[24]. This means that the channel would be able to conduct anions and the authors suggest that its principal role may be in providing an exit pathway for anions, e.g. phosphate and lactate, from the cell. The major evidence favouring this postulation is that two chloride-channel blockers, DPC and SITS, were effective in decreasing the open probability of the channel, but that the sodium-channel blocker, amiloride, which reversibly blocked the apical Na^+ channel, was without effect[24].

---

*Goldman showed that the magnitude of current flow through an ionic conductance depends upon the concentration of the permeating species. In the case of a K channel bathed with low extracellular K and high intracellular K, one would have predicted larger outward currents than inward[3].

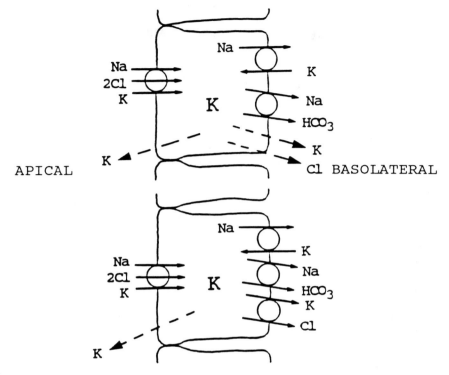

Figure 6.5 Model of dominant conductances and transport processes in the two cell types of the diluting segment, HBC cells (upper) and LBC cells (lower). Broken lines indicate conductances. See text for details

DILUTING SEGMENT

Cell model

The thick ascending limb of the mammal and the early distal tubule of the amphibian share many transport properties and are known as the diluting segments[25,26] (Figure 6.5). These segments have low water permeabilities and reabsorb NaCl against steep electrochemical gradients resulting in a dilute tubular fluid. There is a lumen positive transepithelial potential which is abolished by the loop diuretic, furosemide. The energy for transport is provided for by the Na^+, K^+-ATPase which maintains a low intracellular sodium activity. Sodium enters the cell from the tubule lumen down this sodium chemical gradient together with a K^+ ion and two Cl^- ions on an electroneutral carrier.

129

The apical membrane contains a potassium conductance which allows for recycling of potassium between the cell and the tubule lumen; this recycling of potassium is necessary for the continued reabsorption of sodium chloride since K^+ is required as a substrate for the co-transporter and the tubular delivery of K^+ is less than the amount of NaCl reabsorbed. This K^+ conductance varies with changes in the mineralocorticoid and acid–base status of the animal, in both cases mediated by an intracellular acidification[27].

The basolateral membrane of both amphibians and mammals contains a chloride conductance; it is the series arrangement of an apical potassium conductance and basolateral chloride conductance that leads to the observed lumen positive potential. Morphologically, there are two cell types in both the amphibian[28] and the mammalian[29] diluting segments. This heterogeneity is also seen in the electrical properties of the cells, but only with regard to the basolateral membrane; these cells were termed by Guggino[30] high basolateral conductance (HBC) and low basolateral conductance (LBC) cells. In the HBC cells, the basolateral membrane contains large K^+ and Cl^- conductances in parallel, whereas, in the LBC cells, an electroneutral KCl co-transporter is present. In the rabbit, it appears that the cells are principally of the LBC type[25], lacking a basolateral K^+ conductance.

In addition to these major transport pathways, there is an electrogenic $NaHCO_3$ co-transporter on the basolateral membrane of the amphibian diluting segments cells which is similar to that described for the proximal tubule[31]. This transporter would be expected to be important in maintaining the intracellular pH during chronic K-loading when a normally silent apical Na^+/H^+ exchanger becomes activated[27].

Channels

Two types of K^+ channel have been described in the apical membrane of the amphibian early distal tubule. One of these is a large conductance calcium-activated channel; the other is a channel which is apparently composed of four equally conductive subunits (MBC or multi-barrelled channel)[32,33]. There have been no published reports of channels in the apical membrane of the intact tubule of the mammalian diluting segment, although calcium activated channels have been discovered in the apical membrane of cultured cells derived from both rabbit and chick kidneys[34,35]. Cation-selective channels have been recorded in the basolateral membrane of mouse tubules[36].

Apical channels

The calcium-activated channel of *Amphiuma* (an amphibian) early distal tubule has a large conductance (120 pS with symmetrical 100 mmol L^{-1} K^+ solutions), is highly selective for K^+ over Na^+ and is inhibited by barium from either the cytoplasmic or extracellular face[37]. In the cell-attached condition, at the resting membrane potential, the open probability is immeasurably low; channel openings do not become evident until the channel is depolarized by some 100 mV. This may well argue against the channels having any physiological role since one would never expect the apical membrane potential to alter by this amount. However, I would like to offer a number of reasons why I believe the channel may contribute to the apical K^+ conductance, but maybe only after hormonal stimulation:

(1) The regulation of the channel appears to be multifactorial, with intracellular Ca, ATP and pH all being possible regulators *in vivo*[37,38];
(2) Similar channels in the cultured mammalian thick ascending limb cells are similarly closed under resting conditions but can be stimulated by the addition of ADH or forskolin[35];
(3) The channels are present at a high density on the cell membrane (the probability of finding a channel in a patch is around 0.5).

Following excision of the patch into the bath solution which has a high (1 mmol L^{-1}) calcium concentration, the open probability increases to almost 1 (Figure 6.6). Hyperpolarizing the patch causes the open probability to decrease, with an e-fold (2.3-fold) change in open probability for a 32 mV change in membrane potential (Figure 6.6). The calcium sensitivity of the channel can be best studied using excised inside-out patches. At calcium concentrations above 10^{-6} mol L^{-1}, the channel is predominantly open. Lowering the calcium activity below this level causes an abrupt decrease in channel activity[32]. These properties are typical of large-conductance calcium-activated potassium channels[15].

The channel is also sensitive to changes in the bath (cytoplasmic) concentration of ATP in inside-out patches with an apparent K_d of around 5 mmol L^{-1} in the presence of 1 mol L^{-1} calcium[38]. The inhibition can be overcome by raising the total calcium concentration to 10 mol L^{-1} whilst maintaining the ATP concentration constant. ATP apparently works by altering the calcium sensitivity of the channel by about two orders of magnitude, since the calcium activity in the 5 mol l^{-1} ATP, 1 mmol L^{-1} Ca solution was about 5.10^{-5} mol L^{-1} and gave an inhibition consistent with a calcium activity between 10^{-7} and 10^{-6} mol L^{-1} Ca. This effect is not due to phosphorylation of the channel since the inhibition was still evident in the absence of magnesium (which is normally required as a substrate for ATP-mediated phosphorylation) and a non-hydrolysable analogue of ATP, AMP-PNP, was equipotent in inhibiting the channel[38].

It was generally thought that the properties of the calcium-activated potassium channels (CAKs) of epithelia were the same as those in excitable tissues. However, we were unable to inhibit CAKs from smooth muscle, even using an ATP concentration of 10 mmol L^{-1}, and we tentatively proposed that this may be a distinguishing feature of epithelial CAKs[38].

The CAKs are highly selective for K^+ over sodium, but they do allow the permeation of other alkali cations, with the relative permeability sequence (determined from the shift in reversal potential under bi-ionic conditions):

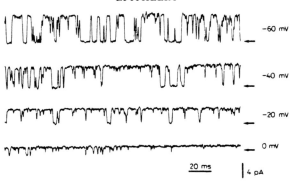

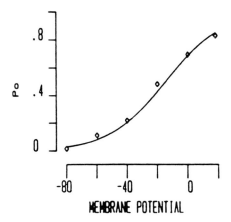

Figure 6.6 Voltage-dependence of calcium activated potassium channel in an excised, inside-out patch from the apical membrane of *Amphiuma* early distal tubule. (Reproduced with kind permission of *Federal Proceedings* and S. Karger AG, Basel). Upper trace: single-channel currents at four different clamp potentials. Arrows indicate closed channel currents, openings are upwards in these traces. Note the high open probability at 0 mV due to the presence of a high (1 mmol L^{-1}) calcium concentration in the bath solution[32]. Lower trace: open probability as a function of membrane potential. As the patch is depolarized the open probability increases[37].

$$K^+ > Rb^+ > NH_4^+ > Na^+$$

With all of the substituting ions, the conductance was reduced, but, unlike the CAK of the rat cortical collecting tubule (CCT, see later), significant amounts of current were carried by Rb^+ (Figure 6.7).

The multibarrelled channel (MBC) of the *Amphiuma* diluting segments is apparently made up of four parallel, equiconductive subunits, each of which gates independently of the others[33]. In addition to these gates, there is a main gate which acts to expose or occlude all of the subunits in unison (Figure 6.8). The channel was a high open probability in the cell-attached condition (p_0 subunits = 0.9, P_0 main gate = 0.8). At first sight, the single

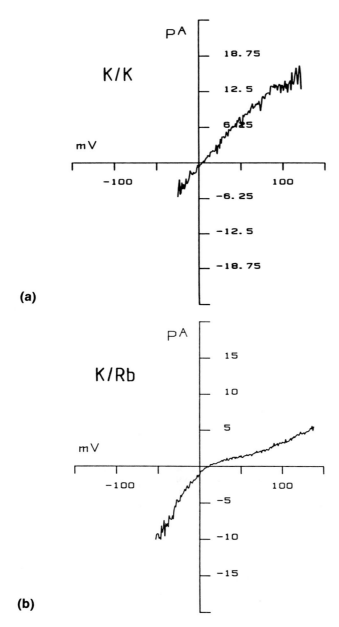

(a)

(b)

Figure 6.7 Open channel current/voltage relationships for calcium-activated channels from apical membrane of *Amphiuma* early distal tubule in excised, inside-out patches. **a.** 100 mmol L^{-1} potassium on both sides of the patch. **b.** 100 mmol L^{-1} rubidium substituted for potassium in the bath solution. Note the smaller but measurable outward (positive) current carried by rubidium and that the current axis is smaller than **a.**

channel records seem to be derived from more than one channel but the frequent closings to the zero current level reveal the presence of a common gating mechanism which we termed the main gate. When the main gate is open, the transitions between the subunits can be observed. If there were four equiconductive subunits in the patch, then we would expect the subunit currents to be integer multiples of the single subunit current. The dotted lines in Figure 6.8 are drawn at intervals of a quarter of the open channel current and one can see currents corresponding to 2, 3 and 4 of the subunits being open simultaneously. I/V curves of the channel currents show an individual subunit conductance of 7.9 pS, giving an overall single channel conductance of 31.6 pS (100 mmol L^{-1} K^{+} in pipette, cell-attached) The channel displays inward rectification, with the outward conductance appearing to be around a quarter of the inward conductance. More quantitative evidence favouring the above model is given by the amplitude histogram (Figure 6.8). If the patch contained four channels, each gating independently of the others, then the amplitude histogram should be described by the binomial distribution. This is obviously so for the subunits (right hand peaks), where the binomial distribution, shown by the dotted lines, well describes the data for a subunit open probability of 0.9. However, the probability of four channels being simultaneously closed with an open probability of 0.9 is only 9.10^{-5}, whereas the experimentally determined probability of the channel being closed was 0.22 – this is the major evidence in favour of the main gate. Kinetic analysis showed that the main gate did not interfere with the opening and closing rates of the subunit gates[33], underlining the independence of these two gating phenomena.

The channels are blocked by barium from the extracellular face in a dose- and concentration-dependent manner (K. Kawahara, M. Hunter and G. Giebisch, unpublished observations). Barium induces full channel closures without causing a change in the gating of the subunits, indicating that barium blocks at the main gate level and is denied access to the subunits.

Unfortunately, the MBCs disappear from the patch very quickly upon excision from the cell and so we have not yet been able to determine which intracellular factors are responsible for their regulation. Certainly, the presence of two types of gate would afford the possibility of regulation of the channel by several factors; for example, the subunit gates may be sensitive to intracellular pH and the main gate to phosphorylation. We can account for all of the conductance of the apical membrane of the tubule given the density of the channels (which can exceed 50 channels per patch) and the high open probability in the cell-attached condition. It is also worth noting that the author has never seen both of these channels in the same patch. This raises the possibility that the distribution of K^{+} channels may reflect the different cell types within the tubule, i.e. the CAKs may be present on one of the cell types and the MBCs on the other.

Using the whole cell clamp technique, we investigated the whole cell potassium conductance of single isolated early distal tubule cells from the frog[10]. By using calcium-free solutions in the pipette solution we hoped to minimise the contribution of the CAKs to the whole cell current, and by using predominantly K-gluconate solutions in both the pipette and the bath,

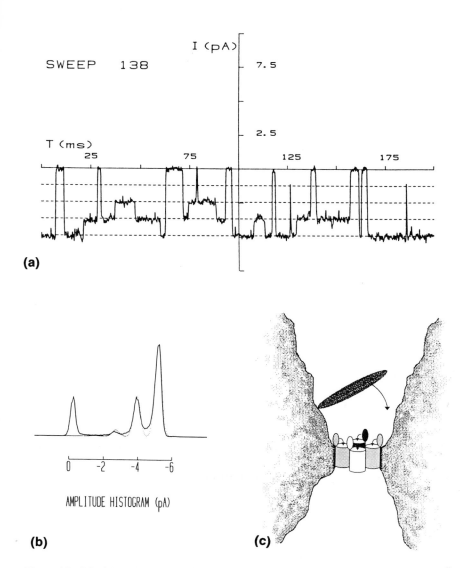

Figure 6.8 Multi-barrelled channel from apical membrane of *Amphiuma* early distal tubule[33]. (Reproduced with kind permission of *Nature (London)*). **a**. single-channel currents measured cell-attached with 100 mmol L[-1] KCl in the pipette and an applied potential of 140 mV, channel openings are downwards in this trace. Dotted lines are drawn at intervals of 1/4 of the open channel current. **b**. amplitude histogram of the above channel. The large peak at the left corresponds to the closed channel current. The peaks at the right correspond to two, three and four channels being open. Dotted line indicates predicted binomial distribution for four independently gating channels, each having an open probability of 0.903. **c**. cartoon showing proposed structure of multi-barrelled channel

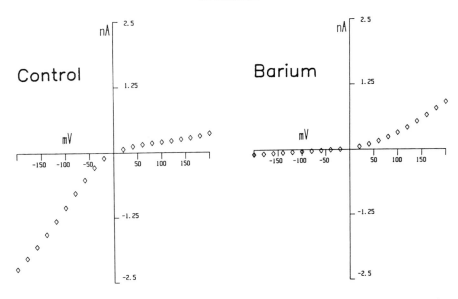

Figure 6.9 Whole cell K⁺ current/voltage relationships for frog early distal tubule cells in absence (left) and presence (right) of 10 mmol L⁻¹ barium[10]. In control conditions, the current shows inward rectification. The inward current is almost totally inhibited by the K-channel blocker, barium. (Reproduced with kind permission of the *American Journal of Physiology*)

we measured almost solely K^+ currents. In these experiments we measure the apical K^+ currents from the LBC cells and the apical and basolateral K^+ currents from the HBC cells.

The whole cell K^+ conductance is inward rectifying, with the inward current beng 4-fold larger than the outward current (the same as the apical MBC). This inward current was reduced by 92% upon the addition of 10 mmol L^{-1} barium to the bath solution (Figure 6.9). By using pipette solutions of varying pHs, we found that the cell K^+ conductance was dependent upon intracellular pH over the range 7.2 to 7.6. This very steep dependence of K^+ conductance on intracellular pH is the same as that recently demonstrated using fused cells of frog kidney[39]. On the other hand, acidification of the extracellular solution by 1.4 pH units was without effect upon the cell conductance[38]. These results agree well with the previous finding that, in isolated perfused tubules of *Amphiuma*, the apical K^+ conductance was reduced following acidification of the bath solution (both iso- and hyper-capnic) but not of the luminal perfusate, indicating that it was a change in the *intracellular* pH which mediated the change in K^+ conductance[40].

In summary there are two types of K^+ channel in the apical membrane of the amphibian diluting segment, only one of which is open at the resting membrane potential under the experimental conditions used so far. This channel has unusual gating characteristics which indicate that it is made up of four smaller channels in parallel.

Basolateral channels

An unexpected finding has been the identification of cation selective channels which do not discriminate between sodium or potassium in the basolateral membrane of cortical thick ascending limbs of mouse kidney[36]. The channels have an ohmic conductance of 27 pS in the presence of 140 mmol L^{-1} sodium and/or potassium and appear to discriminate perfectly between cations and anions. There is no clear dependence of channel activity on voltage, but the channels are activated by raising the bath calcium concentration with excised inside-out patches. However, like the calcium-activated K$^+$ channels described above, these channels were usually inactive in cell-attached patches, and only became active upon excision into bathing media containing higher calcium levels than those of the cytosol. Given the absence of a significant basolateral cation conductance in the intact epithelium[25], it is thus unlikely that these channels are open under normal conditions.

COLLECTING TUBULE

Cell model

The prime ion transport functions of the cortical collecting tubule are (a) to provide a regulated reabsorption of sodium and secretion of potassium, both of these processes being controlled by the mineralocorticoid hormone, aldosterone, and (b) acidification of the urine[41,42]. Electrophysiological and morphological results show that there are at least two cell types in the collecting tubule of amphibians and the cortical collecting tubule of mammals; the principal and the intercalated cells[28,43-45] (Figure 6.10).

The principal cells are characterized by an apical Na$^+$ conductance that is sensitive to the diuretic amiloride[46]. In mammals[46], but not in the amphibian (or at least, *Amphiuma*)[43], there is a parallel apical K$^+$ conductance, which allows for the secretion of K$^+$ by the tubule and which is inhibited by the K$^+$-channel blocker, barium. There is also evidence for electroneutral KCl symport across the apical membrane, providing an additional route for K$^+$ secretion[47]. The basolateral membrane contains both K$^+$ and Cl$^-$ conductances[48]. The energy for transcellular transport is provided by the Na$^+$, K$^+$-ATPase located in the basolateral membrane. This acts to maintain sodium below, and potassium above, electrochemical equilibrium, thus providing the driving forces favouring both sodium entry and potassium secretion across the apical membrane. It is the transepithelial sodium transport which is largely responsible for the observed lumen negative transepithelial potential.

The intercalated cells are involved in proton secretion and the apical membrane contains proton pumps but no measurable conductance, resulting in the cells having apical to basolateral resistance ratios indistinguishable from unity[44]. The basolateral membrane has a large conductance for chloride (or anions) but is one of the few cell types thought to be lacking a

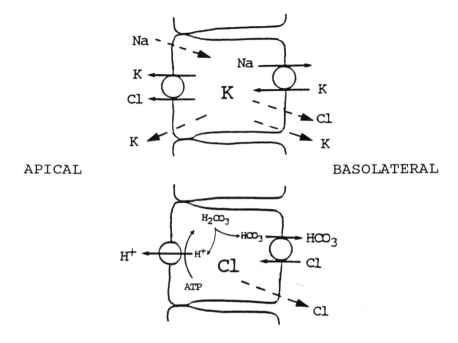

Figure 6.10 Model of dominant conductances and transport processes in the two cell types of the collecting tubule, principal cell (upper) and intercalated cell (lower). Broken lines indicate conductances. See text for details

Na^+, K^+-ATPase. To my knowledge, no channel recordings have been made from intercalated cells.

Apical channels

In all, three types of channel have been discovered in the apical membrane of the mammalian cortical collecting tubule, two potassium channels and an amiloride-sensitive sodium channel.

Apical K^+ channels

The CAK channel is present as well as a much lower conductance K^+ channel.

The CAK is highly selective for potassium over sodium with a selectivity ratio (K^+:Na^+) of over 40:1[49] and a conductance of around 100 pS[7,50]. The channels are activated by micromolar amounts of calcium and inhibited by barium in a dose-dependent manner[47]. Kinetic analysis of the channel records showed that there were at least two closed and two open states[49].

With a bath calcium concentration of 1 mmol L^{-1}, the arithmetic mean open time was 22.1 ms and the closed time was 3.6 ms. Lowering the calcium activity of the bath to 10^{-6} ml L^{-1} reduced the mean open time to 2.4 ms but caused no change in the closed time. This indicates that raising the cytosolic calcium activity maintains the channel in the open state once it has opened, but does not predispose the channel to open any quicker while it is in the closed state.

Depolarization increases the open probability in both cell-attached and excised patches[49]. This dependence of open probability on voltage has a sensitivity of an e-fold change in open probability for a 32 mV change in membrane potential. The voltage sensitivity of the channels is also affected by the calcium activity, such that the open probability versus membrane potential curve is shifted to the right when the calcium concentration is lowered[50]. This relationship has lead to the postulation that depolarization may cause a locally elevated concentration of calcium ions close to the calcium binding site, effectively altering the binding of calcium to the activation site[51]. However, a number of findings argue against this possibility:

(1) A K^+ channel from the basolateral membrane of rabbit urinary bladder shows the same voltage sensitivity as that of the rabbit cortical collecting tubule, but is insensitive to the calcium concentration[52], and

(2) Application of n-bromoacetamide, a protein-specific reactive agent, to the cytosolic face of inside-out patches containing CAKs is able to remove the calcium-sensitive component of channel activation whilst leaving the voltage sensitivity unchanged[53].

The second type of K^+ channel of the CCT has a much smaller conductance (9 pS) than that of the CAKs (100 pS), and, for the remainder of this chapter, shall be referred to as LCK (low conductance K^+ channels)[6,54]. These channels are insensitive to changes in the bath calcium concentration in inside-out patches and are also insensitive to potential over the range ±40 mV of the resting membrane potential[54]. Thus, the LCKs are markedly different from the CAKs in a number of respects. The conductance of 9 pS is that for outward current flow in the cell-attached condition. In the presence of symmetrical 140 mmol L^{-1} K^+ solutions, the channels show inward rectification, i.e. the inward current is greater than the outward currents for a given driving force, with a conductance of 25 pS.

A number of arguments have recently been put forward by Frindt and Palmer favouring the LCKs as those responsible for the K^+ conductance of the apical membrane of the CCT in the physiological condition[54]:

(1) The open probability of the CAKs is very low in the cell-attached condition unless the membrane is strongly depolarized whereas the LCKs have an open probability of around 0.9;

(2) The conductance of the CAKs to rubidium is low, whilst that of the LCKs is high, yet rubidium is known to substitute for potassium in secretion by the CCT; and

CONTROL

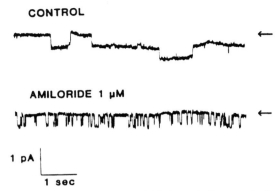

AMILORIDE 1 μM

1 pA

1 sec

Figure 6.11 Single-channel currents through Na⁺ channels in an excised, outside-out patch from rat collecting tubule[55] in the absence (upper trace) and presence (lower trace) of the sodium-channel blocker, amiloride. Arrows indicate closed channel current; openings are downwards in this trace. (Reproduced with kind permission of *Federal Proceedings*)

    (3)   The CAKs were very sensitive to TEA when applied to the extracellular face, whereas TEA was without effect upon the electrical properties of isolated perfused CCTs.

These arguments beg the question of the role of the CAKs in the CCT: why are they present at all? If the channels are closed, then they obviously cannot contribute to the apical K^+ conductance. It may be that the channels are not being stimulated under the conditions in which they have been studied. Indeed, in a cultured cell line from chick kidney, forskolin and ADH stimulated CAKs to open[35]. The results of such experiments have not yet been reported for CAKs from the CCT (nor, to my knowledge, have they been performed).

Na⁺ channels

Sodium channels have been described in the apical membrane of the rat, but not the rabbit, CCT. The channels have a small conductance (5 pS) in the cell-attached condition[55]. Both the single-channel current and the conductance saturate as functions of the bathing sodium concentration[56], with an apparent K_m of ~20 mmol L^{-1}. The channel conducts lithium ions slightly better than sodium. Like the macroscopic tubular sodium conductance, the channels are sensitive to micromolar concentrations of the diuretic, amiloride, which causes a flickering of the channel between open and blocked states (Figure 6.11). The effect of amiloride demonstrates sidedness, such that it is only effective when applied to the extracellular face of the channel.

The channels are only mildly sensitive to voltage, requiring a 100 mV change in membrane potential for an e-fold change in open probability, suggesting that membrane potential is an unlikely physiological regulator of the channel*. Cytosolic pH may be an effective regulator of channel activity,

---

In vivo, at least long-term changes in the macroscopic Na conductance appear to be mediated by changes in the channel number, rather than the open probability (macroscopic current is the product of the single-channel current, open probability and number of channels

since reducing the pH of the fluid bathing the cytosolic face of excised patches from 7.4 to 6.4 caused an eight-fold reduction in open probability[55]. This effect of acidification was rapid and reversible and was brought about by both a reduction in the mean open times and an increase in mean closed time of the channel.

It has been postulated for some time that the amiloride-sensitive sodium channel of epithelia was down regulated by raising the intracellular calcium level[57]. This regulation can be seen at the single-channel level. In cell-attached patches, addition of the calcium ionophore, ionomycin, in the presence of high bath calcium leads to a ten-fold decrease in channel open probability. However, raising the calcium activity of the fluid bathing the cytosolic face of excised sodium channels has no effect upon channel activity[55]. Thus, although raising the cytosolic calcium does reduce the channel open probability, calcium does not regulate the channel activity directly but presumably acts as an intracellular second messenger.

CONCLUSIONS

We have now entered a new era of renal electrophysiology in which we are able to measure ion flow across the membranes of the cells at the molecular level. This is providing us with very precise information about the regulation and biophysical properties of membrane conductances and of the channels through which the ions flow. Although we must take extreme care in projecting the single-channel findings to those of the epithelium, we are now in a position to begin to make predictions of the response of a particular cell to certain manoeuvres; for example, we would predict that a rise in the intracellular calcium would cause an increase in the apical potassium conductance of all of the nephron segments. In those cases where more than one channel type is present in a particular membrane, we need to be able to discern the proportion that each of them contributes to the total membrane conductance. With careful combination of flux data, conventional electro-physiology and single-channel studies, we can anticipate large advances in our understanding of the nature, role and regulation of renal and epithelial conductances over the next few years.

Acknowlegements

I am deeply indebted to Gerhard Giebisch who has constantly encouraged not only my work, but that of renal transport in general, and to Brian Cohen who gently guided me through my first patch experiments. I would like to thank all of my colleagues who generously supplied me with reprints and abstracts, and special thanks to those who allowed me the privilege of access to their unpublished work. Thanks also to Stephen Town for help in the preparation of the diagrams and to Drs Barry Argent, Mike Gray and Chris Jones for their comments on the manuscript.

in the membrane). This is suggested by the increased probability of finding channels in patches either before (<3%) or after (~50%) maintenance of rats on a low-sodium diet for a period of 1 week, a condition under which the endogenous levels of plasma aldosterone are elevated[55].

References

1. Seldin, D. W. and Giebisch, G. (1985). *The Kidney: Physiology and Pathophysiology.* (New York: Raven Press)
2. Stein, W. D. (1986). *Transport and Diffusion Across Cell Membranes.* (London: Academic Press)
3. Hille, B. (1984). *Ionic Channels in Excitable Membranes.* (Sunderland, Mass.: Sinauer Associates)
4. Hamill, O. P., Marty, A., Neher, E., Sakmann, B. and Sigworth, F. J. (1981). Improved patch-clamp techniques for high-resolution current recording from cells and cell-free membrane patches. *Pfleugers Arch.*, **391**, 85–100
5. Gögelein, H. and Greger, R. (1984). Single channel recordings from basolateral and apical membranes of renal proximal tubules. *Pleugers Arch.*, **401**, 424–426
6. Koeppen B. M., Beyenbach, K. W. and Helman, S. I. (1984). Single channel currents in renal tubules. *Am. J. Physiol.*, **247**, F380–F384
7. Hunter, M., Lopes, A. G., Boulpaep, E. L. and Giebisch, G. (1984). Single channel recordings of calcium-activated potassium channels in the apical membrane of rabbit cortical collecting tubules. *Proc.Natl. Acad. Sci.*, **81**, 4237–4239
8. Kawahara, K., Hunter, M. and Giebisch, G. (1987). Potassium channels in *Necturus* proximal tubule. *Am. J. Physiol.*, **253**, F488–F494
9. Parent, L., Cardinal J. and Sauvé, R. (1986). Single channel investigation of the basolateral membrane ionic permeability in the rabbit proximal convoluted tubule. *Biophysical J.*, **49**, 159a
10. Hunter, M., Oberleithner, H., Henderson, R. M. and Giebisch, G. (1988). Whole-cell potassium currents in single early distal tubule cells. *Am. J. Physiol.*, **255**, F699–F703
11. Giebisch, G. and Aronson, P. S. (1986). The proximal nephron. In Andreoli, T. E., Hoffman, J. F., Fanestil, D. D. and Schultz, S. G., (eds.) *Physiology of Membrane Disorders.* (New York, London: Plenum Publishing Corp.) pp. 669–700.
12. Kondo, Y. and Fromter, E. (1987). Axial heterogeneity of sodium-bicarbonate cotransport in proximal straight tubule of rabbit kidney. *Pfleugers Arch.*, **410**, 481–486
13. Lang, F., Oberleithner, H. and Giebisch, G. (1986). Electrophysiological heterogeneity of proximal convoluted tubules in *Amphiuma* kidney. *Am. J. Physiol.*, **251**, F1063–F1072
14. Gögelein, H. and Greger, R. (1986). Sodium-selective channels in the apical membrane of rabbit late proximal tubules (pars recta). *Pfleugers Arch.*, **406**, 198–203
15. Latorre, R. and Miller, C. (1983). Conduction and selectivity in potassium channels. *J. Membr. Biol.*, **71**, 11–30
16. Gögelein, H. and Greger, R. (1988). Patch clamp analysis of ionic channels in renal proximal tubules. In *Proceedings of the Xth International Congress of Nephrology*, Vol.1, pp. 159–178. (Cambridge: Cambridge University Press)
17. Kinsella, J. L. and Aronson, P. S. (1981). Amiloride inhibition of the Na/H exchanger in renal microvillus membrane vesicles. *Am. J. Physiol.*, **241**, F374–F379
18. Benos, D. J. (1982). Amiloride: A molecular probe of sodium transport in tissues and cells. *Am. J. Physiol.*, **242**, C131–C145
19. Sackin, H. and Palmer, L. G. (1987). Basolateral potassium channels in renal proximal tubule. *Am. J. Physiol.*, **253**, F476–F487
20. Sackin, H. (1987). Stretch-activated potassium channels in renal proximal tubule. *Am. J. Physiol.*, **253**, F1253–F1262
21. Hunter, M. (1988). Pressure-induced activation of channels in the basolateral membrane of frog proximal tubule cells. *J. Physiol*, **403**, 24P
22. Messner, G., Oberleithner, H. and Lang, F. (1985). The effect of phenylalanine on the electrical properties of proximal tubule cells in the frog kidney. *Pfleugers Arch.*, **404**, 138–144
23. Parent, L., Cardinal, J. and Sauvé R, (1988). Single channel analysis of a K⁺ channel at basolateral membrane of rabbit proximal convoluted tubule. Am. J. Physiol., 254, F 105–F 113
24. Gögelein, H. and Greger, R. (1986). A voltage-dependent ionic channel in the basolateral membrane of late proximal tubules of the rabbit kidney. *Pfleugers Arch.*, **407**, (Suppl.),

S142–S148

25. Greger, R. (1985). Ion transport mechanisms in thick ascending limb of Henle's loop of mammalian nephron. *Physiological Reviews*, **65**, 760–793
26. Guggino, W. B., Oberleithner, H. and Giebisch, G. (1985). The amphibian diluting segment. *Am. J. Physiol.*, **254**, F615–F627
27. Oberleithner, H., Weight, M., Westphale, H.-J. and Wang, W. (1987). Aldosterone activates Na/H exchange and raises cytoplasmic pH in target cells of the amphibian kidney. *Proc. Natl. Acad. Sci.*, **84**, 1464–1468
28. Stanton, B.A., Biemesderfer, D., Stetson, D., Kashgarian, M. and Giebisch, G. (1984). Cellular ultrastructure of *Amphiuma* distal nephron: effects of exposure to potassium. *Am. J. Physiol.*, **247**, C204–C216
29. Allen, F. and Tisher, C. C. (1976). Morphology of the ascending thick limb of Henle. *Kidney Int.*, **9**, 8–22
30. Guggino, W. B. (1986). Functional heterogeneity in the early distal tubule of the *Amphiuma* kidney: evidence for two modes of Cl⁻ and K⁺ transport across the basolateral cell membrane. *Am. J. Physiol.*, **250**, F430–F440
31. Wang, W., Dietl, P. and Oberleithner, H. (1987). Evidence for sodium-dependent rheogenic bicarbonate transport in fused cells of frog distal tubules. *Pfleugers Arch.*, **408**, 291–299
32. Hunter, M., Kawahara, K. and Giebisch, G. (1988). Calcium-activated epithelial potassium channels. *Mineral Electrolyte Metab.*, **14**, 48–57
33. Hunter, M. and Giebisch, G. (1987). Multi-barrelled K channels in renal tubules. *Nature (London)*, **327**, 522–524
34. Guggino, S. E., Guggino, W. B., Green, N. and Sacktor, B. (1987). Calcium-activated K channels in cultured medullary thick ascending limb cells. *Am. J. Physiol.*, **252**, C121–C127
35. Guggino, S. E., Suarez-Isla, B. A., Guggino, W. B. and Sacktor, B. (1985). Forskolin and antidiuretic hormone stimulate a calcium-activated potassium channel in cultured kidney cells. *Am. J. Physiol.*, **249**, F448–F455
36. Teulon, J., Paulais, M. and Bouthier, M. (1987). A calcium-activated cation-selective channel in the basolateral membrane of the cortical thick ascending limb of Henle's loop of the mouse. *Biochim. Biophys. Acta*, **905**, 125–132
37. Hunter, M., Kawahara, K. and Giebisch, G. (1986). Potassium channels along the nephron. *Fed. Proc.*, **45**, 2723–2726
38. Hunter, M. and Giebisch, G. (1988). Calcium-activated K-channels of *Amphiuma* early distal tubule: inhibition by ATP. *Pfleugers Arch.*, **412**, 331–333
39. Oberleithner, H., Kersting, U. and Hunter, M. (1988). Cytoplasmic pH determines K conductance in fused renal epithelial cells. *Proc. Natl. Acad. Sci.*, **85**, 8345–8349
40. Guggino, W. B., Stanton, B. A. and Giebisch, G. (1982). Regulation of apical potassium conductance in the isolated early distal tubule of the *Amphiuma* kidney. *Biophys. J.*, **37**, 338
41. Wright, F. S. and Giebisch, G. (1985). Regulation of potassium excretion. In Seldin, D. W. and Giebisch, G. (eds.) *The Kidney: Physiology and Pathophysiology*. (New York: Raven Press)
42. Steinmetz, P. R. (1985). Epithelial hydrogen ion transport. In Seldin D. W. and Giebisch G. (eds.) *The Kidney: Physiology and Pathophysiology*. (New York: Raven Press)
43. Hunter, M., Horisberger, J-D., Stanton, B. A. and Giebisch, G. (1987). *Amphiuma* collecting tubule I: electrophysiological properties. *Am. J. Physiol.*, **253**, 1263–1272
44. Koeppen, B. M. (1987). Electrophysiological identification of principal and intercalated cells in the rabbit outer medullary collecting duct. *Pfleugers Arch.*, **409**, 138–141
45. Madsen, K. M., and Tisher, C. C. (1986). Structural-functional relationships along the distal nephron. *Am. J. Physiol.*, **248**, F449–F453
46. Koeppen, B. M., Biagi, B. A. and Giebisch, G. (1983). Intracellular microelectrode characterization of the rabbit cortical collecting duct. *Am. J. Physiol.*, **244**, F35–F37
47. Velasquez, H., Wright, F. S. and Good, D. (1982). Luminal influences on potassium secretion: chloride replacement with sulfate. *Am. J. Physiol.*, **242**, F46–F55
48. Sansom, S. C. and O'Neil, R. G. (1986). Effects of mineralocorticoids on transport properties of cortical collecting duct basolateral membrane. *Am. J. Physiol.*, **251**, F743–F757

143

49. Hunter, M., Lopes, A. G., Boulpaep, E. and Giebisch, G. (1986). Regulation of single K⁺-channels from apical membrane of rabbit collecting tubule. *Am. J. Physiol.*, **251**, F725–F733

50. Frindt, G. and Palmer, L. (1987). Calcium-activated potassium channels in the apical membrane of the mammalian cortical collecting tubule and their role in K secretion. *Am. J. Physiol.*,**252**, F458–F467

51. Moczydlowski, E. and Latorre, R. (1983). Gating kinetics of Ca-activated K channels from rat muscle incorporated into planar lipid bilayers: Evidence for two voltage-dependent Ca binding reactions. *J. Gen. Physiol.*, **82**, 511–542

52. Hanrahan, J. W., Alles, W. P. and Lewis, S. A. (1985). Single basolateral K channels are expressed in primary cultures of urinary bladder epithelial cells. *Biophys. J.*, **47**, 11a

53. Cornejo, M., Guggino, S. E. and Guggino, W. B. (1987). Modification of calcium-activated K channels cultured medullary thick ascending limb cells by n-bromoacetamide. *J. Memb. Biol.*, **99**, 147–155

54. Frindt, G. and Palmer, L. (1989). Low conductance K channels in the apical membrane of the rat cortical collecting tubule. *Am. J. Physiol.*, **256**, F143–F151

55. Palmer, L. and Frindt, G. (1986). Epithelial sodium channels: characterization by using the patch clamp technique. *Fed. Proc.*, **45**, 2708–2712

56. Palmer, L. and Frindt, G. (1988). Conductance and gating of epithelial Na channels from rat cortical collecting tubule. *J. Gen. Physiol.*, **92**, 121–138

57. Grinstein, S. and Erlij, D. (1978). Intracellular calcium and the regulation of sodium transport in the frog skin. *Proc. R. Soc. Biol*, **202**, 353–360

M. Hunter, Leeds, April 1989

7
Transport Physiology of Renal and Intestinal Cell Culture Models

M. H. MONTROSE

INTRODUCTION

Over the past several years, the use of tissue culture systems has greatly increased our ability to study various aspects of the physiology of renal and intestinal epithelial cells. In both of these tissues, a variety of different cell types is found in the epithelial layers. In contrast, under optimal conditions, cultured intestinal or renal epithelial cells have presented an opportunity to study a homogeneous cell type in a well-defined environment. In other words, cultured cells offer the chance to simplify analyses of complex phenomena, while improving experimental rigour. Two types of experimental culture systems have emerged. The first is *primary culture*, in which freshly isolated cells are placed in culture and cell proliferation is limited. The second system is the use of *established cultured lines*, in which cells are maintained in culture for extended periods. This chapter will focus on the utility of the latter cell type, as a large number of established cell lines are available (see Table 7.1), and this type of system offers the most experimental flexibility.

Despite the availability of such established cell lines, their application to the study of renal and intestinal physiology should be made judiciously. Prior to working with a cell line, investigators must consider the minimal requirements for a cell to be satisfactory for study. In general, it is the topic of study itself which places requirements on the choice of a model cell system. For some questions concerning the biology and physiology of epithelial cells, it is not always essential to have cells with a defined homology to a specific tissue type. Studies which require the expression of features found in a variety of epithelia (such as epithelial polarity, tight junctional structure, and microvillus membrane structure) often fall into this category. Similarly, some functions may be tissue specific, but not region specific. For example, Cl^- secretion is elicited by increases in cellular cAMP in both small and large intestine[42,43], and may therefore be studied in

Table 7.1 Partial listing of established epithelial cell lines from kidney and intestine

Cell line	Subclones	Originating tissue	References
LLC-PK₁		Whole pig kidney	1
	LLC-PK₁ₐ		1,2*
	D+Sc		3
	D+Rc		3
	-D-		3
	CL4		4
	PKE20		5
OK		Whole opossum kidney	6
	PTH responder		7
MDCK		Whole canine kidney	8
	Low resistance		9,10
	High resistance		9, 10
A6		Whole South African toad kidney	11
BSC-1		Whole African green monkey kidney	12
293		Whole human kidney	13
FrhK-4		Whole rhesus kidney	14
LLC-MK2		Whole rhesus kidney	15
PtK1		Whole marsupial kidney	16
PtK2		Whole marsupial kidney	16
RK₁₃		Whole rabbit kidney	17
LLC-RK1		Whole rabbit kidney	15
NRK-52E		Whole rat kidney	18
MDOK		Whole sheep kidney	19
MDBK		Whole bovine kidney	19
MTAL		Thick ascending limb of rabbit	20
M-mTAL-IC		Thick ascending limb of mouse	21
RC.SV1		Cortex of rabbit	22
RC.SV2		Cortex of rabbit	22
RC.SV3		Cortex of rabbit	22
RCCT-28A		Cortical collecting duct of rabbit	23
A-498		Carcinoma – whole human kidney	24
A-704		Carcinoma – whole human kidney	24
Caki-1		Carcinoma – whole human kidney	25
Caki-2		Carcinoma – whole human kidney	25
FHs74 Int		Human small intestine	26†
Intestine 407		Human small intestine	27**
IEC-6		Rat small intestine	28
IEC-18		Rat ileum	28
HCT-8		Carcinoma – human ileocecal intestine	29††
IA-XsSBr		Carcinoma – rat small intestine	30
GPC-16		Carcinoma – guinea pig colon	31
T₈₄		Carcinoma – human colon	32
Caco-2		Carcinoma – human colon	33
HT-29		Carcinoma – human colon	25
	HT29-18		34
	HT29-C₁		34
	HT29-N₂		34
	HT29-D4		35
	HT29-D9		35
	Cl.27H		36
	HT29-19A		37
	HT29-16E		37

Table 7.1 *Continued*

Cell line	Subclones	Originating tissue	References
SW48		Carcinoma – human colon	38
SW403		Carcinoma – human colon	38
SW480		Carcinoma – human colon	38
SW948		Carcinoma – human colon	38
SW1116		Carcinoma – human colon	38
SW1417		Carcinoma – human colon	38
LS180		Carcinoma – human colon	39
LS 174T		Carcinoma – human colon	39
DLD-1		Carcinoma – human colon	40
HCT-15		Carcinoma – human colon	40
HCT-116		Carcinoma – human colon	41
SW837		Carcinoma – human rectum	38
SW1463		Carcinoma – human rectum	38
HRT-18		Carcinoma – human rectum	29††

*Note that two distinct clones have been named LLC-PK$_{1A}$
†No keratin was detected in this line
**Contamination by HeLa markers sugests that this line is not intestinal
††HCT-8 and HRT-18 may be cross-contaminated and therefore identical cell lines

a variety of intestinal cell models[44-46]. However, in order to use a tissue culture cell to study an aspect of renal or intestinal physiology which is specific for a defined cell type, the cultured cell must express the physiological function of interest. This chapter will review information on the characterization of some renal and intestinal tissue culture systems which have been useful for studies of transport physiology.

CHARACTERIZATION OF CELL CULTURE LINES

The goal of characterizing a tissue culture cell is to define whether it is a good model system for answering a specific question. In early work with epithelial tissue culture cells, one strategy was to use information about the expression of a wide variety of native tissue functions by a tissue culture cell, in order to make a general assignment of homology to certain tissue types. Ironically, the ambiguity of such information collected from native tissue (containing a variety of cell types) is often the reason that tissue culture systems have been sought in the first place. In addition, as one might expect, tissue culture cells do not mimic the performance of native tissue cells in all respects. For these reasons, a more successful strategy has been to ask questions specifically related to the research topic which seek to define whether it is possible to work within the limits of homology between the tissue culture cells and native tissue. Such strategies are discussed below under *Homology to native tissue: transport physiology*, after a discussion of the general features of established cell lines.

Homology to native tissue: general features

A simple question concerning any tissue culture cell line is the definition of the originating cell type which was cultured from the native tissue. Unfortunately, even this question is often difficult to answer because of heterogeneity in the starting material as well as heterogeneity in the cultured cells themselves. In the case of renal culture, many of the established cell lines have been obtained from explants of whole kidney preparations and therefore the nephron segment(s) from which the cultured cells have been derived cannot be immediately defined (see Table 7.1). In contrast, the large versus small intestinal origin is known for intestinal cell lines, but this information is compromised by the presence of different cell types in the original epithelial layer (e.g. villus cells, crypt cells, enteroendocrine cells and goblet cells) which have varying proliferative capacities. As one might anticipate, a number of uncloned cell lines have been identified as functionally and morphologically heterogeneous[3,9,10,34-37]. The assessment of homogeneity of a population can be made on either morphological criteria[3,34,35], reaction with antibodies or other markers of specific proteins[34,47], or can be based on the function of subcloned or single cells[3,9,10,35-37,48,49]. In some cases, even clonal lines have the capability to produce a variety of cell types which must be further subcloned to produce a stable and homogeneous cell type. This is best documented for the case of the cloned HT29-18 cell line (a human colon carcinoma), which can be induced to differentiate into multiple cell types[34]. In addition, for unknown reasons, even clonal lines have the tendency to 'drift' in phenotype and should occasionally be reselected for the functions under study.

In order to define the originating cell type of a tissue culture population, one preferable alternative to whole explant culture is the establishment of immortalized tissue culture lines from a defined epithelial source. This has been performed most frequently using defined nephron segments to create established lines of medullary thick ascending limb[22,23] or collecting duct cells[23]. While the use of such cells allows firm conclusions about cellular origin, it is worth noting that even the establishment of such lines does not guarantee that their physiology remains faithful to the originating tissue[23]. Use of a defined epithelial source has been more difficult in intestinal culture[28], because the intestinal epithelia have a multitude of different cell types in the same epithelial layer[42].

A number of cultured renal and intestinal epithelial cells express general features found in many epithelia. These features include an epithelial morphology (all cells in Table 7.1 satisfy this criterion) as well as formation of tight junctions and polarized expression of plasma membrane proteins. In addition to satisfying these 'minimal requirements' as epithelial cells, functions can also be expressed which are only observed in a subset of epithelial cells. While a cell line may express some functions which suggest homology to a desired cell type, it is probably not correct to justify use of a tissue culture cell as based on the expression of functions unrelated to the topic of study. This may be best illustrated by the work performed to evaluate expression of membrane enzymes by renal and intestinal tissue culture cells. The expression of microvillar membrane hydrolases is a function which is known to be restricted to the early segments of the intestine (small intestine) and the nephron (proximal tubule), probably because these regions are involved in absorption of the enzymatic products. As shown in Table 7.2, a number of both intestinal and renal cell lines express such enzymes, but contrary to expectation, this has not been successfully applied as a criterion for homology to early segments of intestinal or renal tissue. For instance, given the colonic origin of the Caco-2 and HT29 cells[25,33], it appears curious that these cell lines should express such enzymes (see Table 7.2). However, when monoclonal antibodies were used to identify the enzymes in Western blots, it was found that the adult human small intestine expresses hydrolases of a different molecular weight from Caco-2 cells, while fetal colon expresses hydrolases of the same molecular weight as Caco-2 cells[47]. On this basis, it has been suggested that Caco-2 cells be used as models of fetal colon[47]. Similarly, when the expression of both membrane and cytosolic enzymes was directly compared among native renal tissue and two renal cell lines (LLC-PK$_1$ and MDCK), it was not possible to assign a proximal or distal phenotype to the tissue culture cells[53]. The enzymes which were observed suggested that a mixture of both proximal and distal tubular functions were being expressed in both cell lines. While population heterogeneity can contribute to such confusion, an assignment of homology based on such criteria is clearly weakened.

Homology to native tissue: transport physiology

The most practical criterion for defining the utility of any given culture model to study membrane transport is to *examine the complement of proteins and functions expressed by the cells which relate directly to the topic of study*. A compilation of such information is presented in Tables 7.3 and 7.4 for a variety of renal and intestinal cell lines, and the relevance of such information to studies of transport physiology is described below. It should be mentioned that, in most cases, measurements of function (transport function in particular) are made on a number of cells simultaneously. Therefore, the heterogeneity of cellular transport function within a population has only been evaluated infrequently[48,49,134,175]. It should also be mentioned that the listing of cells which is presented in Table 7.1 is only a subset of those which may be appropriate model systems. Just as functional similarities exist between the distal segments of the intestine and kidney and those of the bladder, cells cultured from the toad bladder have been shown to be useful for studies of some aspects of renal function[161].

A major function of the intestinal and renal epithelia which has been studied in cultured cells is the vectorial transport of salts and solutes. The cellular basis for this vectorial transport is the expression of specific membrane proteins within restricted plasma membrane domains (i.e. apical or basolateral membranes). As shown in Table 7.3, several membrane

Table 7.2 Membrane enzymes expressed by renal and intestinal cell lines

Cell line	Membrane enzyme	References
LLC-PK$_1$	γ-Glutamyltransferase	50–53
	Aminopeptidase	52,53
	Alkaline phosphatase	50,53
	Trehalase	52,54
OK	γ-Glutamyltransferase	55
	Leucine aminopeptidase	55
MDCK	γ-Glutamyltransferase	53
	Leucine aminopeptidase	53
RC.SV1	γGlutamyltransferase	22
	Leucine aminopeptidase	22
	Dipeptidase IV	22
HT29	Sucrase–Isomaltase	35,37,56–58
	Dipeptidase IV	56
	Aminopeptidase N	35,56,57
	Maltase	37,57
	Alkaline phosphatase	57
Caco-2	Sucrase-isomaltase	47,59,60
	Maltase	60
	Aminopeptidase N	47
	Dipeptidase IV	47
	Alkaline phosphatase	59
	Trehalase	59
	γ-Glutamyltransferase	60
	Alkaline phosphatase	61

transporters have been identified in cultured renal and intestinal cells which are candidates for expressed epithelial membrane proteins. Two features of this epithelial specialization can be used to help define the homology of transport systems in cultured cells to those found in native tissue. In some cases, the *expression of a transport function is limited to defined epithelial cells.* For example, except for the brush-border membrane of the proximal tubule and small intestine, the Na$^+$-glucose co-transporter has not been detected in the plasma membrane of mammalian tissue cells[162]. For this reason, several groups have studied the transport mechanism and regulation of the Na$^+$-glucose co-transporter expressed in renal or intestinal cell lines as a model of the same transporter in native tissue[60,163–165]. Alternatively, *some transport systems* may be *present in both polarized and non-polarized cells, but have specialized functions and a restricted polarity* in epithelial cells. Thus, either a functional analysis and/or an assessment of polarity can be used to define the epithelial origin of a transporter in cultured epithelia. As shown in Table 7.3, the polarized expression of a number of transporters has been established. For example, an apical Na$^+$/H$^+$ exchanger is responsible for both pH regulation and transepithelial transport in the proximal tubule, whereas a pH regulatory basolateral Na$^+$/H$^+$ exchanger is found in more distal nephron segments. The presence of this transport reaction has

Table 7.3 Membrane transporters expressed by renal and intestinal cell lines

Cell line	Membrane transporter	Polarity*	References
LLC-PK$_1$	Na/K/Cl co-transport	A	62
	Na/glucose co-transport	A	52,63,64
	Na/aspartate co-transport	A	65
	Na conductance	A	66
	Na/phosphate co-transport	A	67
	Na/H exchange		68
		A	54,69,70
		BL	48,70
	Lysine transport	BL	71
	Na/alanine cotransport	BL	72
	Cl/HCO$_3$ exchange		73
	Organic cation transport		74
	Ca^{2+}-ATPase		75
OK	Cl channels		76,77
	K channels		78,77
	Na/proline co-transport	A	55
	Na/glutamate co-transport	A	55
	Na/glucose co-transport	A	55
	Na/phosphate co-transport	A	55
	Na/H exchange		79–81,49
	Na/K/Cl co-transport	A	175
			81
BSC-1	Na/HCO$_3$ co-transport		82,83
	Na/H exchange		83,84
	Cl/HCO$_3$ exchange		83,84
JTC.12P3	Na/phosphate co-transport		85
A6	Cl channel		86
	Na conductance	A	87
MDCK	Na/H exchange		84,88,89
	Na/K/Cl co-transport		90–94
	Cl/HCO$_3$ exchange		84
	Lysine transport	BL	71
	Anion channels		95
RC.SV1	Na/glucose co-transport		22
	Na/H exchange		22
T$_{84}$	Cl channels	A	45,96,97
	Na/K/Cl co-transport	BL	98
	K conductance		99,100
HT29	Cl channels		101,102
	Na/K/Cl co-transport		103
	Na/H exchange		104
Caco-2	Na/glucose co-transport		60

*A = apical; BL = basolateral

been described for many cultured renal and intestinal epithelia, but the polarized distribution of the exchanger is known to vary even among strains of the same parental cell line[48,54,69,70].

A variety of hormones regulates transepithelial transport in renal epithelia, and the action of these hormones has been intensively studied in

Table 7.4 Hormone receptors expressed in renal and intestinal cell lines, and membrane transport processes affected by hormone/receptor interactions

Cell line	Hormone	Regulated transport	References
LLC-PK$_1$	Vasopressin		105–108
	Calcitonin		3
	Vitamin D		109
OK	Calcitonin		85,108
	PTH		110,111
		Na/phosphate	85,112–114
		Na/H exchange	49,79–81, 110,115
	Atrial natriuretic peptide		116
		Na/H exchange	115
BSC-1	Epidermal growth factor	Membrane potential	117
JTC.12	PTH		85,118,119
A6	Vasopressin		87,120
	Aldosterone		
		Na flux	87,121–125
		Na,K-ATPase	126,127
MDCK	Bradykinin		128
	Vasopressin		129
		Na transport	130
	Adrenergic		131
		Cl secretion	94,132
	Purine		
		Cl secretion	133
RC.SV1-3	PTH		22
	ß-Adrenergic		22
	Calcitonin		22
RCCT-28A	Bradykinin		23
	PTH		23
	Vasopressin		23
	α-Adrenergic		23
	Calcitonin		23
T$_{84}$	Cholinergic		
		Cl secretion	46
		Calcium flux	134
	VIP		
		Cl secretion	45,46
	Somatostatin		
		Cl secretion	46
HT29	Insulin		135
	α_2-Adrenergic		136–144
	α_{2A}-Adrenergic		145tb
	Neurotensin		146,147
	Aldosterone		148
	Gastrin		149
	Somatostatin		149

Table 7.4 *continued*

Cell line	Hormone	Regulated transport	References
	VIP		140,144,150–158
		Na/K/Cl co-transport	103
Caco-2	Adrenergic	Cl secretion	159
	VIP	Cl secretion	159,160

cultured renal cells. Studies of isolated renal tubule segments have documented a pattern of response to polypeptide and steroid hormones which is characteristic for different regions of the nephron (i.e. a segmental heterogeneity of hormonal response is observed)[166–170]. Therefore, the presence of hormone receptors and hormonally-mediated changes in transport in a given cell line have been used to support the study of tissue culture cells as models of transport regulation in specific epithelia[161]. Using radio-ligand binding or hormone-induced changes in intracellular regulators to document hormone-receptor binding in renal tissue, it has been observed that parathyroid hormone (PTH) receptors are found in most segments of the nephron other than the thick ascending limb of Henle (TALH)[166,168], and are also expressed in the renal cell lines OK, JTC-12.P3, RC.SV1, and RC.SV2 (from opossum, monkey and rabbit kidney, respectively)[22,79,85,112,118]. Similarly, the receptors for vasopressin (antidiuretic hormone; ADH) and calcitonin are expressed in distal nephron segments[166,169–171] and in the MDCK, LLC-PK$_1$, and RC.SV3 cells (from dog, pig and rabbit kidney, respectively)[22,105–108,129] but are not expressed in high amounts in OK or RC.SV1 cells[22,85].

Although a number of peptide and steroid hormones also regulate intestinal transport, information about hormonal responses has been of lesser utility in defining models of transport regulation for specific intestinal segments. There are two reasons for this state of affairs. The first reason is that the axial heterogeneity of hormonal receptors on intestinal epithelial cells is not known in a number of cases. Thus, although HT29 cells express neurotensin receptors[146,147], there is no clear definition of the expression of these receptors in the intestine. The second reason is that some hormones have been shown to alter both small and large intestinal ion transport[42,43]. The advantage of this latter situation is that study of the transport response to these hormones may produce information relating to transport regulation in numerous segments of the intestine. The best example of the latter case concerns the expression of VIP receptors by colonic cell lines. It has been shown that transport activity is regulated by VIP in at least 3 colonic cell lines T$_{84}$, Caco-2 and HT29 cells) as well as in both the small and large intestine[46,103,159,160].

TRANSPORT PHYSIOLOGY IN DIFFERENT CELL LINES

The next section will provide examples of the types of topics and questions which have been successfully approached using established cell lines.

Cellular pH and volume homoeostasis

The factors contributing to ion homoeostasis in ion-transporting epithelial cells are complex, and encourage the development of simple experimental systems. Like most other cells, epithelial cells perform 'housekeeping' functions of pH and volume regulation. In contrast to other cells, these functions must be performed despite the presence of transcellular ion transport. Since both functions can involve large net fluxes of osmolytes and acid/base equivalents across the plasma membrane, cellular ion homoeostasis can become a balancing act between tissue and cellular requirements. As discussed by others, this requires a careful co-ordination of fluxes across apical and basolateral membranes, as well as the ability to accommodate the signals regulating transepithelial transport[81,172]. Surprisingly, only limited information is available on the pH and volume regulation of intestinal cells (either native or cultured)[173,174]. In contrast, these homoeostatic functions have been examined in a number of renal cell lines.

Cellular pH homoeostasis in renal cell lines

Several cultured cell lines have been used to examine the mechanisms contributing to pH homoeostasis in renal epithelia. In the four renal cell lines examined (BSC-1, LLC-PK$_1$, OK and RC.SV1), a Na$^+$/H$^+$ exchanger is responsible for the vast majority of net acid extrusion after cells are subjected to an acid load in the absence of bicarbonate[22,48,49,68,83]. Despite this similarity, clear differences have been noted in the properties of the exchanger between cells, the most notable being that, in the same assay system (optical measurements of intracellular pH in single cells), LLC-PK$_1$ and OK cells express the exchanger in opposite membranes (basolateral and apical, respectively)[48,175]. As the apical exchanger in the OK cell is also inhibited by PTH[49,79–81,110,115], these data have suggested that OK cells express the renal proximal tubular Ha$^+$/H$^+$ exchanger which is regulated by PTH in native tissue (see *Parathyroid hormone* section under *Signal transduction mechanisms regulating transport*). Other laboratories have identified both apical and basolateral Na$^+$/H$^+$ exchange in the uncloned LLC-PK$_1$ cell, and this may be due to different strains of the parent cell line (which is known to be heterogeneous) or to differences in experimental conditions[48,54,69,70]. Mutant clones have been isolated from the LLC-PK$_1$ cell which specifically express more apical Na$^+$/H$^+$ exchange[70], evidence which supports the expression of apical Na$^+$/H$^+$ exchange by the LLC-PK$_1$ cell under some circumstances.

Regulation of intracellular pH in the presence of CO_2/HCO_3 has been examined in less detail, but two HCO_3-dependent transport systems have been identified. In both the BSC-1 (from monkey kidney) and LLC-PK$_1$/C14 cells, Cl$^-$/HCO$_3$ exchange has been identified[73,83]. In the BSC-1 cell, Na$^+$/HCO$_3$ co-transport has also been identified[82], but is is not known if this latter system is Cl$^-$ dependent. The latter point is important for comparison with the cells of the proximal tubule and thick ascending limb, as Cl$^-$-independent Na$^+$/HCO$_3^-$ co-transport is the only observed HCO$_3^-$-dependent transport system significantly affecting intracellular pH[176,177].

Cellular volume homoeostasis

As described in a preceding paragraph, control of cell volume is an important requirement for ion-transporting epithelial cells. The observation of cell volume regulation, or cell survival in different osmolarities, has been examined in several cultured renal cell lines. Control of cell volume is especially important to the cells of the renal medulla because these cells are required to survive in an environment which has large fluctuations in osmolarity. MDCK cells (commonly used as a model of distal nephron) have been examined in both hyper- and hypo-osmolarity, and only demonstrate rapid volume regulation in conditions of lowered osmolarity[91]. Cells cultured from the papillary epithelium (GRP-MAL/MTAL)), MDCK, and LLC-PK$_1$ cells have been shown to have a slow adaption to higher osmolarity via accumulation of specific organic solutes[178]. The accumulated solutes are the same as those observed in the renal medulla[178,179], supporting the homology of this response to the native tissue. In the case of the GRP-MAL cells, this was shown to be due to an increased expression of aldoreductase[180].

In the cells of the mammalian proximal tubule, current evidence supports the presence of K$^+$ and anion channels which are responsible for osmolyte loss from swollen cells[181-183]. In the OK cells, evidence suggests that K$^+$ and Cl$^-$ channels are also activated in response to cell swelling[76], and cause loss of cellular osmolytes (with resultant regulatory volume decrease: RVD)[77]. In these cells, RVD was shown to involve the membrane potential; depolarization was noted in patch clamp experiments, and RVD was inhibited by valinomycin (which was shown to hyperpolarize the plasma membrane)[76,77]. These observations suggested that the activation of anion channels must either precede or be greater than the activation of K$^+$ channels in order for RVD to progress in the OK cell.

An important aspect of epithelial pH and volume homoeostasis is to define the interactions of these 'housekeeping' events with transepithelial transport. This has been examined in the OK cell by examining the control of volume regulation versus transepithelial transport[81]. In OK cells, PTH is known to control activity of apical Na$^+$/H$^+$ exchange[47,79-81,110,115,175], and so has been used as a model system to study the PTH regulation of bicarbonate absorption in the proximal tubule[184,185]. In the OK cell, the Na$^+$/H$^+$ exchanger affected by PTH catalysed a large net uptake of osmolytes, but this transporter was not activated in response to the cellular need to perform regulatory volume increase (RVI) after forced cell shrinkage[77,81]. However, the OK cells were observed to perform RVI and RVD, suggesting that the intracellular regulation of volume regulation was intact[81]. Further, PTH had no significant effect on the progress of RVI in OK cells[81]. This suggested that, in this cultured cell model of the proximal tubule, the control of cell volume and transepithelial transport are separate. This represents a simple solution to the problem of co-ordinating control between these tissue and cellular requirements: each transport system only needs to respond appropriately to the regulatory signals directing a single physiological purpose.

Epithelial transport proteins

The use of cultured epithelial cells has offered an opportunity to identify and isolate renal and intestinal transport systems. To date, a variety of membrane proteins has been isolated using tissue culture cells which are either affirmed or candidate membrane transport proteins from the intestinal and/or renal epithelia.

Using antibodies directed against the Na^+,K^+-ATPase, the α and β subunits of the Na^+,K^+-ATPase have been cloned from A6 (*Xenopus laevis* kidney) cells[186]. Immunofluorescent studies have verified the basolateral localization of Na^+,K^+-ATPase in these cells, and the α subunit was found to be of the α-1 isoform which is common in kidney basolateral membranes[186]. These antibodies have also been used to follow the conformational maturation and intracellular transport of the Na^+,K^+-ATPase following induction by aldosterone[126,127].

Also using A6 cells, Benos *et al.* have identified and purified a large protein complex which has been tentatively identified as the epithelial Na^+ channel[187,188]. This protein complex of 730 kDa may be separated into six components by reduction of sulphydryl bonds with mercaptoethanol, but not by high salt or urea. Evidence suggests that amiloride, a potent inhibitor of the Na^+ channel, binds to only one of these six components[187].

Sorscher *et al.*[189] recently used a monospecific polyclonal antibody to identify a DIDS-binding protein in T_{84} (human colon carcinoma) cells. It has been observed previously that disulphonic stilbenes, such as DIDs, are effective inhibitors of Cl^- channels in colonic cells[190]. Since the antibody identifies a single protein on Western blots, and acts at low concentrations to block Cl^- channels[189], evidence suggests that this DIDS-binding protein is a component of the apical Cl^- channel responsible for transepithelial ion secretion. An unusual observation from this group is that the labelled protein is observed primarily in intracellular vesicles prior to exposure of the cell to secretory agonists[189]. This suggests that induction of ion secretion may require insertion of Cl^- channels into the apical membrane.

Attempts to label the renal $Na^+/$phosphate co-transporter with group-specific reagents have identified several candidate membrane proteins of OK (opossum kidney) cells which are labelled by *N*-acetylimidazole[191]. The presence of phosphate can both protect against the inhibition of $Na^+/$ phosphate co-transport caused by *N*-acetylimidazole, and can simultaneously protect against labelling of proteins of molecular weights 31, 53, 104, and 176 kDa. The amount of labelled protein also varied due to exposure to

PTH, which is known to regulate the expression of phosphate transport in this cell line[85,112-114]. In a comparison with work using native tissue, similar labelling strategies applied to membrane vesicles from intestinal tissue suggest that a 130 kDa protein is a candidate for a Na^+/phosphate co-transporter[192].

Most recently, Wu and Lever have used a monoclonal antibody to identify a 75 kDa protein in LLC-PK$_1$ cells which is a candidate for the Na^+-glucose co-transporter[165]. The LLC-PK$_1$ protein appears similar in molecular weight to the Na^+-glucose co-transporter previously cloned by Hediger et al. from native intestinal tissue[193], but awaits further investigation.

Signal transduction mechanisms regulating transport

A large number of hormones affect transepithelial transport in the intestine and kidney. Based on the observation that many tissue culture cells express hormone receptors (see Table 7.4), it is predicted that most cultured cells would be valuable systems for studying hormonal signal transduction. Unfortunately, the list of systems in which hormonal regulation of transport has been demonstrated is shorter (also shown in Table 7.4). In some senses, this is reasssuring: when hormonal regulation of transport is demonstrated, it is more likely to be due to the expression of a complete epithelial function, as it is not expressed in the majority of tissue culture cells. The reasons for the presence of an incomplete hormonal response in a particular cell line have not been examined frequently. In JTC.12P3 cells (derived from kidney cortex), it was observed that Na^+-dependent phosphate transport was expressed, and that PTH receptor occupancy was coupled to cAMP production[85]. However, since no alteration in phosphate transport was observed due to PTH (as would have been predicted for proximal tubular cells), the results indicated a 'defect' in events 'downstream' from cAMP production in the JTC.12P3 cell line.

The ability to control fully the hormonal environment of cultured cells has facilitated work to understand several aspects of hormonal signal transduction. Three systems have reached an advanced stage in the study of hormonal interactions using renal and intestinal tissue culture cells. These are the study of PTH effects on phosphate transport and Na^+/H^+ exchange (OK cells as a model of proximal tubule), aldosterone effect on Na^+ transport (A6 cells as a model of distal nephron), and VIP regulation of transport (T$_{84}$ and HT29 cells as intestinal models).

Parathyroid hormone

In the proximal tubule, PTH has been shown to decrease reabsorption of phosphate and bicarbonate[184,194]. In the native tissue, the mechanisms which have been implicated are a decrease in the rates of apical Na^+/phosphate co-transport, and a decrease in apical Na^+/H^+ exchange[185,195]. Based on the effect of PTH to decrease these same transport reactions in OK cells, several laboratories have used the OK cell as a model of the PTH regulation of transport in the proximal tubule[49,79–81,85,110,112–115]. Various aspects of PTH action have been investigated. Information suggests that two receptors for PTH exist in the OK cell, with differential coupling to adenylate cyclase[111]. This is supported by recent data which suggest a physical separation of these receptors to the apical and basolateral membrane of OK cells (C. Helmle-Kolb, M. H. Montrose and H. Murer, manuscript in preparation). The data suggest that after receptor occupancy, a number of intracellular events occur: increases in intracellular calcium, increases in cAMP and changes in phosphatidylinositol turnover[85,114]. When the PTH concentration dependency of these biochemical events is compared with the PTH concentration dependency for regulation of phosphate transport or Na^+/H^+ exchange, the data suggest that activation of protein kinase C is likely to be the physiological mediator of PTH action at physiological (10^{-12} mol L^{-1}) concentrations of the hormone[111,114,196]. Despite this evidence that diacylglycerol/protein kinase C is responsible for the physiological changes in ion transport, direct activation of protein kinase A (with forskolin or 8-Br,cAMP) can also cause changes in transport, independent of kinase C activation[85,113–115,196]. If submaximal concentrations of phorbol esters and forskolin are combined, no synergistic effect is observed on changes in transport[85,196]. This suggests that the different biochemical pathways are able to regulate transport independently. If saturating doses of the agonists are combined, there is no additional decrease in transport compared with single additions[85,113,196]. This lack of additivity suggests that the same final target is found in all cases (e.g. phosphorylation of the transporters). The data therefore suggest that a number of regulatory cascades can modulate transport independently, but that the final target (either a phosphorylation site on the transporters themselves or a regulatory protein) is common to the different biochemical pathways.

Aldosterone

In the distal tubule of the kidney (and in the large intestine), aldosterone is known to increase Na^+ reabsorption[197]. The response to aldosterone is characterized by an initial latent phase (1–2 h) followed by a graded increase in transepithelial short-circuit current over the next 8–10 h. Using A6 cells (from *Xenopus laevis* kidney) as a model system, it has been possible to characterize the effect of aldosterone in more detail. Unlike peptide hormones, aldosterone and other steroid hormones have no identified plasma membrane receptors, and instead act by binding to a cytosolic receptor following diffusion across the plasma membrane. In subsequent steps, the aldosterone/receptor complex is translocated to the nucleus, and DNA transcription is stimulated. Using A6 cells, it has been shown that binding of the aldosterone/receptor complex to nuclear membranes has both high- and low-affinity components (k_d = 85 pmol L^{-1} and 16 nmol, L^{-1} respectively), but that occupancy of the low-affinity receptor is responsible for the observed changes in transport[125]. This result is curious with respect to native tissue, as it has classically been the high-affinity receptor which is responsible for the mineralocorticoid effects of aldosterone, and the low-affinity receptor which is responsible for the glucocorticoid effects[197]. Despite this discrepancy, in both A6 cells and native tissue, the changes in transport have been shown to be due to increased apical Na^+ uptake by an amiloride-sensitive Na^+ channel, and increased Na^+ efflux via basolateral Na^+,K^+-ATPase[121,122,124–127]. It has also been shown that aldosterone stimulates the rate of synthesis of both subunits of the Na^+,K^+-ATPase by increasing the amount of mRNA produced and that this occurs only during the late phase of aldosterone action[126]. These results imply that biogenesis of the Na^+,K^+-ATPase is not part of the early events which are mediated by aldosterone.

Vasoactive intestinal polypeptide

In both the large and small intestine, VIP is known to cause changes in ion transport, and is considered a neuromodulator (i.e. it is stored in nerve terminals, not enteroendocrine cells). In the native tissue, both anti-absorptive effects and prosecretory effects of VIP have been noted, presumably due to the different cell types which populate the intestinal epithelial layer[42,43]. In tissue culture model systems, only the prosecretory effects have been documented. In T_{84} and Caco-2 cells, addition of VIP leads to ion secretion which has been attributed to increases in the uptake of ions by basolateral $Na^+/K^+/Cl^-$ co-transport and increased Cl^- efflux via apical Cl^- channels[45,46,159,160]. In the T_{84} cells, the prosecretory effect of VIP was antagonized by somatostatin[46]. In HT29 cells, an effect of VIP on $Na^+/K^+/Cl^-$ co-transport has been noted, but further transport effects of the hormone have not been evaluated[103]. Ironically, the cell type which has been the least characterized with respect to transport effects (HT29) has been subjected to the most extensive experiments concerning VIP receptor function. In this cell line, different laboratories report that VIP binds to a membrane receptor with an apparent affinity for VIP of 1–10 nmol L^{-1} and a molecular weight between 45 and 70 kDa[150,152,153,156–158]. One report also suggests that a nuclear receptor (of unknown physiological function) exists for VIP, but the apparent affinity for VIP was much lower than that observed previously[151]. As it has been shown that the concentration dependency of VIP binding to the cell surface is similar to the concentration dependency of cAMP formation, the data suggest that surface receptor occupancy is directly coupled to activation of adenylate cyclase[150,152,156,158]. Binding of VIP to its receptor is antagonized by the fragment of VIP composed of amino acids 10–28 (i.e. amino acids 1–19 were deleted)[155].

In the HT29 cell line, a number of experimental manoeuvres change the number of VIP receptors at the cell surface. Exposure of the cell to VIP itself leads to a down-regulation of VIP receptors with a $t_{1/2}$ of 2–3 min, and removal of extracellular VIP causes an increase in VIP binding activity which does not require protein synthesis[140,154,157,158]. Similarly, exposure of the cells to phorbol esters or α-adrenergic agonists leads to a simultaneous loss of VIP receptors and VIP-stimulated adenylate cyclase activity, without decreases in the adenylate cyclase activation by either forskolin or cholera toxin[140,154,156,158]. These data suggest that internalization of the receptor leads to a loss of coupling between adenylate cyclase and the VIP receptor. Finally, exposure to cycloheximide (an inhibitor of protein synthesis) results in an increase in VIP receptors at the cell surface, but it is unknown if this leads to an increase in VIP-stimulated adenylate cyclase[157]. A recent review discusses these observations in more detail[158]. This evidence supports the concept that the amount of VIP receptor is dynamically regulated by endocytotic events and membrane cycling.

DISEASE MECHANISMS

A variety of intestinal cell lines catalyse Cl^- secretion in response to secretory agonists (e.g. T_{84}, HT29, Caco-2). This ion secretory response has been used as a model system for studying the pathogenesis of diarrhoea, because net water secretion is known to follow passively the net secretion of solute. The T_{84} cell has been used extensively for studies which seek to examine the direct effect of secretagogues on epithelial cells, independent of effects due to submucosal elements which are present in most native tissue preparations. Thus it has been possible to examine the direct effects of enterotoxins, such as *Clostridium difficile* toxin A, which has been shown to alter tight junctional permeability directly[198], and *Escherichia coli* heat-stable toxin (STa), which alters cellular cGMP levels[199]. In the latter case, it was possible to quantify high-affinity binding sites for STa and to correlate cellular binding with intracellular cGMP accumulation[200]. The cell line has also been used to examine the direct effects of prostaglandins and immune mediators on intestinal cells[201], independent of the cells below the epithelial layer which are responsible for the synthesis of such compounds.

In a number of studies, the T_{84} cell line has been used extensively to define the membrane transport mechanisms mediating ion secretion in response to agonists which alter intracellular calcium, cAMP and cGMP[45,46,98,100,199,201]. Use of this cell line has defined that the regulatory cascades causing changes in these second messengers can operate independently, such that there is no additivity *at the level of second messenger concentration* between agonists which specifically affect only one second messenger system[100,199]. In contrast, it has been demonstrated that there are synergistic effects *on transport* when agonists of different second messenger systems are combined[100]. It appears that agents which increase cAMP and cGMP cause activation of both apical Cl^- channels and basolateral K^+ channels, whereas increases in cellular calcium activate additional K^+ channels to cause a larger transport effect[100].

The T_{84} cell was also central to the recent identification and cloning of the gene defective in cystic fibrosis[202]. This protein is expressed in the T_{84} cell line, presumably as a component of anion transport regulation[202]. This is logical as the function of the intestinal epithelium is affected in cystic fibrosis[203] and since Cl^- channels in the T_{84} cell line are regulated by the second messengers which display altered function in the disease[97]. It is anticipated that the cell line will be of further use as the function and regulation of expression of this protein are examined. Curiously, although no defect in renal function has been identified in cystic fibrosis, current evidence suggests that the mRNA encoded by the gene is also expressed in kidney[202]. This implies that renal cell lines may also be appropriate model systems for the study of the defect in cystic fibrosis.

SUMMARY AND FORECAST

Step 1: Supplementing native tissue

The use of tissue culture cells allows one to ask questions which could not be approached using native tissue. The use of tissue culture is imperative in studies of intestinal and renal epithelia because of the multitude of cell types which can coexist in the epithelial layer, and/or the small amount of tissue which may be obtained from an experimental animal or routine clinical work. This point has been noted by a number of investigators who have used tissue culture to improve control of hormonal environments, perform genetic manipulations, and perform biochemical procedures essential for answering physiological questions.

Step 2: Integrating with native tissue

In many studies, the problem of interpreting the answer obtained from tissue culture cells is the compensatory price for the increased power of analysis. Ideally, tissue culture cells will express functions specific for the originating tissue, but the ideal case has never been found. The most successful approach to this problem has been to bracket observations with epithelial parameters so that the use of a tissue culture cell may be justified. For example, it is considered valid to study the regulatory cascade between hormone/receptor interaction and final physiological effect if the investigator has some assurance that the beginning and the end are 'epithelial'. The true power of tissue culture is used when it is possible to alternate between a tissue culture and native tissue system to utilize the power of each by direct comparisons and cross correlations between systems in the same laboratory. As yet, this has been accomplished in relatively few systems. To attain this goal, it should be possible either to stay within the same species when making comparisons (e.g. the aldosterone transport response of A6 cells versus distal tubule of frog kidney), or to use molecular probes (specific antibodies, nucleotide probes) which can be applied across species boundaries.

Step 3: Integrating with human tissue

The use of human tissue culture cells has increased the body of knowledge about human cell physiology. Such human tissue culture systems avoid species variability in function and drug sensitivity, but maintain experimental versatility and rigour. If such a system can be combined with information from human native tissue, as suggested in Step 2, it presents the opportunity to make rapid advances in the understanding of human function and disorders. It is predicted that the intestinal system is most likely to first reach fruition of these goals, since human colon lines are available and biopsies of human colonic mucosa are obtained routinely. This quality of information is currently only established for the cell biology of human intestinal microvillar hydrolases[47], mucins[204], and villin[205] and has yet to reach fruition in studies of membrane transport.

References

1. Hull, R. N., Cherry, W. R., and Weaver, G. W. (1976). The origin and characteristics of a pig kidney cell strain LLC-PK$_1$. *In Vitro*, **12**, 670–677
2. Viniegra, S. and Rabito, C. A. (1988). Development and polarization of the Na/H antiport system during reorganization of LLC-PK$_{1A}$ cells into an epithelial membrane. *J. Biol. Chem.*, **263**(15). 7099–7104
3. Wohlwend, A., Vassali, J-D., Belin, D. and Orci, L. (1986). LLC-PK$_1$ cells: cloning of phenotypically stable subpopulations. *Am. J. Physiol.*, **250**, C682–C687
4. Amsler, K. and Cook, J. S. (1985). Linear relationship of phlorizin-binding capacity and hexose uptake during differentiation in a clone of LLC-PK$_1$ cells. *J. Cell. Physiol.*, **122**(2), 254–258
5. Haggerty, J. G., Agarwal, N., Cragoe Jr., E. J., Adelberg, E. A. and Slayman, C. W. (1988). LLC-PK$_1$ mutant with increased Na$^+$-H$^+$ exchange and decreased sensitivity to amiloride. *Am. J. Physiol.*, **255**, C495–C501
6. Koyama, H., Goodpaste, C., Miller, M. M., Teplitz, R. L. and Riggs, A. D. (1978). Establishment and characterization of a cell line from the American opossum (*Didelphys Virginiana*). *In Vitro*, **14**(3), 239–246
7. Biber, J., Forgo, J. and Murer, H. (1988). Modulation of Na$^+$-P$_i$ co-transport in opossum kidney cells by extracellular phosphate. *Am. J. Physiol*, **255**, C155–C161
8. No original reference seems to exist for the establishment of MDCK cells. A full description of the cells is in the *Cell Lines and Hybridomas Catalogue* of the American Type Culture Collection (1988). The techniques undoubtedly used to establish MDCK cells are described in S. H. Madin, P. C. Andriese and N. B. Darby (1957). The *in vitro* cultivation of tissues of domestic and laboratory animals. *Am. J. Vet. Res.*, **18**, 932–941
9. Barker, G. and Simmons, N. L. (1981). Identification of two strains of cultured canine renal epithelial cells (MDCK cells) which display entirely different physiological properties. *Q. J. Exp. Physiol.*, **66**, 61–72
10. Richardson, J. A. W., Scalera, V. and Simmons, N. L. (1981). Identification of two strains of MDCK cells which resemble separate nephron segments. *Biochim. Biophys. Acta*, **673**, 26–36
11. Rafferty, K. A. (1969). Mass culture of amphibia cells: methods and observations concerning stability of cell type. In Mizell, M. (ed.) *Biology of Amphibian Tumors*, pp. 52–81. (New York: Springer–Verlag)
12. Hopps, H. E., Bernheim, B. C., Nisolak, A., Tjio, J. H. and Smadel, J. E. (1963). Biologic characteristics of a continuous kidney cell line derived from the African Green monkey. *J. Immunol.*, **91**, 416–424
13. Graham, F. L., Smiley, J., Russel, W. C. and Narin, R. (1977). Characteristics of a human cell line transformed by DNA from human adenovirus type 5. *J. Gen. Virol.*, **36**, 59–72
14. Wallace, R. E., Vasington, P. J., Retricciani, J. C., Hopps, H. E., Lorenz, D. E. and Kadanka, Z. (1977). Development and characterisation of cell lines from subhuman primates. *In Vitro*, **8**(5), 333–341
15. Hull, R. N., Cherry, W. R. and Johnson, I. S. (1956). The adaptation and maintenance of mammalian cells to continuous growth in tissue culture. *Anat. Rec.*, **124**, 490 (abstract)
16. Walen, K. H. and Brown, S. W. (1962). Chromosomes in a marsupial (*Potorous tridactylis*) tissue culure. *Nature (London)*, **194**, 406
17. Beale, A. J., Christofinis, G. C. and Furminger, I. G. S. (1963). Rabbit cells susceptible to rubella virus. *Lancet*, **2**, 640–641

18. DeLarco, T. E. and Todaro, G. J. (1978). Epitheloid and fibroblastic rat kidney cell clones: Epidermal growth factor (EGF) receptors and the effect of mouse sarcoma virus transformation. *J. Cell Physiol*, **94**, 335–342

19. Madin, S. H. and Darby Jr., N. B. (1958). Established kidney cell lines of normal adult bovine and ovine origin. *Proc. Soc. Exp. Biol. Med.*, **98**, 574–576

20. Green, N., Algren, A., Hoyer, J., Triche, T. and Burg, M. (1985). Differentiated lines of cells from rabbit renal medullary thick ascending limbs grown on amnion. *Am. J. Physiol.*, **249**, C97–C104

21. Valentich, J. D. and Stokols, M. F. (1986). An established cell line from mouse kidney medullary thick ascending limb. I. Cell culture techniques, morphology, and antigenic expression. *Am. J. Physiol.*, **251**, C299–C311 1986

22. Vandewalle, A., Lelongt, B., Gentieau-Legendre, M., Baudouin, B., Antoine, M., Estrade, S., Chatelet, F., Verroust, P., Cassingena, R. and Ronco, P. (1989). Maintenance of proximal and distal cell functions in SV40-transformed tubular cell lines derived from rabbit kidney cortex. *J. Cell. Physiol.*, **141**(1), 203–221

23. Arend, L. J., Handler, J. S., Rhim, J. S., Gusovsky, F. and Spielman, W. S. (1989). Adenosine-sensitive phosphoinositide turnover in a newly established renal cell line. *Am. J. Physiol*, **256**, F1067–F1074

24. Girard, D. J., Aaronson, S. A., Todaro, G. J., Arnstem, P., Kersey, J. H., Dosik, H. and Parks, W. P. (1973). In vitro cultivation of human tumors: Establishment of cell lines derived from a series of solid tumors. *J. Natl. Cancer Inst.*, **51**(5), 1417–1423

25. Fogh, J. and Trempe, G. (1975). New human tumor cell lines. In Fogh, J. (ed.) *Human Tumor Cells in Vitro*, pp. 115–159. (New York: Plenum Press)

26. Owens, R. B., Smith, H. S., Nelson-Rees, W. A. and Springer, E. L. (1976). Epithelial cell cultures from normal and cancerous human tissues. *J. Natl. Cancer Inst.*, **56**(4), 843–849

27. Henle, G. and Deinhardt, F. (1956). The establishment of strains of human cells in tissue culture. *J. Immunol.*, **79**, 54–59

28. Quaroni, A, Wands, J., Trelstad, R. L. and Isselbacher, K. J. (1979). Epithelioid cell cultures from rat small intestine. Characterization by morphologic and immunologic criteria. *J. Cell Biol.*, **80**, 248–265

29. Tomkins, W. A. F., Watrach, A. M., Schmale, J. D., Schultz, R. M. and Harris, J. A. (1974). Cultural and antigenic properties of newly established cell strains derived from adenocarcinomas of the human colon and rectum. *J. Natl. Cancer Inst.*, **52**, 1101–1110

30. Stevens, R. H., Brooks, G. P., Osborne, J. W., England, C. W. and White, D. W. (1977). Lymphocyte cytotoxicity in X-irradiation-induced adenocarcinoma of the rat small bowel. *J. Natl. Cancer Inst.*, **59**(4), 1315–1319

31. O'Donnell, R. W. and Cockerell, G. L. (1981). Establishment and biological properties of a guinea pig colonic adenocarcinoma cell line induced by *N*-methyl-*N*-nitrosourea. *Cancer Res.*, **41**, 2372–2377

32. Murakami, H. and Masui, H. (1980). Hormonal control of human colon carcinoma cell growth in serum-free medium. *Proc. Natl. Acad. Sci. USA*, **77**, 3464–3468

33. Fogh, J., Wright, W. C. and Loveless, J. D. (1977). Absence of Hela cell contamination in 169 cell lines derived from human tumors. *J. Natl. Cancer Inst.*, **58**, 209–214

34. Huet, C., Sahuquillo-Merino, C., Coudrier, E. and Louvard, D. (1987). Absorptive and mucus-secreting subclones isolated from a multipotent intestinal cell line (HT-29) provide new models for cell polarity and terminal differentiation. *J. Cell Biol.*, **105**, 345–357

35. Fantini, J., Abadie, B., Tirard, A., Remy, L., Ripert, J. P. El Battari, A. and Marvaldi, J. (1986). Spontaneous and induced dome formation by two clonal cell populations derived from a human adenocarcinoma cell line, HT29. *J. Cell Sci.*, **83**, 235–249

36. Laboisse, C. L., Maoret, J-J., Triadou, N. and Augeron, C. (1988). Restoration by polyethylene glycol of characteristics of intestinal differentiation in subpopulations of the human colonic adenocarcinoma cell line HT29. *Cancer Res.*, **48**, 2498–2504

37. Augeron, C. and Laboisse, C. L. (1984). Emergence of permanently differentiated cell clones in a human colonic cancer cell line in culture after treatment with sodium butyrate. *Cancer Res.*, **44**, 3961–3969

38. Leibovitz, A., Stinson, J. C., McCombs, W. B., McCoy, C. E., Mazur, K. C., and Mabry, N. D. (1976). Classification of human colorectal adenocarcinoma cell lines. *Cancer Res.* **36**(12), 4562–4569

39. Tom, B. H., Rutzky, L. P., Jakstys, M. M., Oyasu, R., Kaye, C. I. and Kahan, B. D. (1976). Human colonic adenocarcinoma cells. Establishment and description of a new line. *In Vitro*, **12**, 180–191

40. Dexter, D. L., Barbosa, J. A. and Calabresi, P. (1979). *N,N*-dimethylformamide-induced alteration of cell culture characteristics and loss of tumorigenicity in culture human colon carcinoma cells. *Cancer Res.*, **39**, 1020–1025

41. Brattain, M. G., Brattain, D. E., Fine, W. D., Khaled, F. M., Marks, M. E., Kimball, P. M., Arcolano, L. A. and Danbury, B. H. (1981). Initiation and characterization of cultures of human colonic carcinoma with different biological characteristics utilizing feeder layers of confluent fibroblasts. *Oncodev. Biol. Med*, **2**, 355–366

42. Donowitz, M. and Welsh, M. J. (1987). Regulation of mammalian small intestinal electrolyte secretion. In Johnson, L. R. (ed.) *Physiology of the Gastrointestinal Tract*, 2nd edn., pp. 1351–1388. (New York: Raven Press)

43. Binder, H. J. and Sandle, G. I. (1987). Electrolyte absorption and secretion in the mammalian colon. In Johnson, L. R. (ed.) *Physiology of the Gastrointestinal Tract*, 2nd edn., pp. 1389–1418 (New York: Raven Press)

44. Grasset, E., Pinto, M., Dussaulx, E., Zweibaum, A., and Desjeux, J-F. (1984). Epithelial properties of human colonic carcinoma cell line Caco-2: electrical parameters. *Am. J. Physiol*, **247**, C260–C267

45. Mandel, K. G., Dharmsathaphorn, K. and McRoberts, J. A. (1985). Characterization of a cyclic AMP-activated Cl transport pathway in the apical membrane of a human colonic epithelial cell line. *J. Biol. Chem.*, **261**, 704–712

46. Dharmsathaphorn, K., McRoberts, J. A., Mandel, K. G., Tisdale, L. D. and Masui, H. (1984). A human colonic tumor cell line that maintains vectorial electrolyte transport. *Am. J. Physiol.*, **246**, G204–G208

47. Hauri, H-P., Sterchi, E. E., Bienz, D., Fransen, J. A. M. and Marxer, A. (1985). Expression and intracellular transport of microvillus membrane hydrolases in human intestinal epithelial cells. *J. Cell Biol.*, **101**, 838–851

48. Montrose, M. H., Friedrich, T. and Murer, H. (1987). Measurements of intracellular pH in single LLC-PK$_1$ cells: Recovery from an acid load via basolateral Na$^+$/H$^+$ exchange. *J. Memb. Biol.*, **97**, 63–78

49. Montrose, M. H., Condrau, M. A. and Murer, H. (1989). Flow cytometric analysis of intracellular pH in cultured opossum kidney (OK) cells. *J. Memb. Biol.*, **108**, 31–43.

50. Rabito, C. A., Kreisberg, J. I. and Wight, D. (1984). Alkaline phosphatase and gamma-glutamyl transpeptidase as polarization markers during the organization of LLC-PK$_1$ cells into an epithelial membrane. *J. Biol. Chem.*, **259**(1), 574–582.

51. Sepulveda, F. V., Burton, K. A. and Pearson, J. D. (1982). The development of gamma-glutamyltransferase in a pig renal-epithelial-cell line *in vitro*. *Biochem. J.*, **208**, 509–512

52. Inui, K-I., Saito, H., Takano, M., Okano, T., Kitazawa, S. and Hori, R. (1984). Enzyme activities and sodium-dependent active D-glucose transport in apical membrane vesicles isolated from kidney epithelial cell line (LLC-PK$_1$). *Biochim. Biophys. Acta*, **769**, 514–518

53. Gstraunthalter, G., Pfaller, W. and Kotanko, P. (1985). Biochemical characterization of renal epithelial cell cultures (LLC-PK$_1$ and MDCK). *Am. J. Physiol.*, **248**, F536–F544

54. Moran, A., Biber, J. and Murer, H. (1986). A sodium–hydrogen exchange system in isolated apical membrane from LLC-PK$_1$ epithelia. *Am. J. Physiol.*, **251**, F1003–F1008

55. Malström, K., Stange, G. and Murer, H. (1987). Identification of proximal tubular transport functions in the established kidney cell line, OK. *Biochim. Biophys. Acta*, **902**, 269–277
Note: The name of the first author is misspelled in the paper: it should be Malmström.

56. Trugnan, G., Rousset, M., Chantret, I., Barbat, A. and Zweibaum, A. (1987). The posttranslational processing of sucrase-isomaltase in HT-29 cells is a function of their state of enterocytic differentiation. *J. Cell Biol.*, **104**, 1199–1205

57. Pinto, M. Appay, M-D., Simon-Assmann, P., Chevalier, G., Dracopoli, N., Fogh, J. and Zweibaum, A. (1982). Enterocytic differentiation of cultured human colon cancer cells by replacement of glucose by galactose in the medium. *Biol. Cell*, **44**, 193–196

58. Polak-Charcon, S., Hekmati, M. and Ben-Shaul, Y. (1989). The effect of modifying the culture medium on cell polarity in a human colon carcinoma cell line. *Cell Diff. Dev.*, **26**, 119–129

59. Mohrmann, I., Mohrmann, M., Biber, J. and Murer, H. (1986). Sodium-dependent transport of P_i by an established intestinal epithelial cell line (Caco-2). *Am. J. Physiol.*, **250**, G323–G330

60. Blais, A., Bissonnette, P. and Berteloot, A. (1987). Common characteristics for Na^+-dependent sugar transport in Caco-2 cells and human fetal colon. *J. Memb. Biol.*, **99**, 113–125

61. Gum, J. R., Kam, W. K., Byrd, J. C., Hicks, J. W., Sleisenger, M. H. and Kim, Y. S. (1987). Effects of sodium butyrate on human colonic adenocarcinoma cells – Induction of placental-like alkaline phosphatase. *J. Biol. Chem.*, **262**, 1092–1097

62. Brown, C. D. A. and Murer, H. (1985). Characterization of $Na^+/K^+/2Cl$ cotransport system in the apical membrane of a renal epithelial cell line (LLC-PK$_1$). *J. Memb. Biol.*, **87**, 131–139

63. Rabito, C. A. (1981). Localization of the Na-sugar cotransport system in a kidney epithelial cell line (LLC-PK$_1$). *Biochim. Biophys. Acta*, **649**, 286–296

64. Misfeldt, D. S. (1981). Transepithelial transport in cell culture: C-glucose transport by a pig kidney cell line (LLC-PK$_1$). *J. Memb. Biol.*, **59**, 13–18

65. Rabito, C. A. and Karish, M. V. (1983). Polarized amino acid transport by an epithelial cell line of renal origin (LLC-PK$_1$). *J. Biol. Chem.*, **258**(4), 2543–2547

66. Cantiello, H. F., Scott, J. A. and Rabito, C. A. (1987). Conductive Na transport in an epithelial cell line (LLC-PK$_1$) with characteristics of proximal tubular cells. *Am. J. Physiol.*, **252**, F590–F597

67. Rabito, C. A. (1983). Phosphate uptake by a kidney cell line (LLC-PK$_1$). *Am. J. Physiol.*, **245**, F22–F31

68. Montrose, M. H. and Murer, H. (1986). Regulation of intracellular pH in LLC-PK$_1$ cells by Na^+/H^+ exchange. *J. Memb. Biol.*, **93**, 33–42

69. Cantiello, H. F., Scott, J. A. and Rabito, C. A. (1986). Polarized distribution of the Na^+/H^+ exchange system in a renal cell line (LLC-PK$_1$) with characteristics of proximal tubular cells. *J. Biol. Chem.*, **261**(7), 3252–3258

70. Haggerty, J. G., Agarwal, N., Reilly, R. F., Adelberg, E. A. and Salyman, C. W. (1988). Pharmacologically different Na^+/H^+ antiporters on the apical and basolateral surfaces of cultured porcine kidney cells (LLC-PK$_1$). *Proc. Natl. Acad. Sci. USA.*, **85**, 6797–6801

71. Sepulveda, F. V. and Pearson, J. D. (1985). Cationic amino acid transport by two renal epithelial cell lines: LLC-PK$_1$ and MDCK cells. *J. Cell. Physiol.*, **123**, 144–150

72. Sepulveda, F. V. and Pearson, J. D. (1984). Localization of alanine uptake by cultured renal epithelial cells (LLC-PK$_1$) to the basolateral membrane. *J. Cell. Physiol.*, **118**, 211–217

73. Chaillet, J. R., Amsler, K. and Boron, W. F. (1986). Optical measurements of intracellular pH in single LLC-PK$_1$ cells: Demonstration of Cl-HCO$_3$ exchange. *Proc. Natl. Acad. Sci.*, **83**, 522–526

74. Fauth, C., Rossier, B. and Roch-Ramel, F. (1988). Transport of tetraethylammonium by a kidney epithelial cell line (LLC-PK$_1$). *Am. J. Physiol.*, **254**, F351–F357

75. Parys, J. B., De Smedt, H. and Broghgraef, R. (1986). Calcium transport systems in the LLC-PK$_1$ renal epithelial established cell line. *Biochim. Biophys. Acta*, **888**, 70–81

76. Ubl, J., Murer, H. Kolb, H. A. (1988). Ion channels activated by osmotic and mechanical stress in membranes of opossum kidney cells. *J. Memb. Biol.*, **104**(3), 223–232

77. Knoblauch, C., Montrose, M. H. and Murer, H. (1989). Regulatory volume decrease by cultured renal cells. *Am J. Physiol.*, **256**, C252–C259

78. Ubl, J., Murer, H. and Kolb, H. A. (1988). Hypotonic shock evokes opening of Ca^{2+}-activated K channels in opossum kidney cells. *Pfleugers Arch.*, **412**(5), 551–553

79. Pollock, A. S., Warnock, D. G. and Strewler, G. J. (1986). Parathyroid hormone inhibition of Na-H antiporter activity in a cultured renal cell line. *Am. J. Physiol*, **250**, F217–F225

80. Miller, R. T. and Pollock, A. S. (1987). Modification of the internal pH sensitivity of the Na/H antiporter by parathyroid hormone in a cultured renal cell line. *J. Biol. Chem.*, **262**(19), 9115–9120

81. Montrose, M. H., Knoblauch, C. and Murer, H. (1988). Separate control of regulatory volume increase and Na^+/H^+ exchange by cultured renal cells. *Am. J. Physiol.*, **255**, C76–C85

82. Jentsch, T. J., Schill, B. S., Schwartz, P., Matthes, H., Keller, S. K. and Wiederholt, M. (1985). Kidney epithelial cells of monkey origin (BSC-1) express a sodium bicarbonate cotransport. *J. Biol. Chem.*, **260**, 15552–15560

83. Jentsch, T. J., Janicke, I., Sorgenfrei, D., Keller, S. K. and Wiederholt, M. (1986). The regulation of intracellular pH in monkey kidney epithelial cells (BSC-1). *J. Biol. Chem.*, **261**(26), 12120–12127

84. Reinersen, K. V., Tonnessen, T. I., Jacobsen, J., Sandvig, K. and Olsnes, S. (1988). Role of chloride/bicarbonate antiport in the control of cytosolic pH. *J. Biol. Chem.*, **263**(23), 11117–11125

85. Malmström K. and Murer, H. (1986). Parathyroid hormone inhibits phosphate transport in OK cells but not in LLC-PK$_1$ and JTC-12.P3 cells. *Am J. Physiol.*, **251**, C23–C31

86. Nelson, D. J., Tang, J. M. and Palmer, L. G. (1984). Single-channel recordings of apical membrane chloride conductance in A6 epithelial cells. *J. Memb. Biol.*, **80**, 81–89

87. Perkins, F. M. and Handler, J. S. (1981). Transport properties of toad kidney epithelia in culture. *Am. J. Physiol.*, **241**, C154–C159

88. Rindler, M. J., Taub, M. and Saier, M. H. (1979). Uptake of ^{22}Na$^+$ by cultured dog kidney cells (MDCK). *J. Biol. Chem.*, **254**, 11431–11439

89. Rindler, M. J. and Saier, M. H. Jr. (1981). Evidence for Na/H antiport in cultured dog kidney cells (MDCK). *J. Biol. Chem.*, **256**(21), 10820–10825

90. Rindler, M. J., McRoberts, J. A. and Saier, M. H. (1982). (Na$^+$/K$^+$)-cotransport in the Madin–Darby canine kidney cell line. *J. Biol. Chem.*, **257**, 2254–2259

91. Simmons, N. L. (1984). Epithelial cell volume regulation in hyptonic fluids: Studies using a model tissue culture renal epithelial cell system. *Q. J. Exp. Physiol.*, **69**, 83–95

92. McRoberts, J. A. Erlinger, S., Rindler, M. J. and Saier, M. H. Jr. (1982). Furosemide-sensitive salt transport in the Madin–Darby canine kidney cell line. *J. Biol. Chem.*, **257**(5), 2260–2266

93. Aiton, J. F., Brown, C. D. A., Ogden, P. and Simmons, N. L. (1982). K$^+$ transport in 'tight' epithelial monolayers of MDCK cells. *J. Membr. Biol.*, **65**, 99–109

94. Brown, C. D. A. and Simmons, N. L. (1981). Catecholamine-stimulation of Cl$^-$ secretion in MDCK cell epithelium. *Biochim. Biophys. Acta.*, **649**, 427–435

95. Kolb, H. A., Brown, C. D. A. and Murer, H. (1985). Identification of a voltage-dependent anion channel in the apical membrane of a Cl$^-$-secretory epithelium (MDCK). *Pleugers Arch.*, **403**(3), 262–265

96. Worrell, R. T., Butt, A. G., Cliff, W. H. and Frizzell, R. A. (1989). A volume-sensitive chloride conductance in human colonic cell line T$_{84}$. *Am. J. Physiol.*, **256**, C1111–C1119

97. Halm, D.R., Rechkemmer, G. R., Schoumacher, R. A. and Frizzell, R. A. (1988). Apical membrane chloride channels in a colonic cell line activated by secretory agonists. *Am. J. Physiol.*, **254**, C505–C511

98. Dharmsathaphorn, K., Mandel, K. G., Masui, H. and McRoberts, J. A. (1985). Vasocative intestinal polypeptide-induced chloride secretion by a colonic epithelial cell line. Direct participation of a basolaterally localized Na$^+$, K$^+$, Cl$^-$ cotransport system. *J. Clin. Invest.*, **75**(2), 462–471

99. McRoberts, J. A., Beuerlein, G. and Dharmsathaphorn, K. (1985). Cyclic AMP and Ca^{2+}-activated K$^+$ transport in a human colonic epithelial cell line. *J. Biol. Chem.*, **260**(26), 14163–14172

100. Cartwright, C. A., McRoberts, J. A. Mandel, K. G. and Dharmsathaphorn, K. (1985). Synergistic action of cyclic adenosine monophosphate- and calcium-mediated chloride secretion in a colonic epithelial cell line. *J. Clin. Invest.*, **76**, 1837–1842

101. Hayslett, J. P., Gogelein, H., Kunzelmann, K. and Greger, R. (1987). Characteristics of apical chloride channels in human colon cells (HT29). *Pfleugers Arch.*, **410**(4–5), 487–495

102. Dreinhöfer, J., Gögelein, H. and Greger, R. (1988). Blocking kinetics of Cl$^-$ channels in colonic carcinoma cells (HT29) as revealed by 5-nitro-2-(3-phenylpropylamino) benzoic acid (NPPB). *Biochim. Biophys. Acta*, **946**, 135–142

103. Kim, H. D., Tsai, Y. S., Franklin, C. C. and Turner, J. T. (1988). Characterization of Na$^+$/K$^+$/Cl$^-$ cotransport in cultured HT29 human colonic adenocarcinoma cells. *Biochim. Biophys. Acta*, **946**, 397–404

104. Cantiello, H. F. and Lanier, S. M. (1989). α2-Adrenergic receptors and the Na'/H' exchanger in the intestinal epithelial cell line, HT-29. *J. Biol. Chem.*, **246** (27), 16000–16007

105. Skorecki, K. L., Verkmann, A. S., Jung, C. Y. and Ausiello, D. A. (1986). Evidence for vasopressin activation of adenylate cyclase by subunit dissociation. *Am. J. Physiol.*, **250**, C115–C123

106. Goldring, S. R., Dayer, J. M., Ausiello, D. A. and Krane, S. M. (1978). A cell strain cultured from porcine kidney increases cyclic AMP content upon exposure to calcitonin or vasopressin. *Biochem. Biophys. Res. Commun.*, **83**, 434–440

107. Tang, M-J. and Weinberg, J. M. (1986). Vasopressin-induced increases of cytosolic calcium in LLC-PK₁ cells. *Am. J. Physiol.*, **251**, F1090–F1095

108. Strewler, G. J. (1984). Release of cAMP from a renal epithelial cell line. *Am. J. Physiol.*, **246**, C224–C230

109. Chandler, J. S., Chandler, S. K., Pike, J. W. and Haussler, M. R. (1984). 1,25-Dihydroxyvitamin D3 induces 25-hydroxyvitamin D3-24-hydroxylase in a cultured monkey kidney cell line (LLC-MK₂) apparently deficient in the high affinity receptor for the hormone. *J. Biol. Chem.*, **259**(4), 2214–2222

110. Teitelbaum, A. P., Nissenson, R. A., Zitzner, L. A. and Simon, K. (1986). Dual regulation of PTH-stimulated adenylate cyclase activity by GTP. *Am. J. Physiol.*, **251**, F858–F864

111. Cole, J. A., Eber, S. L., Poelling, R. E. Thorne, P. K. and Forte, L. R. (1987). A dual mechanism for regulation of kidney phosphate transport by parathyroid hormone. *Am. J. Physiol.*, **253**, E221–E227

112. Caverzasio, J., Rizzoli, R. and Bonjour, J.-P. (1986). Sodium dependent phosphate transport inhibited by parathyroid hormone and cyclic AMP stimulation in an opposum kidney cell line. *J. Biol. Chem.*, **261**, 323–3237

113. Malmström, K., Stange, G. and Murer, H. (1988). Intracellular cascades in the parathyroid-hormone-dependent regulation of Na'/phosphate cotransport in OK cells. *Biochem. J.*, **251**, 207–213

114. Quamme, G., Pfeilschifter, J. Murer, H. (1989). Parathyroid hormone inhibition of Na+/phosphate cotransport in OK cells: Generation of second messengers in the regulatory cascade. *Biochem. Biophys. Res. Commun.*, **158** (3), 951–957

115. Moran, A., Montrose, M. H. and Murer, H. (1988). Regulation of Na'/H' exchange in cultured opossum kidney cells by parathyroid hormone, atrial aatruvietic peptide and cyclic nucleotides. *Biochim. Biophys. Acta*, **969**, 48–56 1988

116. Forte, L. R., Krause, W. J. and Freeman, R. H. (1988). Receptors and cGMP signalling mechanism for E. coli enterotoxin in opossum kidney. *Am. J. Physiol.*, **255**, F1040–F1046

117. Rothenberg, P., Reuss, L. and Glaser, L. (1982). Serum and epidermal growth factor transiently depolarize quiescent BSC-1 epithelial cells. *Proc. Natl. Acad. Sci.*, **79**, 7783–7787

118. Ishizuka, I., Tadano, K., Nagata, N., Niimura, Y. and Nagai, Y. (1978). Hormone-specific responses and biosynthesis of sulpholipids in cell lines derived from mammalian kidneys. *Biochim. Biophys. Acta*, **541**, 467–482

119. Takuwa, Y. and Ogata, E. (1985). Characterization of Na-dependent phosphate uptake in cultured kidney cells (JTC-12 monkey cells). *Biochem. J.*, **230**, 715–721

120. Lang, M. A., Muller, J., Preston, A. S. and Handler, J. S. (1986). Complete response to vasopressin requires epithelial organization in A6 cells in culture. *Am. J. Physiol.*, **250**, C138–C145

121. Sariban-Sohraby, S., Burg, M. B. and Turner, R. J. (1983). Apical sodium uptake in toad kidney epithelial cell line A6. *Am. J. Physiol.*, **245**, C167–C171

122. Sariban-Sohraby, S., Burg, M. B and Turner, R. J. (1984). Aldosterone-stimulated sodium uptake by apical membrane vesicles from A6 cells. *J. Biol. Chem.*, **259**(18), 11221–11225

123. Sariban-Sohraby, S., Burg, M. B., Wiesmann, W. P., Chiang, P. K. and Johnson, J. P. (1984). Methylation increases sodium transport into A6 apical membrane vesicles: possible mode of aldosterone action. *Science*, **225**, 745–746

124. Handler, J. S., Preston, A. S., Perkins, F. M. and Matsumura, M. (1982). The effect of adrenal steroid hormones on epithelial formed in culture by A6 cells. *Ann. N. Y. Acad. Sci.*, **453**, 442–454

125. Watlington, C. O., Perkins, F. M., Munson, P. J. and Handler, J. S. (1982). Aldosterone and corticosterone binding and effects on Na$^+$ transport in cultured kidney cells. *Am. J. Physiol.*, **242**, F610–F619

126. Verrey, F., Schaerer, E., Zoerkler, P., Paccolat, M. P., Kraehenbuhl, J.-P. and Rossier, B. C. (1987). Regulation by aldosterone of Na,K-ATPase and mRNAs, protein synthesis, sodium transport in cultured kidney cells. *J. Cell Biol.*, **104**(5), 1231–1237

127. Paccolat, M. P., Geering, K., Gaeggeler, H. P. and Rossier, B. C. (1987). Aldosterone regulation of Na$^+$ transport and Na$^+$, K$^+$-ATPase in A6 cells: role of growth conditions. *Am. J. Physiol.*, **252**, C468–C476

128. Hassid, A. (1981). Transport-active renal tubular epithelial cells (MDCK and LLC-PK$_1$) in culture. Prostaglandin biosynthesis and its regulation by peptide hormones and ionophore. *Prostaglandins*, **21**, 985–1001

129. Mita, S., Takeda, K., Nakane, M., Yasuda, H., Kawashima, K., Yamada, M. and Endo, H. (1980). Responses of a dog kidney cell line (MDCK) to antidiuretic hormone (ADH) with special reference to activation of protein kinase. *Exp. Cell Res.*, **130**, 169–173

130. Reznik, V. M., Shaprio, R. J. and Mendoza, S. A. (1985). Vasopressin Stimulates DNA synthesis and ion transport in quiescent epithelial cells. *Am. J. Physiol.*, **249**, C267–C270

131. Meier, K. E., Snavely, M. D., Brown, S. L., Brown, J. H. and Insel, P. A. (1983). Alpha 1- and beta 2-adrenergic receptor expression in the Madin–Darby canine kidney epithelial cell line. *J. Cell Biol.*, **97**(2), 405–415

132. Rugg, E. L. and Simmons, N. L. (1984). Control of cultured epithelial (MDCK) cell transport function: Identification of a alpha-adrenoceptor coupled to adenylate cyclase. *Q. J. Exp. Physiol.*, **69**, 339–353

133. Simmons, N. L. (1981). Identification of a Purine (P2) receptor linked to ion transport in a cultured renal (MDCK) epithelium. *Br. J. Pharmacol.*, **73**, 379–384 1981

134. Reinlib, L., Mikkelsen, R., Zahniser, D., Dharmsathaphorn, D. and Donowitz, M. (1989). Carbachol-induced cytosolic free Ca^{2+} increases in T$_{84}$ colonic cells seen by microfluorimetry *Am. J. Physiol.* **257**, G950–G960

135. Franklin, C. C., Chin, P. C., Turner, J. T. and Kim, H. D. (1988). Insulin regulation of glucose metabolism in HT29 colonic adenocarcinoma cells: Activation of glycolysis without augmentation of glucose transport. *Biochim. Biophys. Acta*, **972**, 60–68

136. Carpene, C., Paris, H., Cortinovis, C., Viallard, V. and Murat, J. C. (1983). Characterization of α 2-adrenergic receptors in the human colon adenocarcinoma cell line HT29 in culture by [^{3}H]yohimbine binding. *Gen. Pharmacol.*, **14**, 701–703

137. Senard, J. M., Mauriege, P., Daviaud, D. and Paris, H. (1988). α 2-Adrenoceptor in HT29 human colon adenocarcinoma cell-line: Study of [^{3}H]-adrenaline binding. *Eur. J. Pharmacol.*, **162**, 225–236

138. Bouscarel, B., Cortinovis, C., Carpene, C., Murat, J. C. and Paris, H. (1985). α 2-Adrenoceptors in the HT29 human colon adenocarcinoma cell line: Characterization with [^{3}H]clonidine; effects on cyclic AMP accumulation. *Eur. J. Pharmacol.*, **107**, 223–231

139. Bouscarel, B., Murat, J. C. and Paris, H. (1985). Involvement of guanine nucleotide and sodium in regulation of yohimbine and clonidine binding sites in the HT29 human colon adenocarcinoma cell-line. *Gen. Pharmacol.*, **16**, 641–644

140. Jones, S. B. and Bylund, D. B. (1988). Characterization and possible mechanisms of α 2-adrenergic receptor-mediated sensitization of forskolin-stimulated cyclic AMP production in HT29 cells. *J. Biol. Chem.*, **263**, 14236–14244

141. Jones, S. B., Toews, M. L., Turner, J. T. and Bylund, D. B. (1987). α 2-Adrenergic receptor-mediated sensitization of forskilin-stimulated cyclic AMP production. *Proc. Natl. Acad. Sci.*, **84**, 1294–1298

142. Paris, H., Bouscarel, B. Cortinovis, C. and Murat, J. C. (1985). Growth-related variation of α 2-adrenergic receptivity in the HT29 adenocarcinoma cell-line from human colon. *FEBS Lett.*, **184**, 82–86

143. Paris, H., Galitzky, J. and Senard, J. M. (1989). Interactions of full and partial agonists with HT29 cell α2-adrenoceptor: Comparative study of [^{3}H]UK-14,304 and [^{3}H]clonidine binding. *Mol. Parmacol.*, **35**, 345–354.

144. Denis, C., Paris, H. and Murat, J. C. (1986). Hormonal control of fructose 2, 6-bisphospate concentration in the HT29 human colon adenocarcinoma cell line. *Biochem. J.*, **239**, 531–536

145. Bylund, D. B., Ray-Prenger, C. and Murphy, T. J. (1988). Alpha-2A and alpha-2B adrenergic receptor subtypes: Antagonist binding in tissues and cell lines containing only one subtype. *J. Pharmacol. Exp. Ther.*, **245**, 600–607

146. Kitabgi, P., Rostene, W., Dussaillant, M., Schotte, A., Laduron, P. M. and Vincent, J. P. (1987). Two populations of neurotensin binding sites in murine brain: Discrimination by the antihistamine levocabastine reveals markedly different radioautographic distribution. *Eur. J. Pharmacol.*, **140**, 285–293

147. Amar, S., Kitabgi, P. and Vincent, J. P. (1986). Activation of phosphatidylinositol turnover by neurotensin receptors in the human colonic adenocarcinoma cell line HT29. *FEBS Lett.*, **201**, 31–36

148. Lombes, M., Claire, M., Pinto, M., Michaud, A. and Rafestin-Oblin, M. E. (1984). Aldosterone binding in the human colon carcinoma cell line HT29: Correlation with cell differentiation. *J. Steroid Biochem.*, **20**, 329–333

149. Palmer Smith, J. Solomon, T. E. (1988). Effects of gastrin, proglumide, and somatostatin on growth of human colon cancer. *Gastroenterology*, **95**, 1541–1548

150. Couvineau, A., Rousset, M. and Laburthe, M. (1985). Molecular identification and structural requirement of vasoactive intestinal peptide (VIP) receptors in the human colon adenocarcinoma cell line, HT-29. *Biochem. J.*, **231**, 139–143

151. Omary, M. B. and Kagnoff, M. F. (1987). Identification of nuclear receptors for VIP on a human colonic adenocarcinoma cell line. *Science*, **238**, 1578–1580

152. Chastre, E., Emami, S. and Gespach, C. (1986). VIP receptor activity during HT29-18 cell differentiation and rat intestinal development. *Peptides*, **7**, 113–119

153. Chastre, E., Emami, S., Rosselin, G. and Gespatch, C. (1985). Vasoactive intestinal peptide receptor activity and specificity during enterocyte-like differentiation and retrodifferentiation of the human colonic cancerous subclone HT29-18. *FEBS Lett.*, **188**, 197–204

154. Turner, J. T., Bollinger, D. W. and Toews, M. L. (1988). Vasoactive intestinal peptide receptor/adenylate cyclase system: Differences between agonist- and protein kinase C-mediated desensitization and further evidence for receptor internalization. *J. Pharmacol. Exp. Ther.*, **247**, 417–423

155. Turner, J. T., Jones, S. B. and Bylund, D. B. (1986). A fragment of vasocative intestinal peptide, VIP(10–28), is an antagonist of VIP in the colon carcinoma cell line, HT29. *Peptides*, **7**, 849–854

156. Bozou, J. C., Couvineau, A., Rouyer-Fessard, C., Laburthe, M., Vincent, J. P. and Kitabgi, P. (1987). Phorbol ester induces loss of VIP stimulation of adenylate cyclase and VIP-binding sites in HT29 cells. *FEBS Lett.*, **211**, 151–154

157. Luis, J., Martin, J. M., El Battari, A., Fantini, J., Giannellini, F., Marvaldi, F. and Pichon, J. (1987). Cycloheximide induces accumulation of vasoactive intestinal peptide (VIP) binding sites at the cell surface of a human colonic adenocarcinoma cell line (HT29-D4): Evidence for the presence of an intracellular pool of VIP receptors. *Eur. J. Biochem.*, **167**, 391–396

158. Luis, J., Martin, M. N. El Battari, A., Marvaldi, J. and Pichon, J. (1988). The vasoactive intestinal peptide (VIP) receptor: Recent data and hypothesis. *Biochimie*, **7**, 1311–1322

159. Grasset, E., Bernabeu, J. and Pinto, M. (1985). Epithelial properties of human colonic carcinoma cell line Caco-2: Effect of secretagogues. *Am. J. Physiol.*, **248**, C410–C418

160. Burnham, D. B. and Fondacaro, J. D. (1989). Secretagogue-induced protein phosphorylation and chloride transport in Caco-2 cells. *Am. J. Physiol.*, **256**, G808–G816

161. Handler, J. S. Perkins, F. M. and Johnson, J. P. (1980). Studies of renal cell function cell culture techniques. *Am. J. Physiol.*, **238**, F1–F9

162. Murer, H. and Kinne, R. The use of isolated membrane vesicles to study epithelial transport processes. *J. Memb. Biol.*, **55**, 81–95

163. Moran, A., Davis, L. J. and Turner, R. J. (1988). High affinity phlorizin binding to the LLC-PK$_1$ cells exhibits a sodium:phlorizin stoichiometry of 2:1. *J. Biol. Chem.*, **263**(1), 187–192

164. Moran, A., Handler, J. S. and Hagan, M. (1986). Role of cell replication in regulation of Na-coupled hexose transport in LLC-PK$_1$ epithelial cells. *Am. J. Physiol.*,**250**, C314–318

165. Wu, J. S. and Lever, J. E. (1989). Developmentally regulated 75-kilodalton protein expressed in LLC-PK$_1$ cultures is a component of the renal Na$^+$/glucose cotransport system. *J. Cell. Biochem.*, **40**(1), 83–89

166. Morel, F. and Doucet, A. (1986). Hormonal control of kidney functions at the cell level. *Physiol. Rev.*, **66**(2), 377–446
167. Gnionsahe, A., Claire, M., Koechlin, N., Bonvalet, J. P. and Farman, N. (1989). Aldosterone binding sites along nephron of Xenopus and rabbit. *Am. J. Physiol.*, **257**, R87–R95
168. Niendorf, A., Arps, H., Sieck, M. and Dietel, M. (1987). Immunoreactivity of PTH-binding in intact bovine kidney tissue and cultured cortical kidney cells indicative for specific receptors. *Acta Endocrinol. (Suppl.)*, **281**, 207–211
169. Sexton, P. M., Adam, W. R. Mosely, J. M., Martin, T. J. and Mendelsohn, F. A. (1987). Localization and characterization of renal calcitonin receptors by in vitro autoradiography. *Kidney Int.*, **32**(6), 862–868
170. Soeckel, M. E., Freund-Mercier, M. J., Palacios, J. M., Richard, P. and Porte, A. (1987). Autoradiographic localization of binding sites for oxytocin and vasopressin in the rat kidney. *J. Endocrinol.*, **113**(2), 179–182
171. Tribollet, E., Barberis, C., Dreifuss, J. J. and Jard, S. 1988). Autoradiographic localization of vasopressin and oxytocin binding sites in rat kidney. *Kidney Int.*, **33**(5), 959–965
172. Schultz, S. G. and Hudson, R. L. (1986). How do sodium-absorbing cells do their job and survive? *News Physiol. Sci.*, **1**, 185–189
173. Gore, J. and Hoinard, C. (1988). Na$^+$/H$^+$ exchange in isolated hamster enterocytes. Its major role in intracellular pH regulation. *Gastroenterology*, **97**, 882–887
174. Hoinard, C. and Gore, J. (1988). Cytoplasmic pH in isolated rat enterocytes. Role of Na$^+$/H$^+$ exchanger. *Biochim. Biophys. Acta*, **941**, 111–118
175. Montrose, M. H., and Murer, H. (1990). Polarity and kinetics of Na$^+$/H$^+$ exchange in cultured opossum kidney (OK) cells. *Am. J. Physiol.* **259**, C121–C133
176. Alpern, R. J. and Chambers, M. (1987). Basolateral membrane Cl/HCO$_3$ exchange in the rat proximal convoluted tubule. Na-dependent and -independent modes. *J. Gen. Physiol.*, **89**(4), 581–598
177. Krapf, R. (1988). Basolateral membrane H/OH/HCO$_3$ transport in the rat cortical thick ascending limb. Evidence for an electrogenic Na/HCO$_3$ cotransporter in parallel with a Na/H antiporter. *J. Clin. Invest.*, **82**, 234–241
178. Nakanishi, T., Balaban, R. S. and Burg, M. B. (1988). Survey of osmolytes in renal cell lines *Am. J. Physiol*, **255**, C181–C191
179. Bagnasco, S., Balaban, R., Fales, H. M., Yang, Y. M. and Burg, M. (1986). Predominant osmotically active organic solutes in rat and rabbit renal medullas. *J. Biol. Chem.*, **261**(13), 5872–5877
180. Uchida, S., Garcia-Perez, A., Murphy, H. and Burg, M. (1989). Signal for induction of aldose reductase in renal medullary cells by high external NaCl. *Am. J. Physiol.*, **256**, C614–C620
181. Kirk, K., DiBona, D. and Schafer, J. (1987). Regulatory volume decrease in perfused proximal nephron: Evidence for a dumping of cell K$^+$. *Am. J. Physiol.*, **252**, F933–F942
182. Kirk, K. L., Schafer, J. A. and DiBona, D. (1987). Cell volume regulation in rabbit proximal straight tubule perfused in vitro. *Am. J. Physiol.*, **252**, F922–F932
183. Welling, P. A., Linshaw, M. A. and Sullivan, L. P. (1985). Effect of barium on cell volume regulation in rabbit proximal straight tubules. *Am. J. Physiol.*, **249**, F20–F27
184. Iino, Y. and Burg, M. B. (1979). Effect of parathyroid hormone on bicarbonate absorption by proximal tubules in vitro. *Am. J. Physiol.*, **236**, F378–F391
185. Kahn, A. M. Dolson, G. M. Hise, M. K., Bennett, S. C. and Weinman, E. J. (1985). Parathyroid hormone and dibutryl cAMP inhibit Na–H exchange in renal brush border vesicles. *Am. J. Physiol.*, **248**, F212–F218
186. Verrey, F., Kairouz, P., Schaerer, E., Fuentes, P., Geering, K., Rossier, B. C. and Kraehenbuhl, J. P. (1989). Primary sequence of Xenopus laevis Na$^+$-K$^+$-ATPase and its localization in A6 kidney cells. *Am. J. Physiol.*, **256**, F1034–1043
187. Benos, D. J., Saccomani, G. and Sariban-Sohraby, S. (1987). The epithelial sodium channel. Subunit number and location of the amiloride binding site. *J. Biol. Chem.*, **262**(22), 10613–10618
188. Benos, D. J., Saccomani, G., Brenner, B. M. and Sariban-Sohraby, S. (1986). Purification and characterization of the amiloride-sensitive sodium channel for A6 cultured cells and bovine renal papilla. *Proc. Natl. Acad. Sci. USA*, **83**(22), 8525–8529

170

189. Sorscher, E. J., Bridges, R. J., Frizzell, R. A. and Benos, D. J. (1989). Antibodies against the stilbene binding domain of band-3 identify a candidate Cl channel protein in human colon (T_{84}) cells. *FASEB J.*, **3**(3), A602 (abstract)

190. Bridges, R. J., Worrell, R. T., Frizzell, R. A. and Benos, D. J. (1989). Stilbene disulfonate blockade of colonic secretory Cl channels in planar lipid bilayers. *Am. J. Physiol.*, **256**, C902–C912

191. Wuarin, F., Wu, K., Murer, H. and Biber, J. (1989). The Na^+/P_i-cotransporter of OK cells: reaction and tentative identification with N-acetylimidazole. *Biochim. Biophys. Acta*, **981**(2), 185–192

192. Peerce, B. E. (1989). Identification of the intestinal Na-phosphate cotransporter. *Am. J. Physiol.*, **256**, G645–G652

193. Hediger, M. A. Coady, M. J., Ikeda, T. S. and Wright, E. M. (1987). Expression cloning and cDNA sequencing of the $Na^+/$glucose co-transporter. *Nature (London)*, **330**, 379–381

194. Agus, Z. S., Gardner, L. B., Beck, L. H. and Goldberg, M. (1973). Effect of parathyroid hormone on renal tubular reabsorption of calcium, sodium, and phosphate. *Am. J. Physiol.*, **224**, 1143–1148

195. Ever, C., Murer, H. and Kinne, R. (1978). Effect of parathyrin on the transport properties of isolated renal brush border vesicles. *Biochem. J.*, **172**, 49–56

196. Helmle-Kolb, C., Montrose, M. H., Stange, G. and Murer, H. (1990). Regulation of Na^+/H^+ exchange in opossum kidney cells by parathyroid hormone, cyclic AMP and phorbol esters. *Pleugers Arch.* **415**, 461–470

197. Garty, H. (1986). Mechanisms of aldosterone action in tight epithelia. *J. Memb. Biol.*, **90**, 193–205

198. Hecht, G., Pothoulakis, C., LaMont, J. T. and Madara, J. L. (1988). *Clostridium difficile* toxin A perturbs cytoskeletal structure and tight junction permeability of cultured human intestinal epithelial monolayers. *J. Clin. Invest.*, **82**, 1516–1524

199. Houtt, P. A. Liu, W., McRoberts, J. A., Giannella, R. A. and Dharmsathaphorn, K. (1988). Mechanism of action of *Escherichia coli* heat stable enterotoxin in a human colonic cell line. *J. Clin. Invest.*, **82**, 514–523

200. Guarino, A., Cohen, M., Thompson, M., Dharmsathaphorn, M. and Giannella, R. (1987). T84 cell receptor binding and Guanyl cyclase activation by *Escherichia coli* heat-stable toxin. *Am. J. Physiol.*, **253**, G775–G780

201. Wasserman, S. I., Barrett, K. E. Houtt, P. A. Beuerlein, G., Kagnoff, M. F. and Dharmsathaphorn, K. (1988). Immune-related intestinal Cl- secretion. I. Effect of histamine on the T_{84} cell line. *Am. J. Physiol.*, **254**, C53–C62

202. Riordan, J. R., Rommens, J. M., Kerem, B., Alon, N., Rozmahel, R., Grzelczak, Z., Zielenski, J., Lok, S., Plavsic, N., Chou, J. L., Drumm, M. L., Iannuzzi, M. C., Collins, F. S. and Tsui, L. C. (1989). Identification of the cystic fibrosis gene: cloning and characterization of complementary DNA. *Science*, **245**, 1066–1073

203. Berschneider, H. M., Knowles, M. R., Azizkhan, R. G., Boucher, R. C., Tobey, N. A., Orlando, R. C. and Powell, D. W. (1988). Altered intestinal chloride transport in cystic fibrosis. *FASEB J.*, **2**, 2625–2629

204. Phillips, T. E., Huet, C., Bilbo, P. R., Podolsky, D. K. Louvard, D. and Neutra, M. R. (1988). Human intestinal goblet cells in monolayer culture: characterization of a mucus secreting subclone derived from HT29 colon adenocarcinoma cell line. *Gastroenterology*, **94**, 1390–1403

205. Pringault, E., Arpin, M., Garcia, A., Finidori, J. and Louvard, D. (1986). A human villin cDNA clone to investigate the differentiation of intestinal and kidney cells in vivo and in culture. *EMBO J.*, **5**(12), 3319–3124

8
Established Renal Epithelial Cell Lines: Experimental Panacea or Artifact?

N. L. SIMMONS

INTRODUCTION

Cultured epithelial cell lines derived from renal epithelia have become increasingly popular as convenient *in vitro* models for physiological, biochemical and cell biological studies. The most commonly used cell lines are MDCK, derived from a dog kidney by Madin and Darby, the LLCPK$_1$ line derived from a hog kidney by Hull, and, latterly, OK cells from oppossum. The detailed properties of lines and their similarities with different portions of the nephron are dealt with by M. Montrose in this volume. It is evident that exact correspondence of a cell line to an individual nephron segment is unusual; rather, elements of a set of functional characteristics are retained *in vitro*. The purpose of the present chapter is to address the question whether such partial differentiation and coupling of properties found in diverse nephron segments in a single cell line invalidates their use as functional models, or, as is often assumed by workers in the field, provides some universal experimental panacea. As a protagonist for cultured epithelia for a decade, my own conclusion should never be in doubt; however, caveats derived from work with MDCK cell cultures may be useful to newcomers. I wish to demonstrate the apparent validity of functional studies with MDCK epithelia by describing work over the past 5 years on the action of vasoactive intestinal peptide (VIP) upon renal epithelial cells.

Of central importance to a discussion of cultured epithelial properties is that of cellular heterogeneity in non-clonal cultures and the apparent discrepancies reported between different laboratories using 'identical' cell lines[1,2]. For LLCPK$_1$, for instance, cloning of phenotypically stable sub-populations yields cells responsive to calcitonin, and also unresponsive cells[3]. MDCK cells were originally established from the kidneys of a mongrel dog. The heterogeneous nature of the initial cell population is evident in the emergence of strains of MDCK cells with differing physiological properties. It is now evident that two major phenotypes exist. These 'epithelia' differ in morphology, transepithelial resistance, hormonal responsiveness, antigenic properties, lipid composition etc[1,2]. The existence of such heterogeneity may be of some importance in studies in which 'mutant' lines are selected, if selection systems include 'suicide' selection for transport systems that are not present in one or more of the cell types present (see Popowicz and Simmons[4] for an example of this). For the present purposes, work was conducted on Strain 1 MDCK cells which form 'tight' epithelial layers when reconstructed on permeable matrices[1]. The expression of antigenic determinants, recognized by monoclonal antibodies[5,6], suggests that such strains of MDCK cell line may be useful models of distal nephron function.

STIMULATION OF MDCK CULTURED EPITHELIAL ADENYLATE CYCLASE BY VIP AND RELATED HORMONES

Early studies[7] investigated the stimulation of cAMP accumulation in MDCK cells by various renal-acting hormones and receptor agonists. Prostaglandin E_1, glucagon and arginine vasopressin all gave elevations in cAMP accumulation greater than two-fold. Parathyroid hormone (PTH) and calcitonin were not effective. Later studies on β-adrenoreceptor-mediated increases in adenylate cyclase showed significant strain differences in hormonal responsiveness[8].

Since we had identified a secretory flux of Cl^- stimulated by both α and β-adrenoreceptors in strain 1 MDCK epithelia[9-11], we were interested to see if vasoactive intestinal peptide (VIP) stimulated MDCK adenylate cyclase[12] as this hormone has been implicated in the regulation of salt balance in the secretory organs of sharks[13,14] and birds[15]. Strain 1 MDCK cells do indeed show a specific stimulation of adenylate cyclase by VIP that is not observed with structurally related hormones, such as glucagon, secretin and PHI (Figure 8.1). Strain 2 cells show an entirely different pattern of response, in that adenylate cyclase is unresponsive to VIP, but responsive to glucagon (Figure 8.1).

The response of Strain 1 cells to VIP is only a fraction of that observed with forskolin; whereas only a 2.2±0.4 (SE, n = the number of separate experiments = 5), stimulation over basal values of adenylate cyclase is observed with 1 μmol L^{-1} VIP, a 12.6±1.5 (SE, n = 4) fold stimulation is observed with forskolin. Is such a disparity due to the existence of a subpopulation of cells in Strain 1 responsive to VIP? Table 8.1 shows the effects of VIP added in conjunction with other known hormonal agonists of adenylate cyclase in MDCK cells. In each case, although the response to VIP in conjunction with PGE_1, isoprenaline or vasopressin (AVP) exceeds the responses to PGE_1, isoprenaline and vasopressin alone, the observed additive responses are less than those expected for separate pools of adenylate cyclase. It thus seems likely that the responses mediated by these separate hormone agonists share a similar pool of adenylate cyclase. This conclusion is strengthened by studies of clonal lines of MDCK cells. The clonal line, $CL_8 1_b$, was isolated from Strain 1 cells in Kai Simons labor-

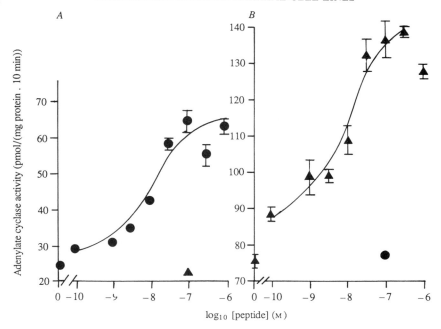

Figure 8.1 **A**   Sensitivity of strain 1 cells to VIP (●), and lack of responsiveness to glucagon (▲). **B**   Sensitivity of strain 2 cells to glucagon (▲) and lack of response to VIP (●). Data are the mean ± SE; $n=3$ for each panel. Half-maximal stimulation of adenylate cyclase to VIP and glucagon were 9.0 ± 2.6 nmol L^{-1} (SD) and 5.8 ± 2.3 nmol L^{-1} respectively. Measurements of adenylate cyclase were made as described by Rugg and Simmons[8]

atory[8] by limiting dilution; the behaviour of this clonal line is similar to the parental strain in its responsiveness to VIP and glucagon. The clonal line DL_{17} differs from Strain 2 in not possessing marked stimulations to either VIP or glucagon[8].

For five separate experiments in which the sensitivity to VIP was determined, the mean half-maximal concentration for adenylate cyclase stimulation was 13.7 ± 6.5 nmol L^{-1} and the slope factor was 1.19 ± 0.14, consistent with the existence of a single class of receptor sites. The sensitivity of Strain 1 cells to VIP is thus similar to that seen in mammalian small intestinal membranes where VIP is involved in the regulation of (Na)Cl and fluid secretion[16]. Half-maximal activation of Strain 2 adenylate cyclase by glucagon was observed at 3.8 ± 1.2 (SE; $n=3$) nmol L^{-1}.

VIP stimulates intracellular cAMP accumulation in intact Strain 1 cells grown in plastic Petri dishes. In the presence of isobutylmethylxanthine (IBMX) to inhibit phosphodiesterase, cAMP levels reach a peak at 4 min, VIP (1 μmol L^{-1}) increases cAMP accumulation from 3.8 ± 0.25 pmol/10^6 (±SD;n = number of replicate cultures, 3) cells to 19.6 ± 2.9 pmol/10^6 cells. Comparison of peak values for cAMP accumulation observed with VIP and other hormonal agonists with data from adenylate cyclase determinations in

175

Table 8.1 Effect of VIP (1 μmol L⁻¹) alone, and in conjunction with maximum concentrations of other hormone agonists, on strain 1 MDCK adenylate cyclase activity. Data are the mean±SE ($n=3$)

	Adenylate cyclase activity (pmol cAMP (mg protein)⁻¹ (10 min)⁻¹		
	No addition	*+ 1 μmol L⁻¹ VIP*	*Expected additive activity with 1 μmol L⁻¹ VIP*
—	26.8 ± 0.9	54.8 ± 2.2†	—
10⁻⁵ mol/L⁻¹ isoprenaline	78.8 ± 1.9	90.0 ± 2.1*	106.8
10⁻⁶ mol/L⁻¹ PGE₁	45.5 ± 2.1	62.5 ± 5.6†	73.5
10⁻⁶ mol/L⁻¹ ADH	34.2 ± 2.1	55.5 ± 1.9*	62.9

Significantly different from values in absence of 1 μmol L⁻¹ VIP: *$p < 0.05$; †$p < 0.01$

cell homogenates show qualitative differences; VIP-dependent accumulations, in cells grown on plastic, are still increasing at 10 μmol L⁻¹ VIP and are smaller than those observed with PGE₁. This most probably indicates considerable restriction of VIP diffusion to the basal–lateral surfaces, even with the use of subconfluent layers. Restriction of amino acid permeation to the basal–lateral cell aspects has been notes in cultures of LLCPK₁ cells grown in plastic Petri dishes[17].

PHYSIOLOGICAL CORRELATES OF VIP ACTION UPON STRAIN 1 CELLS

MDCK cells may be grown upon permeable filter supports. In this way, it is possible to reconstitute a viable epithelium. Figure 8.2 shows an electron micrograph of Strain 1 cells grown on a 0.2 μm pore diameter Millipore filter. The cells are morphologically polarized and exclude La³⁺ penetration across the apical tight junction. Strain 1 cells and the clonal line derived from Strain 1, CL₈1ᵦ, possess epithelial characteristics that are entirely similar, i.e. low transepithelial conductance, and transepithelial ion transport principally of anion (Cl⁻) secretion stimulated by various agonists, such as adrenaline, ATP and prostaglandins[18–20]. The observation of a VIP-responsive adenylate cyclase allows us to look for physiological correlates of VIP stimulation in epithelial layers of MDCK cells grown upon permeable millipore filter supports. Addition of 0.3 μmol L⁻¹ VIP to the basal–lateral bathing solution of short-circuited MDCK epithelial monolayers results in the stimulation of an inwardly directed current (Figure 8.3). For four separate epithelia whose resistance was 2.8±1.7 kΩ.cm², the basal inward short-circuit current (0.82±0.62 μA/cm² SD) is increased at peak values to 2.9±0.3 μA/cm². Glucagon fails to stimulate the short-circuit current. That the VIP-stimulated inward current is most probably due to Cl⁻ secretion (from basal to apical cell aspects) is evident from the data in

176

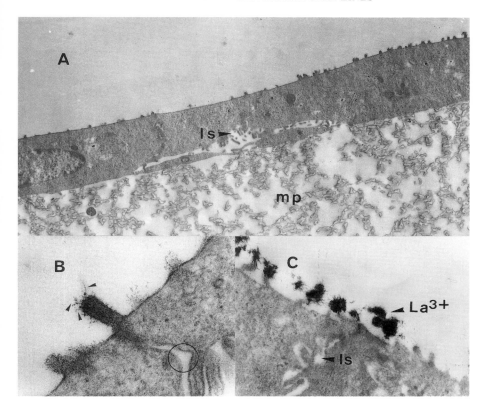

Figure 8.2 **A** Strain 1 MDCK cells grown upon a millipore filter (×10 000). A sparse brush-border is evident; ls = lateral space, mp = millipore filter. **B** High power (×100 000) of the apical tight-junctional region. Note the glycocalyx (arrows) and the lateral space inter-digitations (circled). **C** Lack of penetration of the tight junction by the tracer molecule, La^{3+}, perfused *in situ* in Ussing chambers (×50 000)

Figure 8.3 which demonstrates that the increased inward current is insensitive to replacement of the apical bathing solution Na^+ by choline[+], by the presence of 0.1 mmol L^{-1} amiloride, and by the abolition of the inward current by replacement of the medium Cl^- by NO_3^-. In addition the effect of the Cl-channel blocker, 3-nitro-2-(3phenylpropylamino)-benzoic acid[21] was tested from the apical bathing solution and was found to produce a rapid inhibition of the VIP-stimulated inward current (Figure 8.3). These data are thus consistent with the notion that VIP also stimulates anion (Cl^-) secretion.

It is of interest to compare the magnitude of the VIP-stimulated short-circuit inward current with that of other agonists. Inward currents generated by adrenaline and by exogenous ATP exceed those observed with VIP (or with prostaglandins) often by an order of magnitude (present data[9,10,19]).

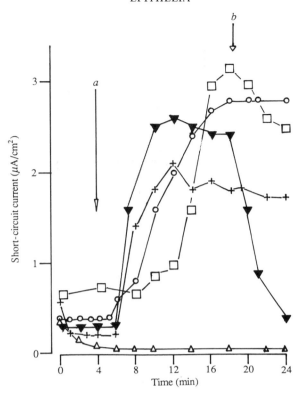

Figure 8.3 Effects of 0.5 μmol L^{-1} VIP upon the short-circuit current of MDCK epithelial layers clamped into Ussing chambers. At arrow *a*, VIP was added to the basal bathing solution to give a final concentration of 0.5 μmol L^{-1} VIP. (O), control data; (□) plus 0.1 mol L^{-1} amiloride added to the apical bathing solution; (+) Na$^+$-free media (choline$^+$ Cl$^-$ replacement, other Na$^+$ salts omitted), apical bathing solution; (△) Cl$^-$-free media, Cl$^-$ salts replaced by NO$_3^-$ salts; (▼) control until arrow at *b* when 100 μmol L^{-1} 3-nitro-2(3-phenylpropylamino)-benzoic acid was added to the apical bathing solution. Data are representative from individual monolayers. Preparation of epithelial monolayers and measurement of short-circuit current are described in references 1 and 19

For stimulation by adrenaline, both α- and β-receptor activation is required to achieve maximal secretory inward currents. β-Adrenoreceptor stimulation resulting in increased intracellular cAMP is thought to activate apical Cl$^-$ conductance via protein phosphorylation[8-10], whereas α-adrenoreceptor activation leads to an increase in cytosolic Ca^{2+}, so increasing a K$^+$ conductance at the basal–lateral membrane[9,11]. Coupled with the operation of an electroneutral ternary co-transporter for Na$^+$, K$^+$ and Cl$^-$ and the Na$^+$/K$^+$/-ATPase, transepthelial secretion is achieved[20]. When pure α-agonists and pure β-agonists are used[9], the following observations are pertinent: first there is synergism of the effects of α- and β-agonists in conjunction[9] and, second, β-agonists do not stimulate a significant activation of K$^+$ efflux across the basal–lateral cell aspects[9,11]. Measurements of membrane potential and of free Ca^{2+} in MDCK cells confirm this separation, namely cAMP activates Cl^- conductance, and Ca^{2+} activates K^+ conductance[22,23]. However, it should be noted that both β-adrenoreceptor activation and cAMP have been reported to lead to a slow sustained rise in intracellular Ca^{2+} in MDCK cells[22,24], suggesting that both increases in cAMP and intracellular Ca^{2+} are involved even in submaximal stimulation of anion secretion. In addition, secretagogues, such as α-adrenoreceptor agonists and adenosine triphosphate may cause endogenous production of prostaglandins[23]. Synergistic activation of inward current may also involve additional intracellular messengers, such as diacylglycerol and inositol phosphates. For VIP activation of anion secretion, it seems clear that primary changes in cAMP and activation of the apical Cl^- conductance are attained without marked synergistic interactions with other second messengers.

ACTION OF VASOACTIVE INTESTINAL PEPTIDE AND GLUCAGON UPON CANINE RENAL ADENYLATE CYCLASE

The possibility suggested by the data from work on MDCK epithelia is that VIP may regulate renal epithelial function by stimulation of the second messenger cAMP. Various studies have looked for effects of VIP upon renal function in several species, including man. Thus, infusions of VIP affect renal haemodynamics and the release of renin[25-28]. In addition to these effects, there is also evidence for direct effects of VIP upon tubular transport independent of renal haemodynamics[25,29].

With these considerations in mind, it was considered worthwhile to turn to *in vitro* studies of renal tissue to substantiate direct renal actions of VIP. We have thus investigated the actions of VIP and glucagon on canine renal adenylate cyclase. Glucagon action upon renal adenylate cyclase has been demonstrated and its tubular sites of action mapped along the rat nephron[30]. Table 8.2 demonstrates that both 1 μmol L^{-1} VIP and glucagon produce significant elevations of adenylate cyclase activity above basal values in plasma membranes isolated from canine cortex. Such stimulations are a small fraction of that observed with either forskolin or parathyroid hormone (PTH) (Table 8.2). In plasma membranes isolated from the outer and inner medulla, the stimulation by PTH declines, no significant stimulation being observed by PTH in the inner medullary membranes. This distribution of PTH-stimulated adenylate cyclase is entirely consistent with the mammalian pattern of nephron-segment stimulation by PTH[30]. In contrast to the effect of PTH, the elevation of adenylate cyclase activity observed with both VIP and glucagon is maintained (Table 8.2) in all three regions dissected from the canine kidneys. A dose-dependent stimulation of canine renal adenylate cyclase by VIP is observed. Half-maximal stimulation was estimated at 18.1±14.0 nmol/L^{-1} (±SE; $n=3$), a value comparable to that observed for rabbit small intestine[16]. Detailed analysis of VIP dose–response curves,

Table 8.2 Stimulation of canine renal adenylate cyclase by VIP, glucagon and PTH in plasma membranes isolated from cortex, outer medulla and inner medulla

| | Adenylate cyclase activity | | |
| | (pmol (mg)$^{-1}$ (15 min)$^{-1}$ for basal; fold activity for stimulations) | | |
	Cortex	Outer medulla	Inner medulla
Basal	94.5 ± 23.3 (4) [1.0]	142.8 ± 15.8 (3) [1.0]	181.1 ± 10.0 (2) [1.0]
1 µmol L^{-1} VIP	1.71 ± 0.16(4)	1.57 ± 0.09(3)	1.50*
1 µmol L^{-1} glucagon	1.89 ± 0.18(4)	2.34 ± 0.28(3)	1.47*
1 µmol L^{-1} b(1–34) PTH	7.0 ± 2.8 (3)	1.35 ± 0.04(3)	1.15$^{n.s.}$
100 µmol L^{-1} forskolin	15.40 ± 4.56(4)	7.54 ± 1.52(3)	6.20

Data are expressed relative to basal values, and are mean values ± SE

Figures in parenthesis give n, the number of separate experimental animals, where measurements of adenylate cyclase were made in triplicate

Significantly different from basal values in the two individual experiments: *p <0.01; n.s. = not significant

however, indicates slope factors[31] less than unity (0.45±0.17 SD; n=3). Peptides related to VIP and glucagon were also tested at 1 µmol L^{-1} for their ability to stimulate renal cortical adenylate cyclase activity; whereas both VIP and glucagon gave marked elevations of activity above basal values (283±11 and 178±68 pmol cAMP (mg protein)$^{-1}$ (15 min)$^{-1}$ respectively, SD; n=3), PHI and secretin gave smaller elevations (87±14 and 52±28 pmol cAMP (mg protein)$^{-1}$ (15 min)$^{-1}$, respectively). The VIP-receptor antagonist (4-Cl,D-Phe^6,Leu^{17})-VIP[32] failed to stimulate canine renal cortical adenylate cyclase activity at 10 µmol L^{-1} (0.76±0.08 times basal activity, SD n=3). At 1 µmol L^{-1} VIP, a 1.75±0.37 (SD n=3) fold stimulation of adenylate cyclase activity was reduced to a 1.32±0.09-fold stimulation by 10 µmol L^{-1} (4-Cl,D-Phe^6,Leu^{17})-VIP. In contrast, (4-Cl,D-Phe^6,Leu^{17})-VIP was without inhibitory effect upon the stimulation observed by 1 µmol L^{-1} glucagon; indeed the effect of glucagon was enhanced.

[^{125}I]IODOTYROSYL VIP BINDING TO CANINE CORTICAL PLASMA MEMBRANES

Figure 8.4 shows the effect of increasing concentrations of VIP and glucagon upon their ability to compete with [^{125}I]VIP for binding to renal cortical plasma membranes. Whereas VIP is an effective competitor, glucagon was without effect. Other peptides were also tested for their ability to compete with [^{125}I]VIP; secretin (1 µmol L^{-1}), PHI (1 µmol L^{-1}), (4-

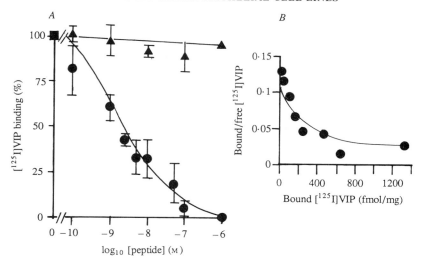

Figure 8.4 **A** Competition of [^{125}I]VIP binding (80 pmol L^{-1}) by unlabelled VIP (●), or glucagon (▲) in canine renal cortical plasma membranes. Data are the mean±SE (*n*=3). Non-specific binding of [^{125}I]VIP was assessed by the inclusion of 1 μmol L^{-1} unlabelled VIP and amounted to 48% of the total binding observed with 80 pmol L^{-1} [^{125}I]VIP. Bound [^{125}I]VIP was separated by rapid filtration and washing. **B** Scatchard plot of specific binding data derived from *A*. A non-linear least-squares fit of binding vs concentration for 2 binding sites of K$_{0.5}$ 2.1 nmol L^{-1} and B$_{max}$ 181 fmol/mg protein, and K$_{0.5}$ 2.2×10^{-1} mmol L^{-1} and B$_{max}$ 6436 fmol/mg protein gives the solid line shown

Cl,DPhe^6,Leu^{17})-VIP (10 μmol L^{-1}) and VIP (1 μmol L^{-1}) gave 96%, 83%, 26% and 35.5% respectively of that binding observed with 80 pmol L^{-1} [^{125}I]VIP alone. Analysis of specific binding data (Figure 8.4b) indicates that the Scatchard curve is markedly curvilinear, suggesting that [^{125}I]VIP interacts with more than a single set of sites in canine renal cortical plasma membranes. Taken together with the data from measurements of adenylate cyclase, this data is consistent with the existence of a VIP receptor coupled to adenylate cyclase present in canine kidney and distinct from effector systems responsive to glucagon.

Thus, in membranes isolated from canine renal cortex, separate responses to the structurally related peptides, VIP and glucagon, can only be inferred from the use of a receptor antagonist to VIP and by [^{125}I]VIP binding, whereas separate responses to glucagon and VIP are segregated into separate strains of the cultured MDCK dog kidney cell line.

These data showing that VIP may stimulate adenylate cyclase in plasma membranes isolated from canine kidneys, extend previous observations made in rabbits[33] and cats[34], and suggest that VIP may indeed have a role as a renal-acting hormone in several mammalian species. As with rabbit and cat, half-maximal stimulation of adenylate cyclase occurs in a dose range that far exceeds the measured values of plasma VIP in mammals. For this reason, it is likely that VIP is released from renal nerves (see below). Further work needs to clarify the intrarenal locus of VIP-sensitive adenylate cyclase in dogs, and to determine the precise physiological actions of VIP.

THE TUBULAR LOCUS OF VIP-STIMULATED ADENYLATE CYCLASE: STUDIES USING MICRODISSECTED NEPHRON SEGMENTS

Data which demonstrate VIP stimulation of adenylate cyclase in purified plasma membranes do not differentiate between effects on vascular rather than tubular elements. In rabbit, separations of glomeruli by magnetic iron oxide trapped in preperfusions, enabled us to establish that VIP stimulated glomerular adenylate cyclase[33]. Whether other elements of the renal vascular tree are also responsive to VIP is not yet known. In addition, adenylate cyclase activity in tubular material from cortex, and in fractionated medullary tubules, was stimulated by VIP[33]. We have been extremely fortunate to be able to extend such observations by determining the action of VIP upon adenylate cyclase activity in defined microdissected segments of the rabbit kidney tubule in collaborative studies conducted in Professor Morel's laboratory in Paris[35-37].

Figure 8.5 shows the distribution of the basal and VIP-stimulated values of adenylate cyclase along the rabbit nephron. The pattern of basal activities (in the presence of 10 μmol L^{-1} GTP) shows low values in most nephron segments except in the distal tubule (especially DCTg) and the cortical collecting tubule.

Highly significant elevations of adenylate cyclase by 1 μmol L^{-1} VIP in the presence of 10 μmol L^{-1} GTP were observed in DCTb, OMCTo and OMCTi ($p < 0.005$ for all 3 segments). A clear gradient of adenylate cyclase responsiveness to 1 μmol L^{-1} VIP exists along the collecting duct since the cortical portion (CCT) was unresponsive whilst only a 2.5-fold stimulation over basal values was observed in the OMCTo compared with the 7.8-fold stimulation observed on OMCTi segments. The absence of any effect obtained with VIP in CCT was substantiated by the large stimulation of adenylate cyclase by arginine vasopressin (AVP) (0.1 μmol L^{-1}) observed in parallel measurements (7.3-fold; $n=7$ experiments). The stimulation by VIP in DCTg, though being of quantitative importance second only to DCTb for all segments studied, represents only a 2.0-fold stimulation over the high basal values. In 2 of the 8 experiments conducted, no significant increase over basal values was observed.

In TDL, 1 μmol L^{-1} VIP produced a significant stimulation of adenylate cyclase activity ($p < 0.05$) in two of the six experiments conducted. When present, this stimulation was of limited magnitude. For the pooled experimental data, VIP action did not reach a statistically significant level (basal 24.3±4.9 (SE) fmol cAMP mm^{-1} (30 min)$^{-1}$, $n=6$; VIP 49.5±11.1; $0.05 < p < 0.1$). In CAL, a significant increase in adenylate cyclase activity was observed in 3 out of the 4 rabbits tested, the mean value of adenylate cyclase rising from 13.8±1.1 fmol cAMP mmol^{-1} (30 min)$^{-1}$ (SE, $n=4$) to only 25.1±1.5 ($p < 0.001$) in the presence of 1 μmol L^{-1} VIP. No effect of VIP was observed in the other nephron segments.

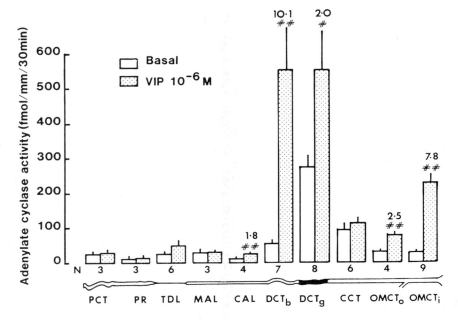

Figure 8.5 Distribution of VIP-sensitive adenylate cyclase along the rabbit nephron. The ten different segments tested are illustrated schematically; abbreviations used are: PCT, proximal convoluted tubule; PR, pars recta; TDL, thin descending limb; MAL, medullary thick ascending limb; CAL, cortical thick ascending limb; DCTb, the 'bright' portion of the distal convoluted tubule; DCTg, the granular portion of the distal convoluted tubule; CCT, the cortical collecting tubule; OMCT, outer medullary collecting tubules from the outer (o) and inner (i) stripes. Data are the mean±SE for n separate animals. Figures above the error bars denote the mean-fold stimulation over basal values in these segments. $p < 0.05$

The magnitude of the VIP action on adenylate cyclase in CAL, DCTb, DCTg and OMCTi to that produced, in the same animals, by other hormone agonists known to stimulate adenylate cyclase in these segments was determined[38-40,64]. The effect of VIP upon adenylate cyclase in individual responsive segments was substantial, representing 25% of the response to salmon calcitonin in DCTb, 26% of that to parathyroid hormone in DCTg and 38% of that to AVP in OMCTi. In contrast for CAL, the VIP effect represented only 4.4% of that produced by calcitonin.

The dose dependence of VIP stimulation was investigated in the main responsive segments, i.e. DCTb and OMCTi. For three separate dose-response relationships obtained with OMCTi, the mean value of half-maximal stimulation of adenylate cyclase by VIP was 25.9±10.1 nmol L^{-1} (SD). For DCTb, half-maximal stimulations of adenylate cyclase by VIP were observed at 17 and 20 nmol L^{-1}. This is similar to that value for renal cortical plasma membranes (see above and reference 33) and for rabbit intestinal basolateral membranes and enterocytes[16]. Significant elevations of activity above basal values were observed at 10 nmol L^{-1} in both DCTb and OMCTi. Marked effects of VIP upon renal function in rabbits are seen when plasma infusions of VIP raise plasma levels to the range of 0.8–2 nmol L^{-1} [25].

The effects of related peptides to VIP on adenylate cyclase were tested on OMCTi. VIP is the most effective of the peptides tested whilst PHI, secretin and glucagon have less or no effect (all tested at 1 µmol L^{-1}). PHI gave a dose-dependent stimulation, lower fold stimulations being observed in 3 experiments at 0.1 µmol L^{-1} PHI (1.9-fold) compared with 1.0 µmol L^{-1} PHI (2-9-fold). In 4 separate experiments with rabbit OMCTi, the response of adenylate cyclase to 1 µmol L^{-1} PHI was 47% of the response to 1 µmol L^{-1} VIP.

In assessing whether the observed effects on adenylate cyclase represent a likely physiological action of the hormone *in vivo*, rather than a pharmacological effect, several criteria should be satisfied: first a distinct pattern of stimulation of adenylate cyclase along the nephron should be evident; secondly, the magnitude and dose dependence of adenylate cyclase stimulation should accord with that observed in other tissues or with known actions of VIP in the target tissue; and, finally, specificity should be displayed with respect to structurally-similar hormones. As outlined above, these criteria are indeed satisfied.

IS VASOACTIVE INTESTINAL PEPTIDE RELEASED FROM RENAL NERVES?

As already mentioned, the dose–response relationship of VIP activation of renal (or tubular) adenylate cyclase renders it extremely unlikely that circulating plasma levels of VIP functionally activate the receptor in renal tissue. It should be recalled that the reported circulating plasma VIP concentrations are only of the order of 10 pmol L^{-1} [25,41], i.e. a value about 1000-fold lower than those producing marked renal effects[25]. VIP activates adenylate cyclase which is functionally contained within the basal–lateral membranes of renal epithelial cells[33] (see also the action of VIP on MDCK epithelia from the basal surfaces, above). Concentration of plasma-derived VIP to functionally active levels within the kidney in the interstitium must be considered an extremely unlikely possibility. What then is the cellular origin of VIP which activates renal adenylate cyclase in cortical (DCTb, DCTg) as well as medullary segments (OMCTi, OMCTo)? A physiological action of VIP almost certainly requires intrarenal VIP release, most probably from nerves.

Several studies have addressed the question of VIP immunoreactivity in kidneys from several species, but not the rabbit. Positive findings have been reported in rat, guinea pig, dog, pig and tupaia kidneys[42–44]. The occurrence of VIP in the kidney of mammals thus appears to be a general phenomenon. In broad terms, VIP immunoreactivity parallels the renal arterial system, forming a perivascular plexus[44]. Earlier studies in dog reported similar findings, VIP-positive immunoreactivity being localized in the region of arcuate and interlobular arteries and, in some cases, in proximity to glomeruli[42]. There was no report of outer medullary VIP-positive immunoreactivity. These findings of VIP immunoreactivity in dog have, however, been questioned by others[28]. In rat, VIP-positive immunoreactivity was considered sparse and associated mainly with vascular elements[43]. Clearly, it is necessary to look for VIP immunoreactivity in rabbit kidney, special attention being focussed on the interstitium that adjoins the tubular segments sensitive to VIP, identified by our own work on microdissected segments.

SPECIES VARIATION IN THE RESPONSE OF RENAL ADENYLATE CYCLASE TO VIP

Species variation in the relative effects of VIP-family peptides upon renal adenylate cyclase, is evident. In dog, rabbit and cat renal membranes, stimulations are observed. However, in rat isolated plasma membranes, we have found little or no stimulation by VIP but pronounced effects of glucagon. In studies of isolated microdissected segments we have also been able to confirm previous observations on rat and rabbit OMCTi which demonstrated an absence of marked actions of VIP on adenylate cyclase in the rat[30,35] and the absence of a marked glucagon effect in the rabbit[35,45]. In the cat[65], the tubular locus of VIP action is similar to that found in the rabbit but, in other responsive species, it is not yet known if the tubular distribution matches that observed in rabbits. In the dog and rat, a VIP effect in the outer medullary collecting tubule has been reported[66].

Taken together, the absence of effects of VIP on renal adenylate cyclase but the presence of the effect of VIP infusions in isolated perfused rat kidney[29] are therefore consistent with a primary action on the renal vasculature and secondary actions on the tubules.

Recently, we have been able to demonstrate a substantial VIP stimulation of adenylate cyclase in plasma membranes prepared from human renal cortex. Tissue samples were obtained from nephrectomy samples from patients undergoing surgery for hypernephroma and ureteric tumours but which were morphologically normal[46]. This work substantiates earlier reports of effects of VIP infusions upon renal function in normal volunteers[47].

THE PHYSIOLOGICAL ROLE OF VIP IN THE KIDNEY

In our own studies on VIP-stimulated adenylate cyclase in rabbit kidney, we can identify both vascular and tubular sites of action. Regulation of renin secretion by VIP has received considerable attention, though a physiological role, at least in dogs, awaits elucidation[28]. The physiological consequences of VIP-stimulated adenylate cyclase in each responsive tubule segment in the rabbit have not yet been defined.

Dimaline, Peart and Unwin[25] have investigated the effects of plasma infusions of VIP upon renal function in conscious rabbits. In these studies, secondary responses arising from reflex sympathetic activation following peripheral vasodilatation are evident. With high doses of VIP infusion giving plasma VIP concentrations in the order of $0.8-2$ nmol L^{-1}, there were VIP-dependent increases in plasma renin and increased fractional excretions of Na^+, K^+ and Cl^-. In addition, there was a fall in urinary pH, consistent with a regulation of H^+ secretion.

Recent experiments carried out in rabbit isolated OMCTi have demonstrated that 8-bromocyclic AMP and forskolin increase H^+ secretion[48]. On the basis of data available in the literature[49], the authors proposed that the cellular mechanism involved in the stimulation of proton secretion was an increase in the chloride conductance of the basolateral membrane. It is well known that urinary acidification occurs in the intercalated cells (type A) which, under normal conditions, predominate in the outer medulla[50]. The localization of this granular cell type corresponds exactly to that of the VIP-sensitive adenylate cyclase along the collecting tubule, as shown in our own study[35] (Figure 8.5). We therefore suggest, as a working hypothesis, that VIP may participate in the regulation of proton secretion, via cAMP[48] in the rabbit. Such a hypothesis is related to the VIP-dependent stimulation of Cl^- secretion in MDCK epithelial layers[12] and T84 cells[51] as well as in the isolated shark rectal gland[13,14]. VIP has also been proposed to stimulate a Cl^-–HCO_3^- exchange in the turtle urinary bladder[52]. Experiments on isolated perfused OMCTi are of course necessary to substantiate and characterize the possible role of VIP in urinary acidification. The role of VIP in regulation of ion transport in the other responsive segments (DCTb and DCTg) is unknown. It should be stressed that physiological studies in these sgements are few and are complicated by the difficulties encountered in microdissection and microperfusion. There is cellular heterogeneity in DCTg and responses to VIP and PTH may not be associated with identical cells. In DCTb, there is cellular homogeneity at the morphological (EM) level and both calcitonin and VIP may stimulate the same pool of adenylate cyclase. Structural studies have shown marked compensatory changes in the DCTb and DCTg when animals are salt loaded[53]. With high-Na and low-K diets, there are marked increases in basolateral membrane area and Na^+/K^+-ATPase activity in the DCT cell (DCTb). Such an alteration is thought to be due to increased Na^+ load at the DCT and increased NaCl absorbtion[53]. An alternative suggestion, equally compatible with the physiological condition, is that of salt secretion (as is observed with MDCK epithelium).

In the periphery, VIP is usually contained in cholinergic nerves. Whether this is the case in renal nerves is unclear. It has recently been reported that MDCK cells possess receptors for bradykinin[54,55] and acetylcholine[56] that produce activation of K^+ conductance by release of Ca^{2+} from intracellular stores or by increased Ca^{2+} influx into the cell. It will thus be of some interest to examine the effects of such agonists upon inward current (Cl^- secretion) when used in conjunction with such agonists as VIP or prostaglandins that activate intracellular cAMP accumulations. Such interactions between kinin and cAMP-stimulated Cl secretion (inward current) have indeed been observed in primary cultures of pig renal papillary collecting tubule cells[57].

Renal infusions of prostaglandins[58-60], kinins[61,62] and VIP[25] are invariably associated with increased urinary Na^+, Cl^- and K^+ losses. It is interesting to speculate that such stimuli may produce such natriuresis and kaliuresis, in part, by a renal tubular mechanism similar to that observed in MDCK cultured epithelium.

The interactions between 'natriuretic' stimuli at the level of the MDCK epithelial cell emphasize the potential difficulty in interpreting renal perfusion data solely from the effects of single agonists where synergistic effects may be of considerable importance and where intrarenal hormones may exert a background 'tone'.

CONCLUDING COMMENTS

The precedents for VIP as a neurohormonal activator of NaCl secretion (balance) in vertebrates are several; the most well-known examples are in the shark rectal gland[13,14] and in the salt glands of birds[15]. In addition, secretory NaCl and volume flow is characteristic of renal tubules in the shark, flounder and killifish[63]. The cellular mechanisms that underly such secretions are similar to that observed in MDCK epithelium. It would indeed appear perverse if evolutionary processes had stripped mammalian kidneys of a potentially useful mechanism for excreting salt in excess of requirements.

It is evident that VIP regulation of secretion in MDCK epithelium has suggested avenues of enquiry that have established VIP as a renal acting neurohormonal agent in mammals. Though we cannot draw exact parallels, it might well be the case that the cultured 'model' might indeed be a very faithful analogue. Panacea or artifact, the jury is still out!

Acknowledgements

Work in the author's laboratory has been supported by the Wellcome Trust, the National Kidney Research Fund and the Medical Research Council (UK). I wish to thank the Cambridge University Press, Elsevier and Springer Verlag for reproduction of published material.

References

1. Barker, G. and Simmons, N. L. (1981). Identification of two strains of cultured canine renal epithelial cells (MDCK cells) which display entirely different properties. *Q. J. Exp. Physiol.*, **66**, 61–72
2. Richardson, J. C. W. and Simmons, N. L. (1981). Identification of two strains of MDCK cells which resemble separate nephron tubule segments. *Biochim. Biophys. Acta*, **673**, 26–36
3. Wohlend, A., Vassalli, J. D., Belin, D. and Orci, L. (1986). LLCPK₁ cells: cloning of phenotypically stable subpopulations. *Am. J. Physiol.*, **250**, C682–C687
4. Popowicz, P. and Simmons, N. L. (1988). Bumetanide binding and inhibition of Na⁺K⁺Cl⁻ cotransport: Demonstration of specificity by the use of MDCK cells deficient in cotransport activity. *Q. J. Exp. Physiol.*, **73**, 193–202
5. Herzlinger, D. A., Easton, T. G. and Ujakian, G. K. (1982). The MDCK epithelial cell-line expresses a cell surface antigen of the kidney distal tubule. *J. Cell Biol.*, **93**, 269–277
6. Ojakian, G. K., Romain, R. E. and Herz, R. E. (1987). A distal nephron glycoprotein has

different cell surface distributions on MDCK cell sublines. *Am. J. Physiol.,* **253**, C433–C443

7. Rindler, M. J., Chuman, L. M., Schaffer, L. and Saier, M. M. (1979). Retention of differentiated properties in an established dog kidney epithelial cell-line (MDCK). *J. Cell Biol.,* **81**, 635–648

8. Rugg, E. L. and Simmons, N. L. (1984). Control of cultured epithelial (MDCK) cell transport function: Identification of a beta adrenoreceptor coupled to adenylate cyclase. *Q. J. Exp. Physiol.,* **69**, 339–353

9. Brown, C. D. A. (1982). The retention of differentiated function in cell-culture: A study of the MDCK cell-line. Chap. 3 pp. 53–77. *PhD Thesis,* University of St Andrews

10. Brown, C. D. A. and Simmons, N. L. (1981). Catecholamine stimulation of Cl secretion in MDCK cell epithelium. *Biochim Biophys. Acta,* **649**, 427–435

11. Brown, C. D. A. and Simmons, N. L. (1982). K transport in tight epithelial monolayers of MDCK cells: Evidence for a calcium activated channel. *Biochim. Biophys. Acta,* **690**, 95–105

12. Rugg, E. L. and Simmons, N. L. (1986). Vasoactive intestinal peptide stimulation of adenylate cyclase and transepithelial transport in the cultured renal epithelial cell system (MDCK). *Renal Physiol.,* **9**, 72

13. Stoff, J. S., Rosa, R., Hallac, R., Silva, P. and Epstein, M. H. (1979). Hormonal regulation of active chloride transport in the dogfish rectal gland. *Am. J. Physiol.,* **237**, F138–F144

14. Stoff, J. S., Silva, P. and Epstein, F. H. (1982). Effect of VIP on active chloride transport in the shark rectal gland. In Said, S. I. (ed.) *Vasoactive Intestinal Peptide,* p. 223. (New York: Raven Press)

15. Gerstberger, R. (1988). Functional vasoactive intestinal polypeptide (VIP)-system in salt glands of Peking duck. *Cell Tiss. Res.,* **252**, 39–48

16. Dharmsathaphorn, K., Harms, V., Yamashiro, D. J., Hughes, R. J., Binder, H. J. and Wright, E. M. (1983). Preferential binding of vasoactive intestinal peptide to the basolateral membrane of rabbit and rat enterocytes. *J. Clin. Invest.,* **71**, 27–35

17. Sepulveda, F. V. and Pearson, J. D. (1984). Localisation of alanine uptake by cultured renal epithelial cells (LLCPK₁) to the basolateral membrane. *J. Cell. Physiol.,* **118**, 211–217

18. Simmons, N. L. (1981). The action of prostaglandins upon a cultured epithelium of canine origin (MDCK). *J. Physiol.,* **310**, 28–29P

19. Simmons, N. L. (1981). Stimulation of Cl secretion by exogenous ATP in cultured MDCK epithelial monolayers. *Biochim. Biophys. Acta,* **646**, 231–242

20. Simmons, N. L., Rugg, E. L. and Tivey, D. R. (1985). Madin Darby canine kidney cells: An in vitro model for intestinal and renal epithelial transport function. *Mol. Physiol.,* **8**, 23–34

21. Wangemann, P., Wittner, M., DiStefano, A., Englert, H. C., Lang, H. J., Schlatter, E. and Greger, R. (1986). Cl-channel blockers in the thick ascending limb of the loop of Henle. Structure activity relationship. *Pfleugers Arch.,* **407**, S128–141

22. Bruer, W. V., Mack, E. and Rothstein, A. (1988). Activation of K and Cl channels by Ca and cAMP in dissociated kidney epithelial (MDCK) cells. *Pfleugers Arch.,* **411**, 450–455

23. Paulmichl, M., Defregger, M. and Lang, F. (1986). Effects of epinephrine on electrical properties of MDCK cells. *Pfleugers Arch.,* **406**, 367–371

24. Chase, H. S. and Wong, S. M. E. (1988). Isoproterenol and cAMP increase intracellular free Ca in MDCK cells. *Am. J. Physiol.,* **254**, F374–F378

25. Dimaline, R., Peart, W. S. and Unwin, R. J. (1983). Effects of vasoactive intestinal polypeptide (VIP) on renal function and plasma renin activity in the conscious rabbit. *J. Physiol.,* **344**, 379–388

26. Porter, J. P. and Ganong, W. F. (1982). Relation of vasoactive intestinal peptide to renin secretion. In Said, S. L. (ed.) *Vasoactive Intestinal Peptide,* pp. 285–297. (New York: Raven Press)

27. Porter, J. P., Said, S. and Ganong, W. F. (1983). Vasoactive intestinal peptide stimulates renin secretion in vitro: evidence for a direct effect on renal juxtaglomerular cells. *Neuroendocrinology,* **36**, 404–408

28. Porter, J. P., Thrasher, T. N., Said, S. I. and Ganong, W. I. (1985). Vasoactive intestinal polypeptide in the regulation of renin secretion. *Am. J. Physiol.,* **249**, F84–89

29. Rosa, R. M., Silva, P., Stoff, J. S. and Epstein, F. H. (1985). Effect of vasoactive intestinal

peptide on isolated perfused rat kidney. *Am. J. Physiol.*, **249**, E494–E497

30. Morel, F. (1981). Sites of hormone action in the mammalian nephron. *Am. J. Physiol.*, **240**, F159–164

31. Munson, P. J. and Rodbard, D. (1980). Ligand: A versatile computerised approach for characterisation of ligand binding systems. *Anal. Biochem.*, **107**, 220–239

32. Pandol, S. J., Dharmsathaphorn, K., Schoeffield, M. S., Vale, W. and Rivier, J. (1986). Vasoactive intestinal polypeptide receptor antagonist. *Am. J. Physiol.*, **250**, G553–G557

33. Griffiths, N. M. and Simmons, N. L. (1987). Vasoactive intestinal polypeptide regulation of rabbit renal adenylate cyclase activity in vitro. *J. Physiol.*, **387**, 1–17P

34. Griffiths, N. M., Rivier, J. and Simmons, N. L. (1988). Vasoactive intestinal peptide stimulation of feline renal adenylate cyclase. *Ann. NY Acad. Sci.*, **527**, 640–643

35. Griffiths, N. M., Chabardes, D., Imbert-Teboul, M., Siaume-Perez, S., Morel, F. and Simmons, N. L. (1988). Distribution of vasoactive intestinal peptide-sensitive adenylate cyclase activity along the rabbit nephron. *Pfleugers Arch.*, **412**, 363–368

36. Imbert, M., Chabardes, D., Montegut, M., Clique, A. and Morel, F. (1975a). Adenylate cyclase activity along the rabbit nephron as measured in single isolated segments. *Pfleugers Arch.*, **354**, 213–228

37. Morel, F., Chabardes, D. and Imbert-Teboul, M. (1978). Methodology for enzymatic studies of isolated tubular segments: Adenylate cyclase. In Martinez-Maldonado, M. (ed.) *Methods in Pharmacology*, Vol. 4b, Chap. 12, p. 297. (New York and London: Plenum Press)

38. Chabardes, D., Imbert, M., Clique, A., Montegut, M. and Morel, F. (1975). PTH sensitive adenyl cyclase activity in different segments of the rabbit nephron. *Pfleugers Arch.*, **354**, 229–239

39. Chabardes, D., Imbert-Teboul, M., Montegut, M., Clique, A. and Morel, F. (1975). Catecholamine sensitive adenylate cyclase activity in different segments of the rabbit nephron. *Pfleugers Arch.*, **361**, 9–15

40. Imbert, M., Chabardes, D., Montegut, M., Clique, A. and Morel, F. (1975b). Vasopressin dependent adenylate cyclase in single segments of rabbit kidney tubule. *Pfleugers Arch.*, **357**, 172–186

41. Pandian, M. R., Harvat, A. and Said, S. I. (1982). Radioimmunoassay of VIP in blood and tissues. In Said, S. L. (ed.) *Vasoactive Intestinal Polypeptide*, p. 35. (New York: Raven Press)

42. Barajas, L., Sokolski, K. N. and Lechago, J. (1983). Vasoactive intestinal polypeptide-immunoreactive nerves in the kidney. *Neurosci. Lett.*, **43**, 263–269

43. Knight, D. S., Beal, J. A., Yuan, Z. P. and Fournet, T. S. (1987). Vasoactive intestinal peptide-immunoreactive nerves in the rat kidney. *Anat. Rec.*, **219**, 193–203

44. Reinecke, M. and Forssmann, W. G. (1988). Neuropeptide (neuropeptide Y, neurotensin, vasoactive polypeptide, substance P, calcitonin gene-related peptide, somatostatin) immunohistochemistry and ultrastructure of renal nerves. *Histochemistry*, **89**, 1–9

45. Bailly, C., Imbert-Teboul, M., Chabardes, D., Hus-Citharel, A., Montegut, M., Clique, A. and Morel, F. (1980). The distal nephron of rat kidney: A target site for glucagon. *Proc. Natl. Acad. Sci. USA*, **77**, 3422–3424

46. Charlton, B., Neal, D. E. and Simmons, N. L. (1990). Vasoactive intestinal peptide stimulation of human renal adenylate cyclase in vitro. *J. Physiol.*, **423**, 475–484

47. Calam, J., Dimaline, R., Peart, W. S., Singh, J. and Unwin, R. J. (1983). Effect of vasoactive intestinal polypeptide on renal function in man. *J. Physiol.*, **345**, 469–476

48. Hays, S., Kokko, J. P. and Jacobson, H. R. (1986). Hormonal regulation of proton secretion in rabbit medullary duct. *J. Clin. Invest.*, **78**, 1279–1286

49. Stone, D. K., Seldin, D. W., Kokko, J. P. and Jacobsen, H. R. (1983). Anion dependence of rabbit medullary collecting duct acidification. *J. Clin. Invest.*, **71**, 1505–1508

50. Madsen, K. M. and Tisher, C. C. (1986). Structural–functional relationships along the distal nephron. *Am. J. Physiol.*, **250**, F1–F15

51. Mandel, K. G., McRoberts, J. A., Beuerlin, G., Foster, E. S. and Dharmsathaphorn, K. (1986). Ba^{2+} inhibition of VIP and A23187-stimulated Cl secretion by T84 cell monolayers. *Am. J. Physiol.*, **250**, C486–C494

52. Durham, J. H., Matons, C. and Brodsky, W. A. (1987). Vasoactive intestinal peptide

stimulates alkali secretion in turtle urinary bladder. *Am. J. Physiol.,* **252**, C428–C435

53. Kaissling, B. (1982). Structural aspects of adaptive changes in renal electrolyte excretion. *Am. J. Physiol.,* **243**, F211–F226

54. Lopez-Burillo, S., O'Brien, J. A., Ilundain, A., Wreggett, K. A. and Sepulveda, F. V. (1988). Activation of a basal–lateral membrane K permeability by bradykinin in MDCK cells. *Biochim. Biophys. Acta,* **939**, 335–342

55. Paulmichl, M., Friedrich, F. and Lang, F. (1987). Effects of bradykinin on electrical properties of Madin Darby canine kidney epitheloid cells. *Pfleugers Arch.,* **408**, 408–418

56. Lang, F., Klotz, L. and Paulmichl, M. (1988). Effect of acetylcholine on electrical properties of subconfluent Madin Darby canine kidney cells. *Biochim. Biophys. Acta,* **941**, 217–224

57. Cuthbert, A. W., George, A. M. and MacVinish, L. J. (1985). Kinin effects on electrogenic ion transport in primary cultures of pig renal papillary collecting tubule cells. *Am. J. Physiol.,* **249**, F439–F447

58. Haylor, J. (1980). Prostaglandin synthesis and renal function in man. *J. Physiol.,* **298**, 383–396

59. Haylor, J. and Lote, C. J. (1980). Renal function in conscious rats after indomethacin. Evidence for a tubular action of endogenous prostaglandins. *J. Physiol.,* **298**, 371–381

60. Levenson, D. J., Simmons, C. E. and Brenner, B. (1982). Archadonic acid metabolism, prostaglandins and the kidney. *Am. J. Med.,* **72**, 354–374

61. Margolius, H. S. (1984). The kallikrein–kinin system and the kidney. *Ann. Rev. Physiol.,* **46**, 309–326

62. Mills, I. H. and Obika, O. F. L. (1977). Increased urinary kallikrein excretion during prostaglandin E_1 infusion in anaesthetised dogs and its relation to natriuresis and diuresis. *J. Physiol.,* **273**, 459–474

63. Beyenbach, K. W. (1986). Secretory NaCl and volume flow in renal tubules. *Am. J. Physiol.,* **250**, R753–R763

64. Chabardes, D., Imbert-Teboul, M., Montegut, M., Clique, A. and Morel, F. (1976). Distribution of calcitonin-sensitive adenylate cyclase along the rabbit kidney tubule. *Proc. Natl. Acad. Sci. USA,* **73**, 3608–3612

65. Griffiths, N. M. and Simmons, N. L. (1990). Localisation and characterisation of functional vasoactive intestinal peptide receptors in feline kidney. *Pfleugers Arch.,* **416**, 80–87

66. Edwards, R. (1988). Distribution of vasoactive intestinal-peptide sensitive adenylate cyclase activity along the nephron. *Eur. J. Pharmacol.,* **157**, 227–230

Section IV
RESPIRATORY EPITHELIA

Introduction

The respiratory section of the book contains four chapters, the first of which by Peter Jeffery presents an overview of the structure, putative functions and proliferative potential of the principal cell types comprising the conductive and respiratory epithelia of the mammalian lung. There is also a discussion of submucosal glands and autonomic nerves. A recurrent theme of the chapter is a consideration of the effects of pathological change associated with asthma, carcinoma and, particularly, cystic fibrosis (CF). The basic lesion in CF is described and the subsequent effects on cell structure and function, autonomic control, mucociliary clearance and lung infection are discussed.

The chapter by Michael Van Scott, James Yankaskas and Richard Boucher presents a model for transepithelial ion transport in the airway and compares the ion-transporting properties of cultured canine and human airway epithelial cells with those of intact tissues. Parameters compared include transepithelial potential difference (V_t), short-circuit current (I_{sc}), conductance (G_t), apical and basolateral membrane potential (V_a and V_b) and fractional resistance (fR_a). In the dog, regional differences are considered when the bioelectric properties of cultured tracheal and bronchial epithelial cells are compared, together with their responses to amiloride and isoproterenol (isoprenaline). In the human case, the authors compare the bioelectric properties of normal nasal epithelia, both freshly-excised and cultured, with those from patients with cystic fibrosis (CF). Cultured human airway epithelial cells display similar bioelectric properties to those of intact tissue with the exception that I_{sc} and G_t are lower in culture. Marked differences are observed between normal and CF nasal epithelia in V_t, I_{sc} and V_a, and these differences are perpetuated in culture. The authors conclude by reviewing some of the recent advances in the establishment of cell cultures from airway epithelium, in particular the establishment of cell lines and the culture of Clara cells. Cell lines are proving of particular value in cystic fibrosis research.

The chapter by John Gatzy, Elaine Krochmal and Stephen Ballard compares solute transport across fetal and adult airway epithelia and addresses the problems in studying transport in alveoli. Beginning with a review of alveolar structure, the authors identify several approaches designed to achieve a greater understanding of alveolar function in the face of obvious practical difficulties associated with the tissue architecture.

The relationship between Cl^- and Na^+ movement and fluid transport across pulmonary epithelium is compared in fetal, prenatal, neonatal and adult epithelia. In the liquid-filled fetal lung, secretion of fluid into the lumen is driven largely by Cl^- secretion whereas the fluid-absorptive process, which begins to supplant liquid secretion shortly before birth, is associated with active Na^+ absorption. The developing capacity for absorption spreads from distal to proximal structures. The authors describe studies of alveolar epithelium in fluorocarbon-blocked adult lobes where they conclude that Na^+ is actively transported but that this does not make a major contribution to fluid absorption. An opportunity to study maturation of the alveolar epithelium is provided by submersion culture of explants from different regions of the fetal lung. The cells in culture secrete Cl^--rich fluid at a rate similar to that estimated for native epithelium and this persists in spite of maturing cell morphology.

Jeffrey Smith addresses the biophysics and regulation of ion channels in cultured airway epithelial cells. He also includes useful introductory sections on mucociliary clearance and electrophysiological theory. The transport of K^+, Na^+ and Cl^- provide the main emphasis of the chapter, with particular focus on Na^+ and Cl^- whose transport is abnormal in CF airway epithelium. Na^+ transport is enhanced, the mechanism of which remains unknown. Cl^- shows reduced permeability in CF and the normal activation of apical Cl^- channels by protein kinase A- and protein kinase C-dependent mechanisms is defective. Non-physiological means of activating Cl^- channels, by membrane depolarization, increased ambient temperature and exposure of the cytosolic membrane surface to trypsin, are preserved in CF, suggesting that the channel conductive properties are patent and that the abnormality lies in regulation. The chapter concludes with a discussion of the CF gene product (CFRT), whose recent discovery opens up new approaches to the study of channel regulation and permits the identification of those cell populations expressing the gene.

See note added in proof on p. 288.

9
Form and Function of Mammalian Airway Epithelia

P. K. JEFFERY

INTRODUCTION

The normal epithelia lining the airways of the lung consist of many morphologically distinct cell types often with distinct, but sometimes overlapping, functions (see Table 9.1). In disease: (1) the integrity of the epithelium may be compromised such as in asthma, leading to a hyper-responsive airway; (2) in carcinoma, malignant transformation of specific epithelial cells may give rise to tumours of differing histological phenotype, each associated with different response to treatment; and (3) inappropriate alteration in the movement of ions and macromolecules can lead to altered transepithelial water flow and drying of airway secretions with increased susceptibility to colonization by bacteria. The possible aberrations of epithelial function are many and for the purposes of the present chapter emphasis will be placed on the normal and, by way of example, the epithelial abnormality in cystic fibrosis, a condition in which there has been much recent progress.

Cystic fibrosis (CF) is the commonest fatal inherited disease in Caucasian populations and affects approximately 1 in 2000 births. The lungs, sweat and salivary glands and pancreas are the body organs classically involved but the extent of involvement of the lung and its progressive deterioration, due to repeated infection, are critical to the eventual fatal course of the disease. The epithelia of several body organs in CF have, in common, defects of electrolyte transport[1,2], i.e. defective permeability to the chloride ion (Cl^-) and less commonly alterations of sodium (Na^+) absorption, both of which are discussed in Chapter 10 by van Scott et al. The defects of epithelial ion translocation which are mirrored by altered electrical potential across epithelia are thought to be responsible for the low water content of fluids from airways, pancreas, uterine cervix and intestine. The low water content, in turn, may explain the presence of the viscid obstructing and chronically infected secretions which are associated subsequently with lung

Table 9.1 Summary of Epithelial cells and their putative functions

Epithelium	Function/s	Submucosal glands	Function/s	Nerve	Function/s
Ciliated	Moves mucus Secretes mucosubstance Controls periciliary fluid and ions	Serous	Mucus-secreting (neutral) Secretion of ions and fluid Secretory piece (component)	NEB	Chemo/mechanoreceptor Modulation of -growth -vessel and bronchial tone -mucous secretion
Mucous	Mucus-secreting (acidic) Absorptive Proliferative		Lactoferrin Lysozyme Small M.W. antiprotease	Nerve terminals	Sensory bronchoconstrictor -cough -secretion -hyperpnoea
Serous	Mucus-secreting (neutral) Secretory piece (component) Lipid Periciliary fluid Proliferative	Mucous	Mucus-secreting (acidic) Proliferation		Motor -secretion -ciliary rate -modifies endocrine response
Clara	Surfactant-hypophase Secretes ions Small M.W. antiprotease	Oncocyte	Ionic + water modulation Degenerate acinar cell		

Cell type	Function
Myo-epithelial	Expulsion of mucus
Endocrine	Secretoregulatory / Vasoregulatory
Intra-acinar nerve	Secretoregulatory
Lymphocyte	Immunoresponsive
DCG/endocrine	Secretes amines (5-HT) / Peptides (bombesin)
Basal	Proliferative
Lymphocyte	Immuno-responsive
Mast cell/globular leukocyte	Releases inflammatory mediators / Transports immunoglobulins
Special-type Indeterminate	Function(s) unknown
Brush (airway and alveolus)	Function unknown
Type I alveolar	Protective
Type II alveolar	Secretes surfactant, ions and fluid / Proliferative

destruction and the fatal outcome in severely affected individuals.

Early functional studies of the airway epithelial defect *in vivo* began when Knowles *et al.*[3] observed that the negative potential difference (mV) across airway epithelia was significantly higher in CF patients than in normal subjects or disease controls (i.e. CF, -60 mV, vs normal, -30 mV). An abnormal response to Cl⁻ superfusion *in vivo* and decreased Cl⁻ fluxes *in vitro* indicated a decreased airway epithelial permeability to the Cl⁻ ion in CF[4,5]. Subsequent functional studies of primary airway epithelial cultures showed abnormalities similar to those observed in the native tissue: cultures derived from CF tracheal and nasal epithelia were Cl⁻ impermeable and failed to secrete Cl⁻ ion when stimulated with the β-agonist, isoproterenol[6,7]. These results indicated that the Cl⁻ impermeability was an inherent property of the epithelial cells themselves, which could be retained in culture, and that a circulating 'factor' was not the cause of the ion transport defect. Intracellular microelectrode recordings demonstrated that the Cl⁻ impermeability was localized to the cellular pathway and specifically to the apical cell membrane[6,8]. Patch-clamp studies indicated that the Cl⁻ transport defect was probably due to failure of an outwardly rectifying apical anion channel to respond to phosphorylation by cyclic AMP-dependent protein kinase or protein kinase C[2,9,10]. Recently, the gene mutated in CF patients has been identified[11,12] and in approximately 70% of cases comprises a specific deletion of three base pairs resulting in the omission of a phenylalanine residue (Phe[508]) in a protein thought to be involved in transmembrane ion conductance (hence named the CF transmembrane conductance regulator or CFTR).

The identification of the CFTR gene and study of normal and mutant CFTR proteins may now make it possible to understand better the control of epithelial ion transport and should help to provide a molecular basis for the development of an effective therapy. For example, it has recently been suggested that introduction by aerosol of normal CFTR genes or protein to airway epithelial cells and their progenitors may be an effective form of gene therapy in the future[13]. It is, however, not yet clear whether all of the many cell types which comprise the epithelial lining of the lung are similarly affected and much work is still required to identify those specific cells which express the basic defect.

The purpose of the present chapter is to give an overview of the normal cellular morphology of the lining epithelium of the 'conductive' and 'respiratory' airways of the lung and of the putative function(s) of the various cell types by way of introduction to the detailed functional studies of airway epithelia which follow. The salient morphological changes to airway surface epithelia in asthma, carcinoma and particularly CF patients will also be summarized. The respiratory airways of the lung are only briefly considered herein: for greater detail, the reader is referred elsewhere[14,15].

CONDUCTIVE AIRWAYS

The conductive airways perform functions beyond the conduction of inspired and expired gases, e.g. warming, humidification and cleansing inhaled air of potentially harmful dust particles, gases, bacteria and other living organisms. The more distal respiratory zone is thereby kept free of pollution and infection by airway defence mechanisms which include: (1) nervous reflexes leading to bronchoconstriction and/or cough; (2) ciliary activity and secretion of mucus, lysozyme, lactoferrin and secretory IgA; and (3) cellular immune response and reactions[14].

Figure 9.1 shows that the airway wall comprises epithelial, lymphoid, muscular, vascular and nervous elements interspersed in a pliable connective tissue support arranged as: (1) a lining mucosa of surface epithelium supported by basement membrane and an elastic lamina propria, in which there are bronchial blood vessels, nerve bundles and free cells (including fibroblasts and mononuclear cells), (2) a submucosa in which lie the bulk of the mucus-secreting glands, muscle and cartilage plates and (3) a relatively thin coat of adventitia.

Surface epithelium

Airway epithelia include the surface epithelium which lines all airways and which is continuous with that forming the tubulo-acinar submucosal mucus-secreting glands which develop from the surface. The stratified squamous epithelium lining much of the larynx gives way to one which is pseudo-stratified, ciliated and columnar when the trachea is reached. The term 'pseudostratified' arises due to all cells resting on the basement membrane but not all reaching the airway lumen (Figure 9.2). In man, this type of epithelium persists throughout the major bronchi, becoming simple cuboidal more peripherally. Ciliated cells predominate, interspersed by mucus-secreting (goblet) cells which are found regularly in the tracheobronchial tree but rarely in bronchioles of less than 1 mm diameter.

A variety of cell types is recognized in airway surface epithelium: at least eight different epithelial cell types have now been delineated[16,17]. In addition, cells involved in the immune response and its reactions may migrate through the epithelial basement membrane: some of these remain within the surface epithelium, whereas others are in the process of passing through to the lumenal surface[14]. The terminal processes of sensory nerve fibres whose cell bodies lie deep to the epithelium, pierce the epithelial basement membrane and lie surrounded by epithelial cells (see Figure 9.2) where they initiate airway reflexes, such as bronchoconstriction and cough[18]. These endings may be inappropriately stimulated in conditions such as asthma and chronic bronchitis.

The variety of cell types and their putative functions are listed in Table 9.1. some of these are now considered in more detail.

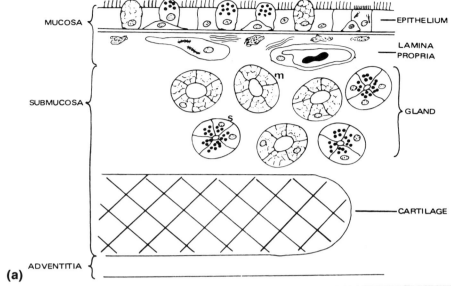

(a)

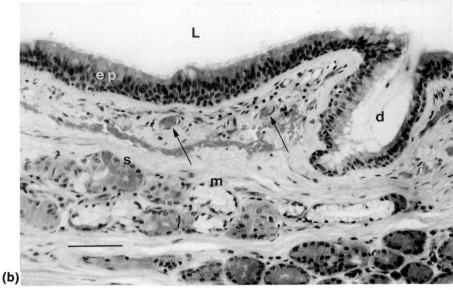

(b)

Figure 9.1 a Diagrammatic representation of bronchial wall comprising (1) a mucosa of surface epithelium with supporting subepithelial connective tissue, bronchial capillaries, nerve bundles and free cells, (2) submucosa consisting of mucous (m) and serous (s) gland acini, smooth muscle and cartilage plates and (3) a thin adventitia. Ions and fluid move across both the surface epithelium and submucosal gland cells. **b** Light micrograph of a histological section through human bronchus stained with haematoxylin and eosin to show some of these features: bronchial lumen (L), ciliated surface epithelium (ep), bronchial capillaries (arrows), mucous (m) and serous (s) gland acini, gland duct (d) opening to surface. (Scale bar = 60 μm)

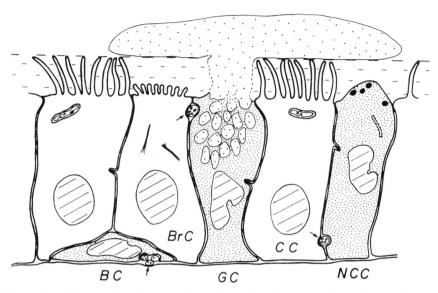

Figure 9.2 Diagram of surface epithelium illustrating the variety of bronchial cell types: ciliated (CC), goblet (GC), basal (BC), non-ciliated serous (NCC), brush (BrC), and intra-epithelial nerve fibres (arrows). Cilia are thought to move the gel-like rafts of mucus by their tips and beat in a periciliary fluid layer of low viscosity whose fluid and ion content is controlled by the epithelial cells

Ciliated cell

The surfaces of ciliated cells (Figures 9.3a and 9.3b) are densely covered by 200–300 cilia per cell, each normally beating at about 1000 times per min with its effective stroke generally in a cranial direction. Where fields of cilia are in continuity there is co-ordination of their beat. Each cell has electron-lucent cytoplasm with an abundance of apically placed mitochrondria and a moderately well-developed Golgi apparatus (Figure 9.3a). Lysosomes and lamellar bodies are often present together with smooth cisternae of endo-plasmic reticulum. The cilia are thought to beat in a periciliary layer of low viscosity, the origin of which is still the subject of speculation[19]. Cilia move the overlying mucous sheet only by their tips, the interaction of the ciliary tips and mucus being facilitated by minute terminal hooklets[20]. Long slender microvilli project between the cilia and are associated with an acidic surface mucosubstance, probably a glycosaminoglycan, which may be an important source of mucosubstance. The rich microvillar border and associated pinocytotic vesicles may play a role in ion translocation and fluid absorption and thereby control the depth of the periciliary fluid layer in which the cilia beat. A quantitative study of ciliated cell microvilli in biopsy specimens has indicated a decreased number of microvilli in CF[21]. Whilst ultra-structural abnormalities of cilia have been clearly shown in genetic con-ditions of ciliary dyskinesia[22], none has been found in CF and the ciliated surface covering, even in chronically infected CF airways, appears to be

201

relatively normal[23,24]. However, mucociliary clearance is depressed and the typically tenacious secretions observed in CF airways resist removal by the normal preparative processes prior to scanning microscopy and remain adherent to cilia (Figure 9.4).

Purification of apical cell membrane vesicles from disaggregated airway epithelial cells and their reconstitution into artificial phospholipid bilayers has recently allowed characterization of ion channels in animal tissues[25]. The application of these techniques to airway cells obtained from CF transplant tissue and to both normal and CF airway epithelial cell lines, immortalized by viral transformation[26,27], will allow the mechanisms of molecular control of apical ion channels and the role of defective CFTR protein(s) to be studied in greater detail. Several studies indicate that the ciliated cell is normally an end-stage cell formed by differentiation and maturation of dividing basal and secretory cell types[28].

Mucous cell

Mucous (goblet), serous, Clara and dense-core granulated (DCG) cells form the normal secretory cell types of the surface epithelium of conductive airways. In human trachea, the normal mean density of surface mucous cells is estimated at between 6000–7000 cells per mm^2 surface epithelium[29]. By electron microscopy, the mucous cell has electron-dense cytoplasm containing electron-lucent, confluent granules of about 800 nm diameter (see Figure 9.3b). Most contain mucin which is acidic due to sialic acid or sulphate groups located at the ends of oligosaccharide side chains, which branch from the protein core of high-molecular-weight glycoprotein[30]. Secretion of the correct amount of mucus with an optimum viscoelastic profile is important in the maintenance of normal mucociliary clearance and this is thought to be defective in CF. Alterations in the predominant histochemical type of mucus have been associated with airway irritation[31] and carcinogenesis[32] but no abnormality specific to CF has been detected[33]. The numbers of mucus-secreting cells increase in chronic bronchitis[34] and, experimentally in animal models of bronchitis, following inhalation of sulphur dioxide[35] or tobacco smoke[36,37]. Their increase in number and extension to the peripheral bronchioles is also a characteristic of small airways disease, particularly in CF[38]. The mucous cell is clearly capable of division and may show stem cell multipotentiality[28]. The solubility and viscosity of mucus varies considerably with ionic strength and divalent cations, such as calcium (Ca^{2+}), cause mucus to form a rigid cross-linked gel

Figure 9.3 a Scanning electron micrograph (SEM) of two partially disaggregated ciliated cells, each with a dense covering of cilia (approximately 200–300 per cell). Long slender apical microvilli, projecting between the bases of the cilia (arrows), are associated with an acidic surface mucosubstance and, together with pinocytotic vesicles, probably play a role in ion and fluid translocation. (Scale bar = 6 μm). **b** Transmission electron micrograph (TEM) of bronchial surface epithelium taken at transplant from a CF patient's (recipient) lung. The apical region of the epithelium appears normal with goblet (GC) and ciliated (CC) cells: the latter have normal mitochondria, numerous cilia and show apical microvilli. Goblet cells contain predominantly clear confluent secretory granules. Areas of 'tight junction' between cells are arrowed. (Scale bar = 1 μm)

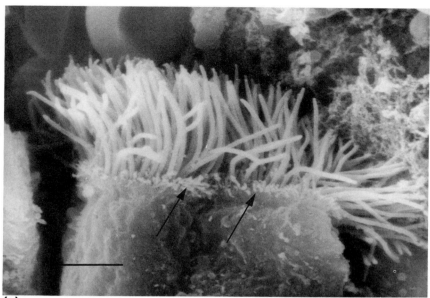

(a)

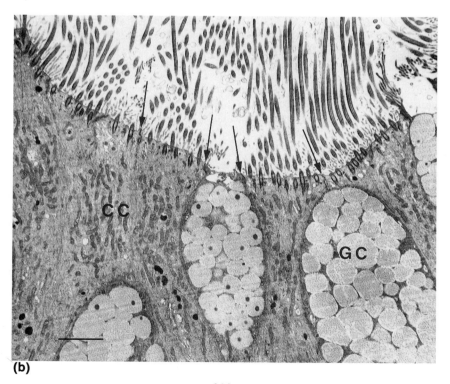

(b)

Figure 9.4 SEM of bronchial surface in freshly processed CF transplant material. The tissue had been rigorously washed in culture medium and processed through alcohols and critical point dried in CO_2. Unlike the normal, secretions with included cells (?macrophages) remained adherent to the highly ciliated surface. (Scale bar = 10 μm)

which may be difficult to clear by mucociliary action or cough. In biopsy studies, mucous cells from patients with CF have been shown to contain significantly raised intracellular (Ca^{2+}) and sulphate (SO_4^{2-}) levels and lower potassium levels than those of patients with chronic bronchitis[39]. The significance of these results is as yet unclear but it is in keeping with the reports of high Ca^{2+} and SO_4^{2-} contents of tracheobronchial secretions[40].

Serous cell

Serous cells have electron-dense cytoplasm, much rough endoplasmic reticulum and, in contrast to mucous cells, discrete electron-dense granules of about 600 nm diameter. Morphologically, serous cells of the surface epithelium resemble those present in the submucosal glands. They have been described in surface epithelium only in the rat, cat, young hamster and fetal humans[41]. Many contain neutral mucin and there is evidence that some may also contain a non-mucoid substance, probably lipid[42].

Clara (non-ciliated bronchiolar) cell

Clara cells in man are restricted in location to the terminal bronchioles where they typically bulge into the airway lumen and contain electron-dense granules of about 500–600 nm diameter, ovoid in man but irregular in most other species. The function of this cell type is as yet undetermined. It may contribute to the production of bronchiolar surfactant[43], a carbohydrate component closely associated with surfactant[44] or an antiprotease[45,46] and is known to have ion-absorbing and secreting properties (see accompanying chapter by van Scott *et al.*). Furthermore, the Clara cell acts as the stem cell of small airways where basal and mucous cells are normally sparse: after irritation or drug administration, both ciliated and mucous cells may develop from the Clara cell subsequent to its division and differentiation[28].

Dense-core granulated (DCG) cell (synonyms: endocrine, Kultchitsky and Feyrter cell)

Argentaffin-positive and argyrophilic cells have been identified within the surface epithelium by light microscopy. By electron microscopy, DCG cells are infrequently found, generally basal in position, but often with a thin cytoplasmic projection reaching the airway lumen[47]. Single cells and clusters of such cells may also be associated with nerve fibres (i.e. so-called neuroepithelial bodies or neurite-receptor complexes)[48]. The cytoplasm of DCG cells usually contains large numbers of small (70–150 nm) spherical granules, each with an electron-dense core surrounded by an electron-lucent halo. Granule subtypes have been described and the cells may contain biogenic amines[49] or peptides, such as bombesin[50] which, when released, may influence vascular and bronchial smooth muscle tone, mucous secretion and ciliary activity. The location of the cell in surface epithelium and its cytoplasmic content make it a prime candidate for sensing hypoxia in the airway lumen. It is likely that, as a consequence, vasoactive substances are released which cause local vasoconstriction and shunting of blood to better ventilated zones of the lung. Whilst hyperplasia of DCG cells has been described in infantile bronchopulmonary dysplasia[51], bronchiectasis associated with small multiple tumours[52] and in bronchi associated with carcinomas of all types[53], no changes have been reported in CF.

Basal cell

It is the presence of a basal cell layer in the large bronchi and trachea which contributes to the pseudostratified appearance of the epithelium (see Figure 9.2). The basal cell has sparse electron-dense cytoplasm, often with bundles of tonofilaments immunoreactive for cytokeratin. The tracheobronchial basal cell is considered by many to be the major stem cell from which the more superficial mucous-secreting and ciliated cells derive[28]. Airway surface epithelium is replaced only slowly with normally less than 1% of cells in division at any one time. However, the mitotic index increases in response to irritation by noxious agents, such as tobacco smoke, and also to sympathomimetic drugs, such as isoprenaline. In these situations, the basal cells play a proliferative role in concert with secretory cells which, as has already been mentioned, also divide.

Many non-ciliated cells fall into the category of 'indeterminate', a classification category comprising a mixture of cells, none of which is clearly classifiable. In addition, normal epithelium may show a number of cells each of which shows features transitional to two or more morphologically well-defined cell types[54]. Determination of the histological phenotype of preneoplastic and neoplastic tissue is relevant to pathogenesis, prognosis and therapy of lung cancer. The patterns of differentiation within a tumour may be multiple and complex and tumour classification can be made by the pathologist only with experience and good judgement and even then not always without difficulty. When deciding the cell of origin of a particular tumour, the assumption is often made that the cell will resemble, in histological phenotype, the main cell type present in the tumour. For example, it is widely assumed that malignant transformation of the epithelial basal cell gives rise to epidermoid carcinoma, of the endocrine (dense-core granulated) cell to small cell anaplastic carcinoma (including oat cell) or the carcinoid tumour. The premise may well turn out to be too restrictive and the pluripotential and proliferative nature of the variety of cell types present in the lung may be seriously underestimated.

Migratory cells

Mononuclear (migratory) cells are present within the normal airway epithelium of humans and many animal species. In health, polymorphonuclear leukocytes are found rarely but are rapidly recruited in inflammatory conditions, such as CF, where they migrate through the surface epithelium to the airway lumen where they are found to have engulfed bacteria (Figures 9.5a and 9.5b). In healthy epithelium, at least two types of mononuclear migratory cell have been distinguished on the basis of morphology and histochemistry: (1) the intraepithelial lymphocyte and (2) the intraepithelial mast cell or closely related 'globular' leukocyte[14]. The intraepithelial lymphocyte may occur singly or in groups organized into a 'lymphoepithelium'. The intraepithelial and subepithelial (so-called 'mucosal') mast cell may be morphologically and functionally distinct from that present in deep connective tissue[55]. Both are, however, capable of releasing mediators of inflammation which, among other things, affect epithelial and vascular permeability. In non-smokers, mast cells form up to 2% of surface epithelial cells but a higher proportion is found in smokers[56] and in smoker's lung subsequently resected for carcinoma[57]. The numbers of distinct types of inflammatory cell have yet to be determined in the lungs of patients with CF. In both mild and severe forms of asthma, there appears to be a tendency to increase in the total number of inflammatory cells in the zone immediately beneath the surface epithelium and there is evidence for their activation[58]. Release of protease from distinct inflammatory cells may disrupt epithelial integrity.

Cell junctions

The epithelial cells of the surface are held together by three types of junction: (1) 'Tight' (zonulae occludens) and 'intermediate' (zonula adherens)

(a)

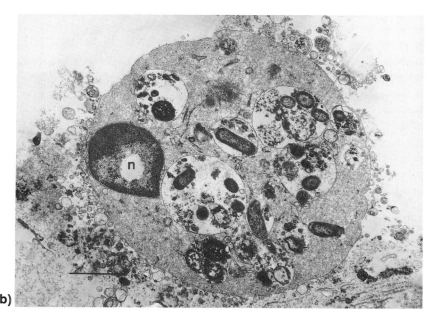

(b)

Figure 9.5 a SEM of CF bronchial lumen showing an accumulation of inflammatory cells (arrow) lying on a ciliated surface (Scale bar = 30 μm). **b** TEM shows bacterial forms (?*Pseudomonas* sp.) (arrows) engulfed by a neutrophil (n), the inflammatory cell type frequently found in lumenal secretions. (Scale bar = 2 μm)

207

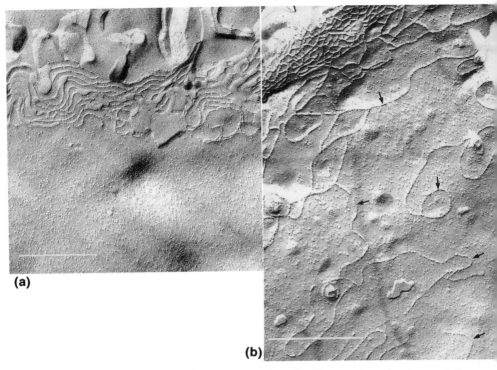

(a)

(b)

Figure 9.6 a Freeze fracture of human bronchial epithelium to show the characteristic appearance of the 'tight junction' joining surface epithelial cells. The junction is composed of interconnecting strands and grooves and extends around the apicolateral surface of the epithelial cells as a 'belt' forming a selectively permeable barrier to the movement of paracellular fluid, macromolecules and ions. (Scale bar = 0.5 μm). **b** There are indications in CF of a 'proliferation' of junctional strands (arrows) below regions of 'normal' tight junction: the proliferation is often extensive and may extend towards the basolateral surface of the plasma membrane. (Scale bar = 1 μm). From reference 64

junctions which together form attachment belts or 'terminal bars' extending around the apicolateral surface of cells and are shown to advantage by freeze–fracture techniques (Figure 9.6a); (2) maculae adherens or desmosomes (i.e. 'spot' junctions), which are frequently found between cells; these may also appear as hemidesmosomes joining the basal lamina and epithelial cells; and (3) nexus or 'gap' junctions which play an important role in cell-to-cell communication.

Normally impermeable, the terminal bar prevents paracellular fluid, macromolecular and ionic movement across the epithelium[59] but, following irritation (e.g. by tobacco smoke or ether), or specific allergen challenge to sensitized animals, the junction appears to become quickly permeable to molecules of 40 000 molecular weight or less placed in the lumen[60-62]. Transport from lumen to submucosa of larger molecules appears to be via a cellular route involving active cell transport[60]. Similar mechanisms are probably involved in the passage from the vascular compartment through the epithelium to the airway lumen of serum components which also contribute to respiratory tract fluid, especially in disease[19]. *In vitro* studies show that the structure of the tight junction is highly labile and that the number of sealing strands may be induced to proliferate rapidly[63]. An abnormal proliferation of tight junctional strands (Figure 9.6b) has also been observed in airway epithelium from patients with CF[64] but the functional relevance of this change in CF is not yet clear. Alterations to tight junctions in asthma may contribute to the increased fragility of the airway epithelium in this condition[58].

Basement membrane

The basement membrane supporting the surface epithelium consists of two morphologically distinct regions: (1) the basal lamina, itself composed of three zones, together made up of type IV collagen, laminin, glycosamino-glycans and fibronectin and (2) a deeper reticular lamina of fine fibrillary type III collagen (i.e. reticulin). In healthy bronchi, the combined thickness of the two zones is about 10 μm but this may increase in chronic inflammatory conditions, including bronchiectasis and asthma[58].

Submucosal glands

The submucosal glands in man are relatively numerous and, in the lower respiratory tract, are found wherever there is supportive cartilage in the airway wall, i.e. from larynx to small bronchi. It has been estimated that some 4000 glands are present in the human trachea[65]. Developing from surface epithelium *in utero*, each gland unit is of the tubulo-alveolar type and, in man, may be composed of four regions whose lumena are continuous: (1) A relatively narrow ciliated duct continuous with the surface epithelium, (2) an expanded collecting duct of cells of indeterminate morphology or of eosinophilic cells (also referred to as 'oncocytes') packed with mitochondria, (3) mucous tubules and mucous acini and (4) serous acini[66,67]. Whilst mucins of both acidic and neutral types are produced, other secretions, such as lysozyme[67], lactoferrin[68], the secretory component of IgA[70], and a low-molecular-weight antiprotease[71] are also produced by the submucosal glands, particularly by serous acini. The movement of ions and water from the vascular compartment to the airway lumen is regulated by both surface epithelium and submucosal glands. In the latter, it is suggested that watery, serous secretions pass from the outermost regions of each gland into the mucous tubules and mix and that the ionic balance of the mixed secretion may be adjusted in the collecting duct before its discharge through the ciliated duct to the bronchial lumen[66]. Discharge is aided by contractile myoepithelial cells which form a basket-like structure around the outer aspects of the acinus.

Both synthesis of intracellular secretion and discharge are influenced by nerves whose terminals lie adjacent to (in humans) or pierce (in cat) the

209

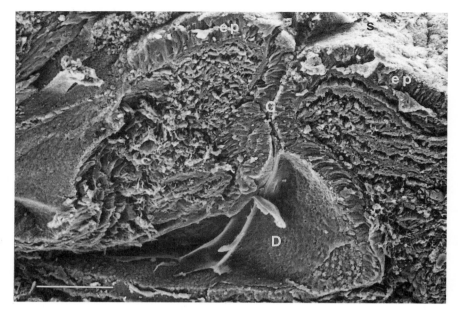

Figure 9.7 SEM of human bronchial mucosa in CF. A fortuitous fracture across the airway wall demonstrates the dilated collecting duct (D) opening onto the surface (S) via the ciliated duct (C) which is continuous with surface epithelium (ep). (Scale bar = 100 μm)

secretory unit. There is evidence that both parasympathetic and sympathetic agonists stimulate secretion although the quantity and quality of the resulting secretion may differ with each[72]. Submucosal glands probably form the major source of tracheobronchial mucus with the balance of contribution by surface epithelium and submucosal glands dependent upon the nature of the stimulus. Submucosal gland mass increases in chronic bronchitis, asthma and cystic fibrosis; the increase being due to cell proliferation within each secretory acinus rather than to an increase in the number of gland units *per se*. Whilst the submucosal glands of the CF patient are structurally normal at birth, one of the earliest changes noted is the dilatation of acinar and duct lumena[23] and this can be seen by scanning electron microscopy (Figure 9.7). A similar lesion is seen in the pancreas and it is suggested that abnormal hydration or solubilization of gland secretions causes lumenal obstruction and impaction, resulting in duct dilatation.

Autonomic nerves

Autonomic control appears to be abnormal in CF with increased sensitivity to α-adrenergic and cholinergic agonists and reduced sensitivity to β-adrenergic agonists[73]. In the trachea and main bronchus, nerve bundles and ganglia are found mainly in the posterior membraneous portion of the airway[74]. On entering the lung, the nerve bundles divide to form distinct peribronchial and perivascular plexuses. The peribronchial plexus further divides to form extra- and endochrondrial (i.e. subepithelial) plexuses. The efferent (motor) innervation has, in general, excitatory and inhibitory components affecting submucosal glands, bronchial smooth muscle and blood vessels. Morphological and physiological studies now indicate that apart from the classic neurotransmitters of the sympathetic (adrenergic) and parasympathetic (cholinergic) nerve supply to lung, a new family of neuropeptide transmitters has been described which have not been thought to belong to the adrenergic nor the cholinergic system (i.e. NANC) but which may co-localize with them. Thus, neural control of bronchial epithelium, glands, muscle and vasculature appears to be more complex than originally proposed[75]. Vasoactive intestinal polypeptide (VIP) is one such neuropeptide and has the properties of inhibiting mucous secretion and stimulating ion and fluid movement across the epithelium and relaxing human bronchial smooth muscle. Reduction of immunoreactivity for VIP has been reported in relation to both sweat glands[76] and bronchial submucosal glands[77] in CF and, recently, complete loss has been reported in the bronchi of patients dying of asthma[78]. Abnormalities of autonomic control have long been implicated in both asthma and CF and, in the latter, recent attention has been focussed on defects in the interaction of intracellular calcium and protein kinase C- and A-dependent mechanisms[10,79,80]. A recent study[81] has also demonstrated a deficiency of B-adrenoceptor binding sites on CF airway epithelia by both membrane binding and autoradiographic techniques. The loss of binding sites is probably secondary to proteolysis by bacterial proteases and release of mediators of inflammation but may nevertheless play a role in the pathogenesis of CF.

Infection

The basic defect in CF in some way facilitates bacterial adherence and subsequent repeated episodes of colonization particularly by *Staphylococcus aureas, Pseudomonas aeruginosa, Klebsiella pneumoniae* and *Haemophilus influenzae*. Colonization with *Pseudomonas* sp., particularly of the mucoid variety, is linked to a decline, both in lung function and in longevity. In a recent study of the ultrastructure of CF airway epithelia in chronically infected lungs obtained freshly at transplant, the author noted the absence of detectable bacterial forms either within the mucosa or directly adhering to epithelial surfaces[24]. Bacteria appeared to be preferentially found in association with the overlying secretions (Figures 9.8a and 9.8b) and engulfed by neurophils (see Figure 9.5b) rather than associated with epithelial cells *per se*. These observations are in accord with previous *in vitro* findings[82] and give support to the hypothesis that loss of mucociliary clearance results in the stagnation of mucus which, with adherent bacteria, forms a focus for persistent infection leading to the release of proteases, inflammation and progressive epithelial and lung damage.

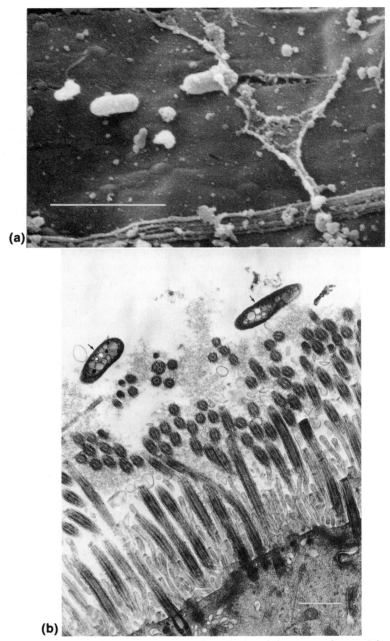

(a)

(b)

Figure 9.8 a SEM of the lumenal aspect of the blanket of airway secretion overlying cilia on which there are two rod-shaped bacterial forms similar to those of *Pseudomonas* sp. cultured in the laboratory. (Scale bar = 3 µm). **b** TEM of ciliary fringe to show bacteria (arrows) associated with epithelial secretions and *not* with apices of the airway cells. (Scale bar = 1 µm)

RESPIRATORY AIRWAYS

The last of the purely conductive airways are known as terminal bronchioles, beyond which are further generations of respiratory bronchioles that both conduct gas and participate in gas exchange. The transitional zone comprises about three generations of respiratory bronchioles after which there are two to nine generations of alveolar ducts, terminating in alveolar sacs (Figures 9.9a).

Adult human lungs contain about 300 million alveoli[83], each alveolus measuring about 250 μm in diameter when expanded. Pores of Kohn (12–13 μm in diameter) are found piercing the alveolar walls, one to seven per alveolus (Figure 9.9b). Expansion of the lung is facilitated by surface tension-reducing lipids secreted by alveolar lining cells and collectively known as pulmonary surfactant: deficiencies of surfactant production are associated with respiratory distress syndromes. (Studies which examine the control of ion and fluid secretion are presented in Chapter 11 by Gatzy *et al.*)

Electron microscopy demonstrates a complete simple squamous epithelium lining all alveoli, which is in continuity with the cuboidal/columnar epithelium of the conductive airways. The alveolar lining epithelium is separated from the underlying capillaries of the interstitium by basement membrane or relatively thick interstitial connective tissue and cells in an alternating pattern. The alveolar epithelium consists of two principal cell types, known as types I and II, or squamous and granular cells, respectively.

Type I cells have few cytoplasmic organelles but have long cytoplasmic extensions (Figure 9.10a), each cell contributing to the lining of more than one alveolus: the processes of each cover up to 5000 μm^2 of the alveolar surface yet measure only about 0.2 μm in thickness[84]. Their function is to provide a complete but thin covering preventing fluid loss while facilitating rapid gas exchange. However, their thinness makes them extremely sensitive to injury. As with conducting airways, type I epithelial cells are connected to each other and to type II cells by tight junctions which provide a selectively permeable barrier to fluid, molecular and ion movement both into and out of the alveolus. In addition, macromolecules and very fine particles may reach the interstitium from the alveolar lumen by pinocytosis from whence they may be removed together with interstitial fluid to the lymphatics of the respiratory and terminal bronchioles.

Type II cells are twice as numerous as the type I cells but due to their cuboidal shape cover only about 7% of the alveolar surface[84] and usually occupy the corners of alveoli (Figures 9.9b and 9.10b). By transmission electron microscopy, their cytoplasm can be seen to include large mitochondria with well-developed cristae, rough endoplasmic reticulum, a Golgi apparatus and characteristic osmiophilic lamellar structures which represent the secretory vacuoles containing pulmonary surfactant. Their role in surfactant secretion has been established by experimental ultrastructural autoradiographic studies tracing the incorporation of surfactant precursors with time[85], and cell separation techniques, enabling pure type II cell suspensions to be studied *in vitro*, confirm that these cells secrete surfactant[86]. Deficiencies of surfactant production are associated with

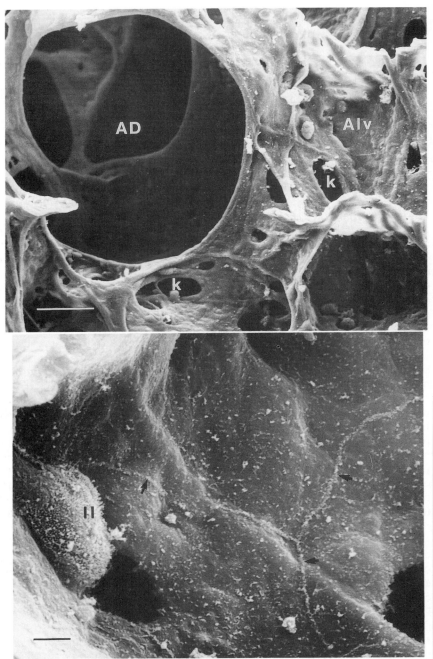

Figure 9.9 a SEM appearance and 'coil spring-like' arrangement of an alveolar duct (AD) with surrounding alveoli (Alv) whose lumena are continuous with that of the AD. Intercommunicating pores of Kohn (k). (Scale bar = 40 μm). **b** SEM image of the alveolar wall demonstrating the junctions of adjacent type I cells (arrows) and a type II cell with apical microvilli in the corner of the alveolus (II) adjacent to a pore of Kohn. (Scale bar = 1 μm)

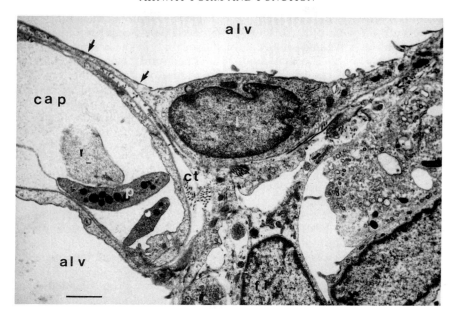

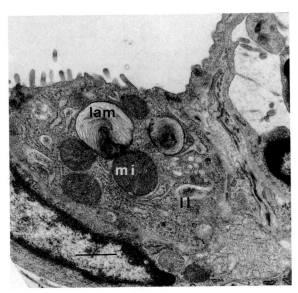

Figure 9.10 **a** TEM of type I alveolar cell (I) with thin cytoplasmic extensions (arrowed) separated from capillary lumen (cap) by either basement membrane alone or interstitial connective tissue (ct). Fibroblast (fi), platelets (p) and alveolar space (alv). (Scale bar = 1 μm). **b** TEM of a type II alveolar cell to show the cuboidal shape, apical microvilli and characteristic osmiophilic lamellar bodies (lam) containing pulmonary surfactant. Mitochondria (mi). (Scale bar = 1 μm)

respiratory distress of the newborn and also occasionally of the adults. Type II cells are the stem cells from which damaged type I cells are replaced by differentiation[28]. Animal studies have shown that, in the normal lung, the turnover time of type II cells is about 25 days and transformation of type II to type I cells may take as little as 2 days. The respiratory zone and its epithelial lining is, therefore, complex in its architecture, and is relatively less accessible to direct study than the conducting airways but plays a substantial and functionally important part in maintenance of a normal fluid lining.

SUMMARY AND CONCLUSION

The present synopsis introduces the reader to the main cell types comprising the airway epithelium, their structure, salient function and proliferative potential. Emphasis has been placed on the conductive airways where much experimental work has focussed but it should be remembered that airway epithelium is a continuum from nares/mouth to the alveolus and that the respiratory airways also have secretory (i.e. producing mucus, ion and fluid) and proliferative potential.

   A number of abnormalities of airway epithelia have been cited and those in CF have been highlighted. A convincing unifying hypothesis for the basic defect of CF is centered on a disorder of epithelial ion and water transport. The most consistent defect in a number of body organs is that of chloride impermeability with, in addition, increased sodium absorption in the respiratory tract. Early dilatation of submucosal gland ducts, indications of proliferation of epithelial tight junctional strands, reductions in the number of ciliated cell microvilli, of VIP immunoreactivity and of epithelial β-adrenoceptor number are all findings which are likely to be secondary to expression of the basic genetic defect(s). Functional studies using mucosal tissues mounted in Ussing chambers or of primary airway epithelial cell and tissue cultures have given and continue to give valuable information regarding the control of ion translocation and mucous secretion, and the effects of pharmacological intervention. Disaggregation of normal and CF airway epithelium and use of immortalized cell lines for reconstitution of their apical membrane vesicles into phospholipid bilayers will allow characterization of single ion channels, a study of their controlling mechanisms and a means by which mutant CFTR proteins can be tested for their effects. An analysis of calcium-dependent mechanisms (known to be altered in CF) may be a particularly worthwhile area of study as it may well provide a rationale for future treatment. Gene therapy is still very much for the future and, in the author's opinion, should be preceded by further studies aimed at localizing gene expression of mutant CFTR protein(s) to the specific cell types which comprise the lung.

Acknowledgements

I thank Dr A. P. R. Brain (King's College London) for his willing participation in these ultrastructural studies and Mr A. Rogers for his help in preparation of the illustrations. I am most grateful to the Cystic Fibrosis Research Trust, the Asthma Research and Cancer Research Campaigns (UK) for their support and Mrs J. Billingham for her patience in preparation of the manuscript.

References

1. Boat, T. E., Welsh, M. J. and Beaudet, A. L. (1989). Cystic fibrosis. In Scriver, C. R., Beaudet, A. L., Sly, W. S., Valle, D. (eds.) *The Metabolic Basis of Inherited Disease*, pp. 2649–2680. (New York: McGraw-Hill)
2. Welsh, M. J. (1987). Electrolyte transport by airway epithelia. *Physiol. Rev.*, 67, 1143–1184
3. Knowles, M., Gatzy, J. T. and Boucher, R. (1981). Increased bioelectric potential difference across respiratory epithelia in cystic fibrosis. *N. Engl. J. Med.*, 305, 1489–1494
4. Knowles, M., Gatzy, J. and Boucher, R. (1983). Relative ion permeability of normal and cystic fibrosis nasal epithelium. *J. Clin. Invest.*, 71, 1410–1417
5. Knowles, M. R., Stutts, M. J., Spock, A., Fischer, N., Gatzy, J. T. and Boucher, R. C. (1983). Abnormal ion permeation through cystic fibrosis respiratory epithelium. *Science*, 221, 1067–1070
6. Widdicombe, J. H., Welsh, M. J. and Finkbeiner, W. E. (1985). Cystic fibrosis decreases the apical membrane chloride permeability of monolayers cultured from cells of tracheal epithelium. *Proc. Natl. Acad. Sci.*, 82, 6167–6171
7. Yankaskas, J. R., Cotton, U. C., Knowles, M. R., Gatzy, J. T. and Boucher, R. C. (1985). Culture of human nasal epithelial cells on collagen matrix supports: a comparison of bioelectric properties of normal and cystic fibrosis epithelia. *Am. Rev. Respir. Dis.*, 132, 1281–1287
8. Cotton, C. U., Stutts, M. J., Knowles, M. R., Gatzy, J. T. and Boucher, R. C. (1987). Abnormal apical cell membrane in cystic fibrosis respiratory epithelium: an in vitro electrophysiologic analysis. *J. Clin. Invest.*, 79, 80–85
9. Schoumacher, R. A., Shoemaker, R. L., Halm, D. R., Tallant, E. A., Wallace, R. W. and Frizzell, R. A. (1987). Phosphorylation fails to activate chloride channels from cystic fibrosis airway cells. *Nature (London)*, 330, 752–754
10. Hwang, T. C., Lu, L., Zeitlin, P. L., Gruenert, D. C., Huganir, R. and Guggino, W. B. (1989). Chloride channels in CF: lack of activation by protein kinase C and cAMP-dependent protein kinase. *Science*, 244, 1351–1353
11. Rommens, J. M., Ianuzzi, M. C., Kerem, B. *et al.* (1989). Identification of the cystic fibrosis gene: chromosome walking and jumping. *Science*, 245, 1059–1065
12. Riordan, J. R., Rommens, J. M., Kerem, B. *et al.* (1989). Identification of the cystic fibrosis gene: cloning and characterization of complementary DNA. *Science*, 245, 1066–1073
13. Editorial (1990). Cystic fibrosis: prospects for screening and therapy. *Lancet*, 335, 79–80
14. Jeffery, P. K. and Corrin, B. (1984). Structural analysis of the respiratory tract. In Bienenstock, J. (ed.) *Immunology of the Lung*, pp. 1–27. (New York: McGraw-Hill)
15. Jeffrey, P. K. (1990). Microscopic structure of the normal lung. In Brewis, R. A. L., Gibson, G. J. O. and Geddes, D. M. (eds.) Texbook of Respiratory Medicine, pp. 57–78. (London: Balliere Tindall)
16. Jeffery, P. K. (1983). Morphology of airway surface epithelial cells and glands. *Am. Rev. Respir. Dis.*, 128, S14–S20
17. Breeze, R. G. and Wheeldon, E. B. (1977). The cells of the pulmonary airways. *Am. Rev. Respir. Dis.*, 116, 705–777
18. Jeffery, P. K. (1986). Innervation of airway epithelium. In Kay, A. B. (ed.) *Asthma: Clinical Pharmacology and Therapeutic Progress*, pp. 376–392. (Oxford: Blackwell)
19. Jeffery, P. K. (1987). The origins of secretions in the lower respiratory tract. *Eur. J. Respir. Dis.*, 71 (suppl. 153), 34–42
20. Jeffery, P. K. and Reid, L. (1975). New observations of rat airway epithelium: a quantitative electron microscopic study. *J. Anat.*, 120, 295–320
21. Gilljam, H., Motakefi, A-M., Robertson, B. and Strandvik, B. (1987). Ultrastructure of bronchial epithelium in adult patients with cystic fibrosis. *Eur. J. Respir. Dis.*, 71, 187–194

22. Greenstone, M., Rutland, A., Dewar, A., MacKay, I. and Cole, P. J. (1988). Primary ciliary dykinesia: physiological and clinical features. *Q. J. Med. (New Series 67)* **253**, 405–430

23. Sturgess, J. (1982). Morphological characteristics of the bronchial mucosa in cystic fibrosis. In Quinton, P. M., Martinez, J. R. and Hopfer, U. (eds.) *Fluid and Electrolyte Abnormalities in Exocrine Glands in Cystic Fibrosis,* pp. 254–270. (San Francisco: San Francisco Press)

24. Jeffery, P. K. and Brain, A. P. R. (1988). Surface morphology of human airway mucosa: normal, carcinoma, or cystic fibrosis. *Scanning Electron Microsc.,* **2**, 553–560

25. Valdivia, H. H., Debinsky, W. P. and Coronado, R. (1988). Reconstitution and phosphorylation of chloride channels from airway epithelium membranes. *Science,* **242**, 1441–1444

26. Gruenert, D. C., Basbaum, C. B., Welsh, M. J., Li, M., Finkbeiner, W. E. and Nadel, J. A. (1988). Characterization of human tracheal epithelial cells transformed by an origin-defective simian virus 40. *Proc. Natl. Acad., Sci.,* **85**, 5951–5955

27. Gruenert, D. C. (1987). Differential properties of human epithelial cells transformed in vitro. *Biotechniques,* **5**, 740–749

28. Ayers, M. and Jeffery, P. K. (1988). Proliferation and differentiation in adult mammalian airway epithelium: a review. *Eur. Respir. J.,* **1**, 58–80

29. Ellefsen, P. and Tos, M. (1972). Goblet cells in the human trachea: quantitative studies of a pathological biopsy material. *Arch. Otolaryngol.,* **95**, 547–555

30. Carlstedt, I. and Sheehan, J. K. (1984). Macromolecular properties and polymeric structure of mucus glycoproteins. In Nugent, J. and O'Connor, M. (eds.) *Mucus and Mucosa. Ciba Foundation Symp. 109,* pp. 157–172. (London: Pitman Medical)

31. Jones, R., Bolduc, P. and Reid, L. (1973). Goblet cell gylcoprotein and tracheal gland hypertrophy in rat airways: the effect of tobacco smoke with or without the anti-inflammatory agent phenylmethyloxydiazole. *Br. J. Exp. Pathol.,* **54**, 229–239

32. Alvarez-Fernandez, E. (1981). Histochemical classification of mucin-producing pulmonary carcinomas based on the qualitative characteristics of the mucin and its relationship to histogenesis. *Histochemistry,* **71**, 117–123

33. Oppenheimer, E. H. (1981). Similarity of the tracheobronchial mucous glands and epithelium in infants with and without cystic fibrosis. *Hum. Pathol.,* **12**, 36–48

34. Reid, L. (1954). Pathology of chronic bronchitis. *Lancet,* **1**, 275–279

35. Lamb, D. and Reid, L. (1968). Mitotic rates, goblet cell increase and histochemical changes in mucus in rat bronchial epithelium during exposure to SO_2. *J. Pathol.,* **96**, 97

36. Jones, R., Bolduc, P. and Reid, L. (1972). Protection of rat bronchial epithelium against tobacco smoke. *Br. Med. J.,* **2**, 142–144

37. Jeffery, P. K. and Reid, L. (1981). The effect of tobacco smoke with or without phenylmethyloxadiazole (PMO) on rat bronchial epithelium: a light electron microscopic study. *J. Pathol.,* **133**, 341–359

38. Esterly, J. R. and Oppenheimer, E. H. (1968). Cystic fibrosis of the pancreas: structural changes in peripheral airways. *Thorax,* **23**, 670–675

39. Roomans, G. M., von Euler, A. M., Muller, R. M. and Gilljam, H. (1986). X-ray microanalysis of goblet cells in bronchial epithelium of patients with cystic fibrosis. *J. Submicrosc. Cytol.,* **18**, 613–615

40. Boat, T. F., Cheng, P. W., Iyer, R. N., Carlson, D. M. and Polony, I. (1976). Human respiratory tract secretions: mucous glycoproteins of non-purulent tracheobronchial secretions, and sputum of patients with bronchitis and cystic fibrosis. *Arch. Biochem. Biophys.,* **177**, 95–104

41. Jeffery, P. K. and Reid, L. (1977). The ultrastructure of the airway lining and its development. In Hodson, W. A. (ed.) *The Development of the Lung,* pp. 87–134. (New York: Marcel Dekker Inc.)

42. Spicer, S. S., Mochizuki, I., Setser, M. E. *et al.* (1980). Complex carbohydrates of rat tracheo-bronchial surface epithelium visualized ultrastructurally. *Am. J. Anat.,* **158**, 93

43. Niden, A. H. (1980). Bronchiolar and large alveolar cell in pulmonary phospholipid metabolism. *Science,* **158**, 1323

44. Gil, J. and Weibel, E. (1971). Extracellular lining of bronchioles after perfusion-fixation of rat lungs for electron microscopy. *Anat. Rec.,* **169**, 185–200

45. de Water, R., Willems, L. N. A., van Muijens, G. N. P., Franken, C., Fransen, J. A. M., Dijkman, J. H. and Kramps, J. A. (1986). Ultrastructural localization of bronchial antileukoprotease in central and peripheral human airways by a gold-labeling technique using monoclonal antibodies. *Am. Rev. Respir. Dis.,* **133**, 882–890

46. Willems, L. N. A., Kramps, J. A., Jeffery, P. K. and Dijkman, J. H. (1988). Antileuko-protease in the developing fetal lung. *Thorax,* **43**, 784–786

47. Bensch, K. G., Corrin, B., Pariente, R. *et al.* (1968). Oat cell carcinoma of the lung: its origin and relationship to bronchial carcinoid. *Cancer,* **22**, 1163

48. Lauweryns, J. M. and Cokelaere, M. (1973). Intrapulmonary neuroepithelial bodies hypoxia-sensitive neuro (chemo-) receptors. *Experientia,* **29**, 1384

49. Lauweryns, J. M., de Bock, V., Verhofstad, A. A. J. and Steinbusch, H. W. M. (1982). Immunohistochemical localization of serotonin in intrapulmonary neuro-epithelial bodies. *Cell Tiss. Res.,* **226**, 215

50. Wharton, J., Polak, J. M., Bloom, S. R. *et al.* (1978). Bombesin-like immunoreactivity in the lung. *Nature (London),* **273**, 769

51. Cutz, E. and Gillan, J. E. (1983). Hyperplasia of bombesin-immunoreactive cells in lungs of infants with chronic pulmonary disease. *Lab. Invest.,* **48**, 4P

52. Tsutsumi, Y., Osamura, Y., Watanabe, K. and Yanaihara, N. (1983). Immunohisto-chemical studies on gastrin-releasing cells in the human lung. *Lab. Invest.,* **48**, 623

53. Gould, V. E., Linnoila, R. I., Memoli, V. A. and Warren, W. H. (1983). Neuroendocrine cells and neuroendocrine neoplasma of the lung. *Pathol. Annu.,* **18**, 287

54. Jeffery, P. K. (1987). Structure and function of adult tracheobronchial epithelium. In McDowell Elizabeth, M. (ed.) *Lung Carcinomas,* pp. 42–73. (London: Churchill Livingstone)

55. Enerback, L. (1986) Mast cell heterogeneity: the evolution of the concept of a specific mucosal mast cell. In Befus, A. D., Bienenstock, J. and Denburg, J. A. (eds.) *Mast Cell Differentiation and Heterogeneity.* (New York: Raven Press)

56. Lamb, D. and Lumsden, A. (1982). Intra-epithelial mast cells in human airway epithelium: evidence for smoking-induced changes in their frequency. *Thorax,* **37**, 334–342

57. McCartney, A., Bull, T. B. and Fox, B. (1985). Is the mast cell associated with chronic bronchitis and cancer? An electron microscopic study. Proceedings of British Thoracic Society. *Thorax,* **40**, 214

58. Jeffrey, P. K., Wandlaw, A., Nelson, F. C., Collins, J. V. and Kay, A. B. (1989). Bronchial biopsies in asthma: an ultrastructural quantitative study and correlation with hyperreactivity. *Am. Rev. Respir. Dis.,* **140**, 1745–1753

59. Madara, J. L. and Dharamsathaphorn, K. (1985). Occluding junction structure–function relationship in a cultured epithelial monolayer. *J. Cell Biol.,* **101**, 2124

60. Richardson, J. B., Bouchard, T. and Ferguson, C. C. (1976). Uptake and transport of exogenous proteins by respiratory epithelium. *Lab. Invest.,* **35**, 307

61. Boucher, R. C., Johnson, J., Inone, S. *et al.* (1980). The effect of cigarette smoking on the permeability of guinea pig airways. *Lab. Invest.,* **43**, 94

62. Hulbert, W. C., Walker, D. C., Jackson, A. and Hogg, J. C. (1981). Airway permeability to horseradish peroxidase in guinea pigs: the repair phase after injury by cigarette smoke. *Am. Rev. Respir. Dis.,* **123**, 320–326

63. Faff, O., Mitreiter, R., Muckter, H., Ben-Shaul, Y. and Bacher, A. (1988). Rapid formation of tight junctions in HT 29 human adenocarcinoma cells by hypertonic salt solutions. *Exp. Cell Res.,* **177**, 60–72

64. Severs, N. J. and Jeffery, P. K. (1988). Freeze fracture demonstration of tight junction abnormalities in airway epithelium of cystic fibrosis patients. *Inst. Phys. Conf.,* Ser. No. 93, Vol. 3, Ch. 5, pp. 141–142. (IOP Publ. Ltd)

65. Tos, M. (1966). Development of the tracheal glands in man. *Acta Pathol. Microbiol. Scand.,* **185**, 1–130

66. Meyrick, B., Sturgess, J. and Reid, L. (1969). Reconstruction of the duct system and secretory tubules of the human bronchial submucosal gland. *Thorax,* **24**, 729

67. Meyrick, B. and Reid, L. (1970). Ultrastructure of cells in the human bronchial submucosal glands. *J. Anat.,* **107**, 201

68. Bowes, D. and Corrin, B. (1977). Ultrastructural immunocytochemical localization of lysozyme in human bronchial glands. *Thorax,* **32**, 163

69. Bowes, D., Clark, A. E. and Corrin, B. (1981). Ultrastructural localisation of lactoferrin and glycoprotein in human bronchial glands. *Thorax*, **36**, 108
70. Brandtzaeg, P. (1974). Mucosal and glandular distribution of immunoglobulin components: immunohistochemistry with a cold ethanol-fixation technique. *Immunology*, **26**, 1101
71. Kramps, J. A., Franken, C., Meijer, C. J. L. M. *et al.* (1981). Localization of low molecular weight protease inhibitor in serous secretory cells of the respiratory tract. *J. Histochem. Cytochem.*, **29**, 712
72. Phipps, R. J., Williams, I. P., Pack, R. J. *et al.* (1982). Adrenergic stimulation of mucus secretion in human bronchi. *Chest*, **81**, 19S
73. Davis, P. B. and Kaliner, M. (1983). Autonomic nervous system abnormalities in cystic fibrosis. *J. Chronic Dis.*, **36**, 269–278
74. Jeffery, P. K. (1982). Bronchial mucosa and its innervation. In Cumming, G. and Bonsignore, G. (eds.) *Cell Biology and the Lung, Vol. 10 Ettore Majorana Life Sciences Series*, pp. 1–32. (New York: Plenum)
75. Barnes, P. J. (1986). State of art: neural control of human airways in health and disease. *Am. Rev. Respir. Dis.*, **134**, 1289–1314
76. Heinz-Erian, P., Rey, R., Flux, M. and Said, S. I. (1985). Deficient vasoactive intestinal peptide innervation in the sweat gland of CF patients. *Science*, **229**, 1407–1408
77. Shipperbottom, C. A., Jeffery, P. K. and Jones, C. J. (1987). Preliminary studies on pulmonary innervation and alterations in cystic fibrosis. *J. Pathol.*, 151/61A
78. Ollerenshaw, S., Jarvis, D., Woolcock, A., Sullivan, C. and Scheiber, T. (1989). Absence of immunoreactive vasoactive intestinal polypeptide in tissue from the lungs of patients with asthma. *N. Engl. J. Med.*, **320**, 1244–1248
79. McPherson, M. A. and Dormer, R. L. (1987). The molecular and biochemical basis of cystic fibrosis. *Biosci. Rep.*, **7**, 167–185
80. Boucher, R. C., Cheng, E. H. C., Paradiso, A. M., Stutts, M. J., Knowles, M. R. and Earp, H. S. (1989). Chloride secretory responses of CF human airway epithelia: preservation of calcium but not protein kinase C- and A-dependent mechanisms. *J. Clin. Invest.*, **84**, 1424–1431
81. Sharma, R. and Jeffery, P. K. (1990). Airway β-adrenoceptor number in cystic fibrosis and asthma. *Clin. Sci.*, **78**, 409–417
82. Ramphal, R. and Pyle, M. (1983). Adherence of mucosal and non-mucoid *Pseudomonas aeruginosa* to acid-injured tracheal epithelium. *Infect. Immun.*, **41**, 345–351
83. Dunnill, M. S. (1962). Postnatal growth of the lung. *Thorax*, **17**, 329
84. Crapo, J. D., Barry, B. E., Cehr, P., Bachofen, M. and Weibel, E. R. (1982). Cell characteristics of the normal lung. *Am. Rev. Respir. Dis.*, **125**, 740
85. Chevalier, G. and Collet, A. J. (1972). In vivo incorporation of choline-^{3}H leucine-^{3}H and galactose-^{3}H in alveolar type II pneumocytes in relation to surfactant synthesis. A quantitative radioautographic study in mouse by electron microscopy. *Anat. Rec.*, **174**, 289
86. Kikkawa, Y., Yoneda, K., Smith, F., Packard, B. and Suzuki, K. (1975). The type II epithelial cells of the lung. II. Chemical composition and phospholipid synthesis. *Lab. Invest.*, **32**, 295

10
Physiological Properties of Cultured Airway Epithelial Cells: Comparison with Intact Epithelial Tissues

M. R. VAN SCOTT, J. R. YANKASKAS AND R. C. BOUCHER

INTRODUCTION

Effective techniques for isolating and culturing airway epithelial cells have been developed in the past 10 years, and systems for studying the physiology of intact epithelial cell barriers in culture have only become common within the past 5 years. Consequently, few physiological properties of the airways epithelia have been studied both in intact and cultured preparations. Ion transport is a primary function of airway epihelia that has been well characterized in both preparations and provides a solid basis for evaluation of cell culture as a tool for studying epithelial cell physiology.

Active ion transport, ion permeabilities, and the resultant bioelectrical properties have been characterized in intact tissue and cell cultures from both dogs and humans. Canine airway epithelia have been particularly useful due to the large data base derived from *in vivo* and *in vitro* studies of intact tissues and quantitative differences in Na^+ and Cl^- transport in the trachea and bronchus of this species. Regional differences in ion transport are less pronounced in human than in canine airways, but the well-documented differences in function of normal and cystic fibrosis tissues are dramatic. These physiological differences may be used to assess the validity of cell cultures as models of normal and diseased epithelial cells *in vivo*.

In this chapter, we examine tissue-specific ion transport properties in cultures of human and canine airway epithelial cells in comparison with those of intact tissues. Summary data from several studies in the same laboratory are assembled into tables which permit comparison of key bioelectric properties from different preparations. To be concise, statistics have been omitted from this chapter, but are available in the original reports. Data from other groups which substantiate the findings are included in the text for comparison. Finally, the significance of these

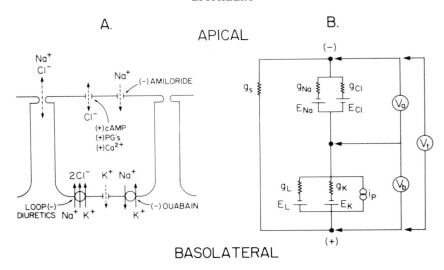

Figure 10.1 Model of ion transport across respiratory epithelial cells (A) and the corresponding electrical analogue (B). Conductance (g); electromotive force (E); extracellular shunt (s); electrogenic pump (P); leak (L); apical, basolateral and transepithelial membrane potential difference (V_a, V_b and V_t, respectively); prostaglandin (PG)

observations to cell physiology today and in the future are considered.

A model commonly used to describe transepithelial ion transport in airways is shown in Figure 10.1a. Two active transport processes, Na^+ and Cl^- secretion dominate net transepithelial ion fluxes. Na^+ enters the cell across the apical membrane through passive conductive channels and is pumped out across the basolateral membrane by a Na^+/K^+ ATPase. The net transcellular flux of Na^+ can be abolished by blocking the conductive channels in the apical membrane with amiloride (10^{-4} mol L^{-1}). Little is known of the factors that control the rate of Na^+ absorption. Chloride enters the cell via a $Na^+/K^+/Cl^-$ co-transporter on the basolateral membrane and exits through passive conductive channels in the apical membrane. Cl^- permeability of the apical membrane and transcellular Cl^- movement is regulated by cAMP and Ca^{2+}. As alluded to above, the absolute rates of active ion transport, the ratio of Na^+ absorption to Cl^- secretion, and the Cl^- secretory response to mediators of different classes vary between airway regions, species and disease states.

As illustrated in Figure 10.1b, the movement of an ion across an epithelial cell membrane can be modelled as an electrical circuit containing a conductance (g) in series with an electromotive force (E) and in parallel with a current source (i). Conductance of an ion is related to the permeability of the membrane for that ion; E is equal and opposite to the equilibrium potential; and i is the current generated during active transport. Current flowing across the basolateral or apical membranes results in a potential difference (V_b or V_a, respectively). Current traversing both the apical and basolateral membranes results in a transepithelial potential difference (V_t)

that is equal to the sum of V_b and V_a. Under open-circuit conditions, transcellular currents return to the basolateral bathing solution through a shunt pathway (g_s) that lies between adjacent epithelial cells.

CANINE TRACHEAL AND BRONCHIAL EPITHELIA

Ion transport in intact tissues

Intact canine tracheal epithelia display a higher transepithelial potential difference (V_t) than bronchial epithelia (−29 mV vs −5 mV *in vivo*)[1]. Studies of freshly excised tissues indicate that short-circuit current (I_{sc}) is similar (70 $\mu A/cm^2$) in both regions, but total tissue conductance (G_t) is greater in the bronchus than in the trachea (7 vs 2 mS/cm^2 *in vitro*[2]). I_{sc} across intact canine tracheal epithelia is accounted for by a combination of net Na^+ absorption and net Cl^- secretion[2]. Na^+ absorption, measured by isotopic techniques, accounts for 45% of the basal I_{sc} across the trachea[2,3]; however, amiloride reduces I_{sc} by only 10–20% in the presence of Cl^--containing solutions. This phenomenon has been explained by an amiloride-induced hyperpolarization of the apical membrane which increases the electro-chemical driving force for Cl^- secretion[4]. In contrast to the trachea, Na^+ absorption across the canine bronchus accounts for 78% of the I_{sc} under basal short-circuit conditions[2].

Regional differences in baseline ion transport are accompanied by differences in responses to agents that stimulate Cl^- secretion. Isoproterenol (10^{-5} mol L^{-1}) increases I_{sc} by 30%, G_t by 15%, and serosal to mucosal flow of Cl^- by 35% in freshly isolated canine tracheal epithelia[4]. The same concentration of isoproterenol has no effect on I_{sc} and Cl^- secretion in canine bronchus, but decreases G_t by 15%[4].

The potential difference profiles reported for tracheal and bronchial epithelial cells are consistent with the patterns of ion transport discussed above. More specifically, the intracellular potential difference measured across the apical membrane (V_a) in the basal state is more negative in tracheal cells than in bronchial cells (−46 mV vs −23 mV)[5,6]. The corresponding basolateral membrane potential differences (V_b) and fractional resistances of the apical membranes ($fR_a = V_a/(V_a + V_b)$ are −64 mV[5] and 0.57[7] for canine trachea and −30 mV and 0.50 for bronchus[6]. Treatment with epinephrine decreases V_a in the trachea to −34 mV[5], has no effect on V_b, and decreases fR_a to 0.38[7]. Effects of epinephrine on intracellular bioelectrical parameters of freshly isolated bronchial cells have not been reported.

Ion transport in cultured epithelial cells

Canine tracheal epithelial cells grown on semipermeable supports form confluent sheets with properties similar to the intact tissue, but the degree to which morphology and function are preserved in culture depends upon the culture system[8-11]. Cells grown in 5% fetal bovine serum (FBS) on collagen-coated polycarbonate filters are flat (i.e. approximately 5 µm tall and 20 µm wide), but exhibit morphological and functional polarity[8,9]. The apical surface exhibits a glycocalyx and numerous microvilli, and is separated from the basolateral membrane by a tight junction. Gap junctions and desmosomes connect adjacent cells[9]. The epithelial sheets develop a conductance[9] of 2.5 mS/cm² and I_{sc} of 5 µA/cm². Agents known to stimulate Cl⁻ secretion in intact tissue (isoproterenol, cAMP, PGE₂, PGF₂ₐ, LTC₄ and LTD₄) also increase I_{sc} in preparations cultured in this system[9].

The morphology and function of canine tracheal epithelial cells cultured in serum-free media on collagen matrices depend upon specific hormones and growth factors added to the media[11]. Insulin or insulin-like growth factor (IGF-1), endothelial cell growth supplement, transferrin, and cholera toxin support proliferation[11,12]. Hydrocortisone and tri-iodothyronine promote morphological differentiation (i.e. cells become cuboidal, 15 µm tall and 10 µm wide, and cilia are present). Cholera toxin increases I_{sc} by 50%. This increase is not sensitive to amiloride and is probably due to increased Cl⁻ secretion. Mean short circuit currents for cells grown in serum-free culture media supplemented with different combinations of the factors listed above range from 20 to 55 µA/cm², with the lowest current observed across preparations cultured in relatively simple media (i.e. Ham's F12 supplemented with insulin, endothelial cell growth supplement and transferrin) and the highest currents observed when the cells are grown in the presence of cholera toxin. Transepithelial conductances range from 1 to 3.3 mS/cm² with lower values being exhibited by preparations grown in relatively simple media (i.e. media supplemented with only insulin, endothelial cell growth supplement and transferrin). In general, the canine tracheal cells in serum-free hormone-supplemented media respond to β-adrenergic stimulation with an increase in I_{sc}, but preparations cultured in the presence of epidermal growth factor (EGF) exhibit a larger increase in current than preparations grown in the absence of this factor.

A microelectrode study[13] of canine tracheal cells cultured on collagen matrices in the presence of insulin, endothelial growth supplement, transferrin, cholera toxin, epidermal growth factor, hydrocortisone, and tri-iodothyronine indicates that the bioelectric profile of cultured cells is similar to the profile observed in freshly excised tissues (Table 10.1). In this study, isoproterenol increased I_{sc} from 47 µA/cm² to 93 µA/cm² and transepithelial conductance from 2.0 mS/cm² to 3.4 mS/cm². The apical membrane depolarized ($V_{a\ pre}$ = -41 mV, $V_{a\ post}$ = -33 mV), the basolateral membrane depolarized slightly ($V_{b\ pre}$ = -60 mV, $V_{b\ post}$ = -57 mV), and fR_a of the apical membrane decreased (0.51 vs 0.31). These observations are consistent with an increased apical membrane Cl⁻ conductance caused by increased intracellular cAMP.

Compared with the tracheal preparations, confluent cultures of canine bronchial epithelial cells in serum-free hormone-supplemented medium on collagen matrices exhibit lower transepithelial potentials (-6 vs -24 mV) and higher conductances (4.9 vs 1.5 mS/cm²), but no significant difference in short circuit currents are noted[13]. V_a and fR_a are similar in cultures of bronchial and tracheal cells (-33 vs -39 mV and 0.54 vs 0.54, respectively),

Table 10.1 Regional differences in the bioelectric properties of cultured canine tracheal and bronchial epithelial cells

	Trachea		Bronchus	
	Baseline	Amiloride	Baseline	Amiloride
Effects of amiloride				
V_t(mV)	-24	-22	-6	-2
I_{sc}(μA/cm^2)	60	54	32	10
G_t(mS/cm^2)	2.0	2.1	4.9	4.5
V_a(mV)	-36	-42	-28	40
V_b(mV)	-61	-64	-34	42
fR_a	0.49	0.54	0.56	0.67
Effects of isoproterenol				
V_t(mV)	-19	-25	-6	-6
I_{sc}(μA/cm^2)	47	93	31	36
G_t(mS/cm^2)	2.0	3.4	4.6	5.3
V_a(mV)	-41	-33	-31	-29
V_b(mV)	-60	-57	-37	-35
fR_a	0.51	0.31	0.57	0.51

See Boucher and Larsen[13] for original data.
Transepithelial potential difference, short circuit current and conductance (V_t, I_{sc}, and G_t, respectively); apical and basolateral membrane potential (V_a, V_b); apical membrane fractional resistance (fR_a); [Amiloride] = 10^{-4} mol L^{-1} in the apical bathing solution, [Isoproterenol] = 10^{-5} mol L^{-1} in the basolateral bathing solution

but V_b is lower in cultured bronchial cells (-49 vs -62 mV). As mentioned above, isoproterenol has a pronounced effect on cultured tracheal cells. In contrast, isoproterenol has a minimal effect on bronchial cells in culture (i.e. no significant difference in V_t, transepithelial resistance (R_t), I_{sc}, V_a, or fR_a was noted)[13]. These observations may reflect a decreased electrochemical gradient for Cl$^-$ across the apical membrane. In summary, data on the transepithelial and intracellular bioelectric characteristics of canine tracheal and bronchial airway epithelia indicate that tissue-specific ion transport properties are qualitatively and, under appropriate conditions, quantitatively preserved in culture.

HUMAN AIRWAY EPITHELIAL CELLS

Ion transport in intact tissues

Differences in Na$^+$ and Cl$^-$ transport between epithelia of normal and cystic fibrosis (CF) patients have provided an additional opportunity to examine how well ion transport function is preserved in cultures of airway epithelial cells. In CF, airway epithelia are characterized by decreased Cl$^-$ permeability of the apical membrane and an increased rate of Na$^+$ absorption[14]. Normal human airway epithelia exhibit ion transport characteristics similar to those observed in the canine bronchus. V_t across freshly excised human bronchial epithelia[15] is -6 mV (lumen negative), I_{sc} is 51 μA/cm^2, and G_t is

9 mS/cm^2. Amiloride in the mucosal bath inhibits 50% of the I_{sc}, and isoproterenol induces a small (10%) increase in I_{sc} without a significant change in ion flows[15]. V_t, I_{sc}, and G_t across nasal epithelia excised from normal patients are similar to the canine and human bronchus (i.e. V_t = -6 mV, I_{sc} = 58 μA/cm^2, and G_t = 11 mS/cm^2)[16]. Amiloride inhibits 67% of the I_{sc}, hyperpolarizes the apical membrane ($V_{a\ pre}$ = -29 mV, $V_{a\ post}$ = -38 mV)[16], and does not change fR_a. Isoproterenol has little effect if added to epithelia in the basal state; but when the Na$^+$ conductance of the apical membrane is blocked using amiloride, isoproterenol increases I_{sc} by 100% ($I_{sc\ pre}$ = 20 μA/cm^2, $I_{sc\ post}$ = 46 μA/cm^2), depolarizes the apical membrane ($V_{a\ pre}$ = -38 mV, $V_{a\ post}$ = -28 mV), and decreases fR_a ($fR_{a\ pre}$ = 0.54, $fR_{a\ post}$ = 0.42)[16]. The permeability of the apical cell surface to Cl$^-$ has been demonstrated in the presence of amiloride by substituting gluconate for Cl$^-$ in the apical bathing solution and measuring the resulting depolarization of the apical membrane ($V_{a\ pre}$ = -38 mV, $V_{a\ post}$ = -20 mV, $fR_{a\ pre}$ = 0.57, $fR_{a\ post}$ = 0.66)[16]. Exposure to isoproterenol further depolarizes the apical membrane to -9 mV but does not affect V_b.

Relative to normal epithelia, V_t and I_{sc} across CF nasal epithelia are consistently high ($V_{t\ CF\ polyps}$ = -15 mV, $I_{sc\ CF\ polyps}$ = 170 μA/cm^2)[16]. Amiloride in the mucosal bathing solution blocks 95% of the I_{sc} across freshly excised CF nasal polyps and increases fR_a from 0.51 to 0.69[16]. Isoproterenol has no effect on V_t, I_{sc}, G_t, V_a, V_b, or fR_a across CF epithelia, even in the presence of amiloride. In amiloride-pretreated tissues, replacing Cl$^-$ in the luminal bathing solution with gluconate causes a slight increase in V_t ($V_{t\ pre}$ = -1 mV, $V_{t\ post}$ = -4 mV), but no significant change in V_a or V_b is observed[16]. Isoproterenol has no effect on V_a when added to freshly excised CF nasal epithelial preparations pretreated with amiloride and bathed with a low Cl$^-$ solution on the mucosal side[16].

Ion transport in cell cultures

In general, human airway epithelial cells in culture display bioelectric properties that are similar to those of intact tissue, although two exceptions are evident (Table 10.2): I_{sc} and G_t are lower in cultured cells than in freshly excised human nasal epithelia (Table 10.2 and reference 17). The reasons for these differences are not clear but may involve bacterial products produced in freshly excised tissues[18], direct handling of the epithelia during excision and mounting of fresh tissue, differences in morphology (e.g. cultured cells may have less basolateral membrane per unit of apical surface area resulting in a lower density of transport elements such as the Na$^+$/K$^+$ ATPase), and/or differences in the densities of transport elements (e.g. conductive channels) in the apical membranes.

As in freshly excised tissues, I_{sc} is higher in CF nasal epithelia than in normal nasal epithelia cultured under identical conditions[19]. Amiloride inhibits 85% of the spontaneous transepithelial potential difference across CF cultures compared with 46% in normal cultures (CF $V_{t\ pre}$ = -25 mV, V_t post = -4 mV; Normal $V_{t\ pre}$ = -9 mV, $V_{t\ post}$ = -5 m-v)[19]. Isoproterenol in the

Table 10.2 Bioelectrical properties of human nasal epithelia from normal and cystic fibrosis patients

	Normal			Cystic fibrosis		
	Excised[a]	Cultured[b]	Cell line[c]	Excised[a]	Cultured[b]	Cell line[c]
Baseline						
V_t(mV)	-6	-12	-1	-15	-30	-2
I_{sc}(μ./cm^2)	58	23	13	170	53	19
G_t(mS/cm^2)	11.4	1.9	12.7	12.6	1.6	8.0
V_a(mV)	-29	-28	-25	-11	-17	-25
V_b(mV)	-34	-41	-26	-26	-48	-28
fR_a	0.56	0.50	0.67	0.59	0.68	0.56
Low luminal Cl						
V_t(mV)	-19	-22	-9	-22	-32	-8
I_{sc}(μ./cm^2)						
G_t(mS/cm^2)	7.5	1.4	13.8	9.6	1.4	10.4
V_a(mV)	-1	-16	-9	-4	-15	-25
V_b(mV)	-20	-38	-18	-26	-47	-33
fR_a	0.57	0.66	0.56	0.57	0.73	0.64
Amiloride						
V_t(mV)	-2	-5	-0.4	-1	-1	-0.5
I_{sc}(μ./cm^2)	20	14	8	5	3	6
G_t(mS/cm^2)	11.7	2.0	16	5.8	1.2	11.5
V_a(mV)	-38	-39	-23	-37	-43	-33
V_b(mV)	-39	-44	-24	-38	-44	-33
fR_a	0.54	0.56	0.72	0.74	0.82	0.66
Amiloride + isoproterenol						
V_t(mV)	-4	-7		-1	-1	
I_{sc}(μ./cm^2)	46	22		4	3	
G_t(mS/cm^2)	12.3	2.4		5.8	1.2	
V_a(mV)	-28	-33		-42	-44	
V_b(mV)	-31	-41		-43	-45	
fR_a	0.42	0.43		0.74	0.85	

Transepithelial potential difference, short circuit current, and conductance (V_t, I_{sc}, and G_t, respectively); apical and basolateral membrane potential (V_a and V_b); apical membrane fractional resistance (fR_a); low luminal [Cl] = 3 mmol/L[-1] [Amiloride] = 10^4 mmol/L[-1] [Isoproterenol] = 10^5 mmol/L[-1] for excised preparations and 10^6 mmol/L[-1] for culture preparations.

[a]See Cotton et al.[17]; [b]See Boucher et al.[7]; [c]See Jetten et al.[32]

Table 10.3 Effects of A23187 on cultured human nasal epithelia pretreated with amiloride $(10^{-4} \text{ mol}/\text{L}^{-1})$

	Normal		Cystic fibrosis	
	Baseline	*A23187*	*Baseline*	*A23187*
$V_t(mV)$	-7.4	-7.4	-1.9	-4.3
$I_{sc}(\mu A/cm^2)$	12.6	20.0	1.2	7.0
fR_a	0.54	0.48	0.68	0.59

See Willumsen and Boucher[22] for original data.
Transepithelial potential difference (V_t), short circuit (I_{sc}), apical membrane fractional resistance (fR_a); $[A23187] = 10^{-6} \text{ mol } L^{-1}$

presence of amiloride increases I_{sc} by 50%, depolarizes the apical membrane, and decreases fR_a in cultures of normal cells, but does not affect CF cells (Table 10.2)[19].

Differences in the apical membrane Cl^- permeability of cultured normal and Cf cells have been studied using microelectrode techniques. The apical membrane fractional resistance of cultured CF nasal epithelial cells is higher than in normal cells (Table 10.2). Decreasing the Cl^- concentration of the apical bathing solution increases V_t and decreases V_a in cultured normal cells but does not change V_t or V_a in cultured CF cells. Intracellular Cl^- activity (a_{Cl}) is the same in both cultured preparations under basal conditions $(43 \text{ mmol } L^{-1}$ vs 47 $\text{mmol } L^{-1}$ for normal and CF, respectively)[20]. Replacing Cl^- in the apical bathing solution with gluconate decreases a_{Cl} to 27 $\text{mmol } L^{-1}$ in cultures of normal cells but has no effect on a_{Cl} in CF cultures[20,21].

As mentioned above, β-adrenergic agonists do not increase the Cl^- conductance (g_{Cl}) of the apical cell membrane in CF cultures. However, apical g_{Cl} in CF cells can be at least partially activated by exposing the cells to Ca^{2+} ionophore (A23187). A23187 hyperpolarizes the apical membrane, increases I_{sc}, and decreases fR_a in amiloride-pretreated cultures of CF and normal nasal epithelial cells (Table 10.3). If Cl^- in the apical bathing solution is replaced with gluconate, A23187 lowers a_{Cl} in CF and normal cells[22] to approximately 37 $\text{mmol } L^{-1}$.

A more detailed discussion on ion channel activity and regulation in cultured normal and CF airway epithelial cells can be found in Chapter 13 contributed by Jeffrey J. Smith.

APPLICATION OF CELL CULTURE TO CELL PHYSIOLOGY

Ability to grow functionally differentiated epithelial cell preparations allows physiologists to study preparations developed by cell and molecular biologists. For example, studies of salt and water transport by subpopulations of epithelial cells from remote areas of the respiratory tract (e.g. alveolar type II cells[23,24] and bronchiolar cells[25]) have been conducted. In addition, the functional state of genetically altered cells and cell lines can be assessed.

With regard to the airways, a technique for isolating and partially purifying non-ciliated (Clara) cells from the small airways of laboratory animals was first described in 1981 by Devereux and Fouts[26]. As methods for culturing epithelial cells were developed, the techniques described by Devereux and Fouts were adapted to yield cultures of Clara cells whose function could be evaluated using conventional techniques[25,27]. Clara cells are isolated from the small airways of rabbits by protease digestion, and partially purified by centrifugal elutriation, discontinuous density gradient centrifugation, and differential adherence[25]. When seeded at high density onto crosslinked collagen matrices, the cells form confluent monolayers within 18 h. As in the canine trachea, the choice of culture media affects the function of the cells. V_t and I_{sc} are greater in preparations maintained in serum-free (Ham's F12 supplemented with insulin, endothelial cell growth supplement, transferrin, cholera toxin, hydrocortisone, tri-iodothyronine and epidermal growth factor) versus serum-supplemented (Ham's F12 + 5% FBS) medium ($V_{t \text{ serum-free}} = 25$ mV vs $V_{t \text{ serum}} = 8$ mV; $I_{sc \text{ serum-free}} = 44$ $\mu A/cm^2$ vs $I_{sc \text{ serum}} = 16$ $\mu A/cm^2$)[25,27]. Amiloride in the apical bathing solution inhibits I_{sc} by approximately 70% in both preparations, and ion flux measurements across serum-free cultures indicate that 90% of the I_{sc} can be accounted for by Na^+ absorption ($J^{Na}_{net} = 1.48$ μEq cm^{-2} h^{-1} which equates to 39.6 $\mu A/cm^2$)[27]. Therefore, unlike the canine trachea, the increase in short circuit current observed across cultures in serum-free hormone-supplemented medium is due primarily to an increase in Na^+ transport, while Cl^- secretory paths are largely unaffected.

An apparent difference between Clara cells and other airway epithelial cell cultures is that Clara cells demonstrate a large electrically silent non-conductive transepithelial Cl^- pathway. This pathway has not been well characterized and its role in salt and water homoeostasis in the small airways is unknown. However, cell culture provides an avenue by which necessary experiments can be conducted.

Cell lines derived from respiratory tract carcinomas have been reported[28] and several cell lines developed from normal or cystic fibrosis airway epithelia with oncogenes or viral genes[29-34]. Some of these cell lines are reported to demonstrate exceptional preservation of morphological characteristics[28]. Studies to determine the ion transport activities and other functional properties of these cells are ongoing. Some lines develop transepithelial transport properties (I_{sc} and G_t)[32,33], although these features may be lost at late passages. The matched CF and control cell lines reported by Jetten et al.[32] develop baseline bioelectrical properties and responses to low luminal Cl^- or amiloride similar to those of cultured epithelia (Table 10.2), but with quantitative differences that may be due to the exogenous gene. Based on conventional and chloride-selective microelectrodes and on patch clamp studies, the CF cell line retains the regulatory abnormality that characterizes CF; that is, a reduced apical membrane chloride conductance and apical Cl^- channels that are activated by Ca^{2+} ionophores but not by cAMP-dependent agonists.

The recent identification of the CF gene[35] and the availability of CF and normal human airway epithelial cell lines may facilitate experiments to identify the functions of the CF gene and its normal counterpart. Investigators are actively developing methods for introducing genetic material, such as the normal counterpart of the CF gene, into cells and assaying for the CF or normal phenotype. Such complementation of function would provide insight into the CF gene's functions and may lead to new therapeutic options. As one might expect, a wide variety of approaches is being tested, and an inevitable result of this work will be the refinement of cell physiological techniques.

Current research trends in the field of airway epithelial physiology reflect a blending of physiology with cell biology, immunology, and molecular biology, and subsequent development of novel techniques for studying airway epithelial cell function in health and disease. The development and validation of cell culture systems for studying epithelial cell function *in vitro* has provided a common base on which physiology and the new disciplines can be integrated. Such an integration could be mutually beneficial, revitalizing the classical discipline while simultaneously accentuating the relevance of current biological techniques.

References

1. Boucher, R. C. Jr., Bromberg, P. A. and Gatzy, J. T. (1980). Airway transepithelial electric potential in vivo: species and regional differences. *J. Appl. Physiol.,* **48**, 169–176
2. Boucher, R. C., Stutts, M. J. and Gatzy, J. T. (1981). Regional differences in bioelectric properties and ion flow in excised canine airways. *J. Appl. Physiol.: Respir. Environ. Exercise Physiol.,* **51**, 706–714
3. Al-Bazzaz, F. J. and Zevin, R. (1984). Ion transport and metabolic effects of amiloride in canine tracheal mucosa. *Lung,* **162**, 357–367
4. Boucher, R. C. and Gatzy, J. T. (1982). Regional effects of autonomic agents on ion transport across excised canine airways. *J. Appl. Physiol.: Respir. Environ. Exercise Physiol.,* **52**, 893–901
5. Shorofsky, S. R., Field, M. and Fozzard, H. A. (1986). Changes in intracellular sodium with chloride secretion in dog tracheal epithelium. *Am. J. Physiol. (Cell),* **19**, C646–C650
6. Boucher, R. C., Narvarte, J., Cotton, C., Stutts, M. J., Knowles, M. R., Finn, A. L. and Gatzy, J. T. (1982). Sodium absorption in mammalian airways. In Quinton, P. M., Martinez, J. R. and Hopfer, U. (eds.) *Fluid and electrolyte abnormalities in exocrine glands in cystic fibrosis*, pp. 271–287. (San Francisco: San Francisco Press, Inc.)
7. Welsh, M. J., Smith, P. L. and Frizzell, R. A. (1982). Chloride secretion by canine tracheal epithelium: the cellular electrical potential profile. *J. Membr. Biol.,* **70**, 227–238
8. Coleman, D. L., Tuet, I. K. and Widdicombe, J. H. (1984). Electrical properties of dog tracheal epithelial cells grown in monolayer culture. *Am. J. Physiol.,* **246**, C355–C359
9. Widdicombe, J. H., Coleman, D. L., Finkbeiner, W. E. and Friend, D. S. (1987). Primary cultures of the dog's tracheal epithelium: fine structure, fluid, and electrolyte transport. *Cell Tissue Res.,* **247**, 95–103
10. Welsh, M. J. (1985). Ion transport by primary cultures of canine tracheal epithelium: methodology, morphology, and electrophysiology. *J. Membr. Biol.,* **88**, 149–163
11. Van Scott, M. R., Lee, N. P., Yankaskas, J. R. and Boucher, R. C. (1988). Effect of hormones on growth and function of cultured canine tracheal epithelial cells. *Am. J. Physiol.,* **255**, C237–C245
12. Retsch-Bogart, G. Z., Stiles, A. D., Van Scott, M. R. and Boucher, R. C. (1988). Effects of IGF-I on tracheal epithelial cell proliferation. *Am. Rev. Respir. Dis.,* **137**, 13A
13. Boucher, R. C. and Larsen, E. H. (1988). Comparison of ion transport by cultured secretory and absorptive canine airway epithelia. *Am. J. Physiol.,* **254**, C535–C547
14. Knowles, M. R., Stutts, M. J., Spock, A., Fischer, N., Gatzy, J. T. and Boucher, R. C. (1983). Abnormal ion permeation through cystic fibrosis respiratory epithelium. *Science,* **221**, 1067–1070

15. Knowles, M., Murray, G., Shallal, J., Askin, F., Ranga, V., Gatzy, V. and Boucher, R. (1984). Bioelectric properties and ion flow across excised human bronchi. *J. Appl. Physiol.: Respir. Environ. Exercise Physiol.,* **56**(4), 868–877

16. Cotton, C. U., Stutts, M. J., Knowles, M. R., Gatzy, J. T. and Boucher, R. C. (1987). Abnormal apical cell membrane in cystic fibrosis respiratory epithelium. An in vitro electrophysiological analysis. *J. Clin. Invest.,* **79**, 80–85

17. Boucher, R. C., Cotton, C. U., Gatzy, J. T., Knowles, M. R. and Yankaskas, J. R. (1988). Evidence for reduced Cl⁻ and increased Na⁺ permeability in cystic fibrosis human primary cell cultures. *J. Physiol. (London),* **405**, 77–103

18. Stutts, M. J., Schwab, J. H., Chen, M. G., Knowles, M. R. and Boucher, R. C. (1986). Effects of Pseudomonas aeruginosa on bronchial epithelial ion transport. *Am. Rev. Respir. Dis.,* **134**, 17–21

19. Yankaskas, J. R., Cotton, C. U., Knowles, M. R., Gatzy, J. T. and Boucher, R. C. (1985). Culture of human nasal epithelial cells on collagen matrix supports. *Am. Rev. Respir. Dis.,* **132**, 1281–1287

20. Willumsen, N. J., Davis, C. W. and Boucher, R. C. (1989). Cellular Cl⁻ transport in cultured cystic fibrosis airway epithelium. *Am. J. Physiol. (Cell),* **256**, C1045–C1053

21. Willumsen, N. J., Davis, C. W. and Boucher, R. C. (1989). Intracellular Cl⁻ activity and Cl⁻ pathways in cultured human airway epithelium. *Am. J. Physiol. (Cell),* **256**, C1033–C1044

22. Willumsen, N. J. and Boucher, R. C. (1989). Activation of an apical Cl⁻ conductance by Ca²⁺ ionophores in cystic fibrosis airway epithelia. *Am. J. Physiol.,* **256**, C226–C233

23. Mason, R. J., Williams, M. C., Widdicombe, J. H., Sanders, M. J., Misfeldt, D. S. and Berry, L. C. (1982). Transepithelial transport by pulmonary alveolar type II cells in primary culture. *Proc. Natl. Acad. Sci. USA,* **79**, 6033–6037

24. Goodman, B. E., Fleischer, R. S. and Crandall, E. D. (1983). Evidence for active Na⁺ transport by cultured monolayers of pulmonary alveolar epithelial cells. *Am. J. Physiol.,* **245**, C78–C83

25. Van Scott, M. R., Hester, S. and Boucher, R. C. (1987). Ion transport by rabbit nonciliated bronchiolar epithelial cells (Clara cells) in culture. *Proc. Natl. Acad. Sci. USA,* **84**, 5496–5500

26. Devereux, T. R. and Fouts, J. R. (1980). Isolation and identification of Clara cells from rabbit lung. *In Vitro,* **16**, 958–968

27. Van Scott, M. R., Davis, C. W. and Boucher, R. C. (1989). Na⁺ and Cl⁻ transport across rabbit nonciliated bronchillar epithelial (Clara) cells. *Am. J. Physiol.,* **256**, C893–C901

28. Gazdar, A. F. (1986). Advances in the biology of non-small cell lung cancer. *Chest,* **89**, 277S–283S

29. Yoakum, G. H., Lechner, J. F., Gabrielson, E. W., Korba, B. E., Malan-Shibley, L., Willey, J. C., Valerio, M. G., Shamsuddin, A. M., Trump, B. F. and Harris, C. C. (1985). Transformation of human bronchial epithelial cells transfected by Harvey ras oncogene. *Science,* **227**, 1174–1179

30. Reddel, R. R., Ke, Y., Gerwin, B. I., McMenamin, M. G., Lechner, J. F., Su, R. T., Brash, D. E., Park, J-B., Rhim, J. S. and Harris, C. C. (1988). Transformation of human bronchial epithelial cells by infection with SV40 or adenovirus-12 SV40 hybrid virus, or transfection via strontium phosphate coprecipitation with a plasmid containing SV40 early region genes. *Cancer Res.,* **48**, 1904–1909

31. Gruenert, D. C., Basbaum, C. B., Welsh, M. J., Li, M., Finkbeiner, W. E. and Nadel, J. A. (1988). Characterization of human tracheal epithelial cells transformed by an origin-defective simian virus 40. *Proc. Natl. Acad. Sci. USA,* **85**, 5951–5955

32. Jetten, A. M., Yankaskas, J. R., Stutts, M. J., Willumsen, N. J. and Boucher, R. C. (1989). Persistence of abnormal chloride conductance regulation in transformed cystic fibrosis epithelia. *Science,* **244**, 1472–1475

33. Scholte, B. J., Kansen, M., Hoogeveen, A. T., Willemse, R., Rhim, J. S., Van der Kamp, A. W. M. and Bijman, J. (1989). Immortalization of nasal polyp epithelial cells from cystic fibrosis patients. *Exp. Cell Res.,* **182**, 559–571

34. Buchanan, J. A., Yeger, H., Tabcharani, J. A., Jensen, T. J., Auerbach, W., Hanrahan, J. W., Riordan, J. R. and Buchwald, M. (1990). Transformed sweat gland and nasal epithelial cell lines from control and cystic fibrosis individuals. *J. Cell Sci.,* **95**, 109–123

231

35. Riordan, J. R., Rommens, J. M., Kerem, B-T., Alon, N., Rozmahel, R., Grzelczak, Z., Zielenski, J., Lok, S., Plavsic, N., Chou, J-L., Drumm, M. L., Iannuzzi, M. C., Collins, F. S. and Tsui, L-C. (1989). Identification of the cystic fibrosis gene: Cloning and characterization of complementary DNA. *Science,* **245**, 1066–1073

11
Solution Transport Across Alveolar Epithelia of Fetal and Adult Lungs

J. T. GATZY, E. M. KROCHMAL and S. T. BALLARD

INTRODUCTION AND BACKGROUND

The complex architecture of the alveolar region of the lung has hampered investigation of solution transport by the epithelium that lines the airspace. Alveolar units are small (~80 μm diameter) and joined with other alveoli to form sacs. The sacs are connected to other alveoli by a highly branched duct system and pores of Kohn which allow bulk flow and the passage of particles as large as cells[1]. These properties, coupled with the lack of rigidity of the air-filled adult lung, make it difficult to apply techniques that have been developed to study large sheets of epithelia or the epithelia of small tubules within other organs. Solute and water movement across epithelia in different regions of the gastrointestinal tract[2], renal tubule[3] and, more recently, large airways have been studied extensively. These studies have led to working models that describe the modes and paths of solution flow through the epithelial barrier. This information provides a basis for evaluation of the role that cells disaggregated from the epithelium and maintained or grown in cell culture may have played in the function of the intact epithelial barrier. However, there is little direct information about solute and water transport across the native alveolar epithelium *in vivo* or *in vitro*. Consequently, it is difficult to judge how closely the behavior of cultured barriers that are enriched with an alveolar cell type resembles the function of the native epithelium.

Moreover, there is no widely accepted cultured epithelial barrier composed of the cell type that covers most of the interface between the airspace and interstitium. Type I pneumocytes comprise less than one half of the adult population of cells but cover more than 95% of the surface[1]. These cells are nearly devoid of organelles and are less than 1 μm thick except in the vicinity of the nucleus. Type II pneumocytes line less than 5% of the alveolar surface. These cells are cuboidal and relatively rich in organelles, such as mitochondria, endoplasmic reticulum and lamellar bodies. Even

233

though lamellar bodies have been noted in type I cells[4], there is strong evidence that lamellar bodies which contain pulmonary surfactant and associated proteins are synthesized and organized within the type II cell. Whereas type II cells secrete lamellar bodies into the airspace, the lamellar forms within type I cells are thought to participate in a process of surfactant reuptake and recycling. The secretion or reabsorption of crystalloids with surfactant has not been studied. Injury to the alveolar surface results in a proliferation of type II cells[5]. This observation has spawned the notion that type II cells are the progenitors of type I cells.

Pulmonary alveolar macrophages may also be part of the alveolar lining, but many of these cells can be washed out of the alveolar region by lung lavage. Accordingly, macrophages must be free or loosely attached.

By conventional electron microscopy and freeze–fracture analysis, the junctions that join the lumenal borders of type I and II pneumocytes of the intact alveolus fit the criteria for a morphological 'tight' junction[6]. These junctions, in contrast to junctions between capillary endothelial cells in the same region, exclude relatively large protein probes, such as horseradish peroxidase (molecular radius ~3.5 nm), that are injected into the pulmonary circulation.

Only preparations enriched with type II cells have been routinely grown in culture[7,8]. Without a barrier composed of type I cells, it is difficult to project how the two cell types might contribute to the barrier properties of the native epithelium, even if these properties were well-known. Consequently, information about membrane transport and permeability of type II cells is valuable but cannot be meaningfully integrated, at this time, into a scheme for the native barrier.

Based on this perspective, several important characterizations are necessary before phenomenological evaluation of the pulmonary alveolar epithelium can approach that for other major epithelia. These include:

(1) Assessment of barrier properties of an intact alveolar region that is separated from the adjacent airway epithelia;

(2) Disaggregation and culture of an epithelium representative of the mixed cell population of the native barrier;

(3) Comparison of properties of the mixed cell barrier with properties of the native epithelium;

(4) Culture and characterization of a barrier composed of type I pneumocytes;

(5) Integration of properties of barriers with type I and type II pneumocytes into a simulation of properties of the native barrier.

Even with this information, the importance of integrated transport by the alveolar epithelium of air-filled lung in normal (or patho-) physiology is not obvious. Impressive physical forces are present at the air–epithelial interface and have been proposed as major regulators of alveolar surface liquid volume and composition[9]. The melding of this view with evidence for active ion transport from liquid-filled preparations is a major challenge for future research. The remainder of this chapter is devoted to reviewing how closely our knowledge of alveolar epithelial function approaches projections based on the behaviour of the entire pulmonary epithelium and the ideal characterization described above.

FETAL PULMONARY EPITHELIUM

The lung of the normally developing fetus is liquid filled, and it was shown many years ago that liquid in the developing airspace is the result of liquid secretion rather than aspiration of amniotic fluid[10]. The maximum rate of secretion has been estimated to be 2–3 ml h^{-1} (kg body weight)$^{-1}$ in sheep[11]. The processes that drive liquid into the lumen have been studied in sheep and rabbits by several groups[12-15]. Secretion of Cl$^-$ (with attendent Na$^+$ and water flow) appears to be the major driving force. The Cl$^-$ concentration in fetal lung liquid of dogs and sheep is 22–47% greater than that of plasma[16,17], a finding that cannot be accounted for by Donnan forces. Moreover, the electric potential difference (PD) across the pulmonary epithelium of fetal sheep lung is about 4 mV, lumen negative[13]. The ratio of unidirectional flows of ^{36}Cl$^-$ across the epithelial barrier do not fit the prediction for passive movement, whereas ^{22}Na$^+$ fluxes appear to follow the gradient of electrochemical driving force[18]. Consequently, Cl$^-$ accumulates in lung liquid against a gradient of both electrical and chemical potential. The pattern of permeation of hydrophilic probe molecules of different size can be described[11] as flow through 'pores' with an equivalent radius of about 0.6 nm. This finding, coupled with the low permeability to even small probes, implies that the entire fetal pulmonary epithelium is a barrier comparable to that of other 'tight' epithelia, such as amphibian skin and urinary bladder, mammalian large intestine, and the distal tubule of the renal nephron.

Shortly before birth, liquid secretion slows and is supplanted by an absorptive process which is inhibited when amiloride, a Na$^+$ channel blocker, is added to lung liquid[14]. This observation implicates a role for active Na$^+$ transport in absorption. At the time of reabsorption, concentrations of catecholamines in the maternal and fetal circulation are raised. Injection of isoproterenol, a β-adrenergic agonist, into the fetal circulation induces an absorption that is inhibited by the β-blocker, propranolol[19]. Consequently, interaction of the β-agonist with β-receptors appears to be required for the induction of absorption. We will show later that exposure of excised fetal alveolar and airway epithelia (and some adult airway epithelia) to β-agonists routinely leads to Cl$^-$ secretion (also blocked by propranolol) rather than Na$^+$ absorption. Recent studies with fetal sheep lung *in vivo* have demonstrated that thyroidectomy abolishes the absorptive response to β-agonists[20]. This observation implies that at least one other humoral agent is required to tune the β-receptor–effector system for the absorptive response. From the ouabain-sensitive ^{86}Rb uptake into pulmonary granulocytes (type II cells) disaggregated from rabbit lung, Bland and Boyd[21] inferred that paths of Na$^+$ absorption increased with gestational age.

FETAL AND NEONATAL AIRWAY EPITHELIUM

Because liquid begins to accumulate in the primitive trachea before the development of alveoli[22], it was suspected that airways participate in the production of lung liquid. Clear evidence for this expectation emerged from studies of airways excised from fetal sheep and dogs[17,23,24]. With the transepithelial PD clamped to zero (short-circuited), unidirectional fluxes of $^{22}Na^+$ across the fetal trachea were symmetrical and compatible with passive permeation, whereas the flow of $^{36}Cl^-$ from the submucosa to lumenal bath exceeded the flux in the opposite direction. Net Cl^- flow (secretion) was about equal to the current required to hold the voltage clamp (equivalent short-circuit current). Current, voltage, and Cl^- secretion were stimulated by isoproterenol. Lumenal amiloride did not affect bioelectric properties of the epithelium or ion fluxes but replacement of submucosal Na^+ by choline inhibited the short-circuit current. The evidence suggested that a path for Na^+ absorption had not yet developed and that Na^+-dependent Cl^- secretion dominated active ion flow. This pattern persisted for tracheas excised from neonatal dogs up to one month of age and for week-old neonatal sheep. Although it was difficult to mount smaller fetal airways, such as bronchi, in the chambers and ensure adequate edge seal for an accurate measure of net ion flux, stimulation of transepithelial PD and short-circuit current by isoproterenol in the submucosal bath and the absence of a response to lumenal amiloride suggested that Cl^- was secreted in this region of fetal airways. When bronchi from neonatal (1-month-old) dogs were excised and mounted in chambers, measurements of unidirectional flows of $^{22}Na^+$ revealed a net Na^+ absorption, whereas $^{36}Cl^-$ flows were usually symmetrical under short-circuit conditions and compatible with passive permeation. Isoproterenol induced or increased Cl^- secretion without affecting Na^+ absorption. Amiloride blocked the absorption of Na^+.

The pattern of unidirectional Na^+ and Cl^- flow across excised neonatal sheep trachea indicated that Cl^- secretion dominated active ion transport by tissue from the newborn, but Na^+ absorption was the major net flow across tracheas excised from one-week or older lambs[25].

These studies of developing airways suggest that Na^+ absorption matures well after birth and that this process could not contribute to the clearance of liquid from the airspace of these species before and immediately after birth. In addition, the appearance of Na^+ absorption in bronchi, but not in the trachea, of the one-month-old dogs implies that the capacity for absorption spreads from distal to proximal structures. However, there may be important differences in the time-table for maturation of this function in different species or in the same species reared under different conditions. For example, Olver and Robinson[26] found a small Na^+ absorption but no evidence of Cl^- secretion by tracheas excised from fetal sheep. More important, the PD across nasal epithelium of premature and term human infants *in vivo* was comparable to that of adult subjects and the magnitude of the decrease in PD induced by exposure of the lumenal surface of the neonatal or adult epithelium to amiloride was similar[27]. Since the permeability and transport properties of the nasal epithelium, trachea and

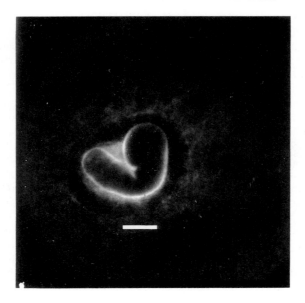

Figure 11.1   Trachea excised from a rat fetus at day 16 of gestation and submerged in culture for two weeks (bar represents 1 mm). Reprinted, with permission, from *Am. J. Physiol.*

large bronchi of human adults are comparable, the amiloride sensitivity of the transepithelial PD of premature infants indicates that Na^+ transport may be much more fully developed in the large airways of humans around the time of birth.

Tracheas have been explanted from fetal rats and held in submersion culture for up to three weeks[28,29]. Under these conditions, the cut ends seal over and liquid is secreted into the lumen (Figure 11.1). After two weeks in culture, most non-epithelial cells have disappeared and the liquid cysts are surrounded by a pseudostratified epithelial monolayer with an associated basement membrane. The majority of the epithelial cells are ciliated (Figure 11.2). Analyses of liquid obtained by micropuncture of cysts or from the responses of ion-selective electrodes during the impalement of cysts are characterized by a Cl^- concentration greater than and Na^+ concentration about equal to that of the culture medium[29]. Since the cyst lumina are about 15 mV negative with respect to the bath (Table 11.1), Cl^- must accumulate against the electrochemical gradient. Further, the response of transepithelial PD to drugs or changes in solution composition that would be expected to affect solute transport suggested that Na^+ was not absorbed (insensitivity to lumenal amiloride) and that Cl^- secretion (inhibited by bumetanide) was stimulated by β-adrenergic agonists (terbutaline) and dependent upon basolateral (bath) Na^+.

When tracheal epithelial cells were disaggregated from fetal rabbits in late gestation and were cultured on planar collagen gel matrices, only 7% of the basal short-circuit current was sensitive to amiloride whereas most of the

Figure 11.2   Cross-section of fetal tracheal explant after two weeks in submersion culture (bar represents 20 μm, arrows point to ciliated cells)

residual current was inhibited by furosemide, a loop diuretic which is likely to block the path of Cl⁻ secretion[30]. The response pattern of cultured epithelium from adult tracheas was similar except that inhibition by amiloride was greater (24%) and furosemide inhibition was smaller (53%) than the values for fetal epithelium.

All studies of fetal airway epithelia in culture lead to the conclusion that a barrier develops with typical morphological cell differentiation. But, ion and volume transport seem to be locked into the fetal pattern. The modes of ion transport, however, are probably not different from those described for the adult airways (Figure 11.3). Inhibition of Cl⁻ secretion by loop diuretics suggests that Cl⁻ enters the cell through the basolateral membrane on a electroneutral, Na^+, (K^+?), Cl⁻ co-transporter which is driven by the trans-membrane Na^+ (and Cl⁻) gradient. Replacement of basolateral Na^+ by choline impedes Cl⁻ entry. Cl⁻ efflux from cell to lumen occurs through Cl⁻ channels in the apical membrane. Evidence for this channel comes from microelectrode studies of tracheal epithelium that was freshly excised from fetal dogs[23] (Figure 11.4). Replacement of Cl⁻ in the lumenal bath by gluconate depolarized the apical membrane. Pretreatment of the preparation with a β-adrenergic agonist appeared to open more channels and effected an even greater voltage response across the apical membrane when lumenal Cl⁻ was replaced.

Na^+ absorption, when it develops in fetal or neonatal airways, exhibits the hallmarks of the pump–leak system that has been described for many epithelia. Na^+ in the lumenal compartment is drawn down the electro-chemical gradient through channels in the apical membrane that can be

238

Table 11.1 Effects of drugs or bath composition on the transepithelial PD across fetal rat pulmonary explants in culture for 14 days

Drug or bath	% Change from basal PD (treated-control*) (mean ± SE)	
	Alveolar buds (basal PD=1.8 mV)	Tracheas (basal PD=12 mV)
Bumetanide (100 μmol L^{-1}, bath)	−284±61†	−35±6†
Terbutaline (10 μmol L^{-1}, bath)	56± 9†	34±6†
Amiloride (100 μmol L^{-1}, lumen)	1±10	−5±4
Ouabain (2.3 mmol L^{-1}, bath)	−101±12†	−77±3†
Na$^\cdot$-free Ringer	−31±23	−78±5†

*Change in PD of 9 explants by a drug or Na$^\cdot$-free (choline) Ringer bath compared with PD change that accompanied incubation of an equal number of explants with drug vehicle or Na$^\cdot$-replete Ringer
†Change significantly different from zero ($p < 0.05$)

blocked by amiloride. Na$^+$ in the cell is pumped out by a transporter that is linked to a cardiac glycoside-sensitive Na$^+$, K$^+$-ATPase in the basolateral membrane. This transporter contributes indirectly to cell negativity by establishing the transmembrane K$^+$ gradient and, perhaps, directly by its own electrogenicity. The low intracellular Na$^+$ concentration provides a driving force for the secondary co-transport of Cl$^-$ and Na$^+$ that was described above. Cardiac glycosides disrupt Cl$^-$ secretion by destroying the transmembrane Na$^+$ gradient.

FETAL ALVEOLAR EPITHELIUM

The most direct evidence for liquid secretion by the epithelium of the fetal alveolar region comes from cultures of explants of fetal rat lung. When alveolar buds of 14-day fetuses are separated from the primitive trachea and submersed in culture medium supplemented with newborn[28] or fetal calf serum[29], each bud forms about 10 liquid-filled cysts (Figure 11.5). The cysts are lined with an epithelial monolayer which is made up of thin (2–4 μm) cells in the septal regions between cysts (Figure 11.6) and of thin cells and cuboidal cells with lamellar bodies in the barrier between the cyst lumena and the bath[29] (Figure 11.6).

Fetal explant volume and cyst liquid increase from close to zero to 900 nl after one week in culture and to nearly 2900 nl after two weeks. The rate of liquid secretion approximates 0.2 μl cm^{-2} h^{-1}. Lamellar bodies are infrequently found in the epithelial cells of one-week buds but are present in about half of the cuboidal cells after two weeks in culture. These observations indicate that the epithelium of the alveolar region, like that of the fetal

239

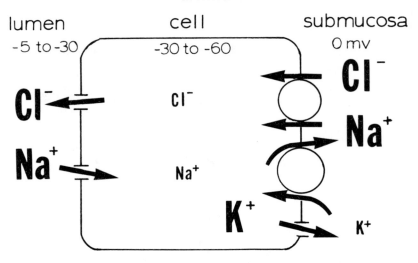

Figure 11.3 Schematic representation of major paths of ion flow into and across airway epithelia at a representative range of transmembrane and of transepithelial PD

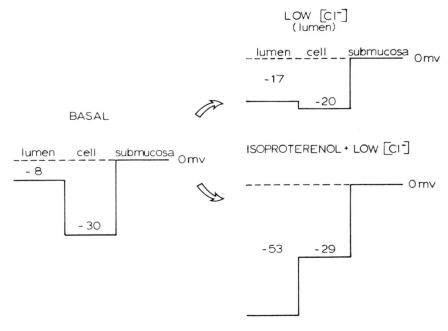

Figure 11.4 Profile of bioelectric PD across the cell membranes of excised fetal canine tracheal epithelium during exposure of the lumenal surface to Krebs Ringer bicarbonate solution (KRB) or a KRB with most (122 of 125 mmol L^{-1}) of the chloride replaced by gluconate[24]. The submucosal bath was the electrical reference (0 mV) and contained KRB or KRB with 10^{-5} mol L^{-1} isoproterenol

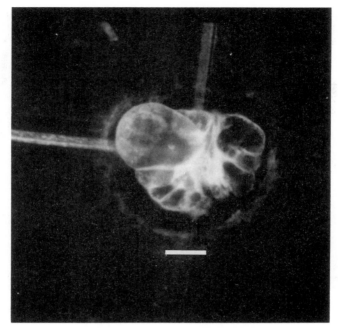

Figure 11.5 Alveolar bud excised from a rat fetus at day 14 of gestation and submerged in culture for two weeks (bar represents 1 mm). Reprinted, with permission, from *Am. J. Physiol.*

trachea in submersion culture, undergoes morphological cell differentiation but maintains a solution transport typical of early fetal life.

Analysis of cyst liquid by microsampling or ion-selective electrodes revealed a Cl^- concentration or activity that was 15–40% greater than that of the bath[29]. Na^+ and K^+ concentrations in alveolar cysts, like those of fetal tracheal cysts, were not significantly different from cation concentrations of the bath. Since the lumen of the alveolar cyst was 1–3 mV negative with respect to the bath, Cl^- was accumulated against a gradient of electro-chemical potential.

The transepithelial PD of alveolar cysts was inhibited by bumetanide or ouabain in the bath, but not by amiloride that was microinjected into the cyst lumen or by replacement of bath Na^+ by choline (Table 11.1). The effects of the loop diuretic on alveolar cyst PD and the insensitivity to amiloride imply a role for Cl^-, but not Na^+, transport in basal epithelial electrogenesis. However, this Cl^- transport, in contrast to the process in tracheal cysts, did not require basolateral (bath) Na^+. The pattern is not unique: exposure of the alveolar epithelium of the bullfrog lung to a Na^+-free medium does not affect Cl^- secretion[31].

The β-adrenergic agonist, terbutaline, stimulated the PD across fetal alveolar cysts. Stimulation was blunted when alveolar buds or tracheas were pretreated with bumetanide (Figure 11.7). These results imply that the

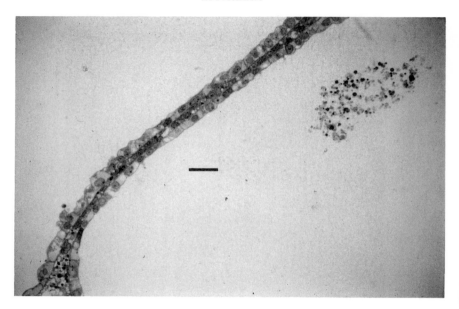

Figure 11.6 Cross-section of septum between two liquid-filled cysts of an alveolar bud after two weeks in submersion culture (bar represents 20 μm)

increase in PD is the result of a stimulation of Cl^- secretion rather than Na^+ absorption.

To summarize, the composition of alveolar cysts and the response of the transepithelial PD to drugs that affect ion transport favour the existence of a Cl^--secretory process which drives solution into the cyst lumen. Only the Na^+ independence of this process distinguishes it from Cl^- secretion by fetal tracheal epithelium.

ADULT PULMONARY EPITHELIUM

Ever since the studies of Courtice and Phipps[32], it has been clear that large volumes of liquid can move across the barrier between the airspace and blood. Kylstra believed that the enormous surface area of the lung could be used as a dialysis membrane[33], but was surprised to learn that solutes moved across the barrier slowly. This relatively low permeability was confirmed for small[36] and large hydrophilic solutes[35]. Taylor and Gaar[36] showed with several hydrophilic solutes that reflection coefficients of the vascular endothelium of the vascular perfused excised canine lungs were considerably smaller than those of the pulmonary epithelium. Matthay and coworkers[37,38] demonstrated that liquid could be cleared, against a gradient of colloid osmotic pressure, from autologous serum that was added to the airspace of lungs of anaesthetized sheep, dogs and rabbits. Absorption was stimulated by terbutaline and inhibited by amiloride that was added to the

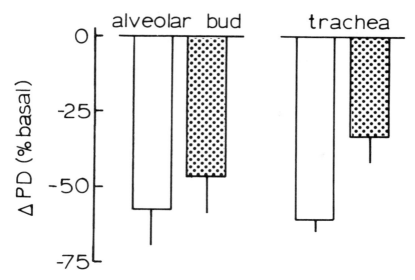

Figure 11.7 Change in PD across liquid-filled cysts of alveolar buds or tracheas in submersion culture induced by 10^{-4} mol L^{-1} bumetanide (open histograms) or by bumetanide and 10^{-5} mol L^{-1} terbutaline (stippled histograms) in the culture medium

airspace liquid. Terbutaline also increased the unidirectional flow of $^{22}Na^+$ but not sucrose from airspace to the perfusate of the excised rat lung[39]. The unidirectional flux of Na^+ was inhibited by a high concentration (10^{-3} mol L^{-1}) of amiloride in the perfusate. Whereas these observations are compatible with active Na^+ absorption, they do not exclude exchange processes (e.g. Na^+/H^+ exchange) or changes in the electrical driving force. Attempts to demonstrate basal volume absorption, stimulation by terbutaline, and inhibition by ouabain or amiloride from changes in weight of the perfused lung were interpreted by the authors as compatible with active Na^+ absorption. However, basal weight changes were erratic and the weight gain induced by amiloride implies more than inhibition of absorption of airspace liquid. Effros et al.[40] found that the glucose concentration of liquid in the airspace of vascular-perfused rat lung declined more rapidly than the concentration in the perfusate. Moreover, reabsorption, as measured by a volume marker, was inhibited when glucose in the airspace liquid was replaced by mannitol. They proposed that glucose transport could generate an osmotic driving force for volume absorption. The most comprehensive studies of liquid clearance from the vascular-perfused rat lung (isolated or cross-circulated in situ) have been carried out by Basset et al.[41-43]. Unidirectional flows of radio-Na^+, -Cl^- and -mannitol into and out of airspace liquid were monitored along with the concentration of labelled albumin, a nearly impermeant volume marker, in airspace liquid. Volume was cleared from the airspace by a process that was inhibited by ouabain in the perfusion, or by amiloride or phloridzin (inhibtor of Na^+–glucose co-transport) in the airspace liquid, but not by phloretin (inhibitor of glucose-facilitated diffusion). Estimates of ion flow that was unassociated with reabsorption suggested that the transepithelial PD was too small ($<$5 mV) to affect flux and that small hydrophilic solutes flowed around cells by unrestricted diffusion. K^+ flux into an ion-free airspace solution with glucose was more rapid than the flow of Na^+ and was inhibited by Ba^{2+} in the airspace solution. Ouabain in the airspace reduced ^{86}Rb flow out of airspace instillate, but Ba^{2+} in the instillate slightly increased ^{86}Rb efflux. Basset and co-workers propose glucose-dependent and independent paths of Na^+ and solution absorption. In addition, they suggest a path for K^+ secretion and absorption, each of which requires ouabain-sensitive K^+ accumulation at one cell surface and a Ba^{2+}-inhibitable leak at the other. Whereas secretion of K^+ by this kind of pathway is common for many epithelia (e.g. renal distal tubule[3], toad urinary bladder treated with amphotericin B[44]), stimulation of ^{86}Rb efflux from the airspace induced by Ba^{2+} is in the wrong direction to support the theory. Regardless of the nature of the transport processes, none of the paths described by any of these investigators can be unequivocally assigned to the alveolar epithelium.

ADULT ALVEOLAR EPITHELIUM

Amphibian lungs are composed of a short airway connected to a single chamber lined mostly with primitive alveolar epithelial cells. This simple architecture makes it easy to mount the lung as a sheet in a traditional flux chamber. Studies of bullfrog lung demonstrated that the epithelium generated a 20 mV (lumen negative) PD and secreted Cl^- by a process that was insensitive to amiloride, was inhibited by loop diuretics, but did not require Na^+ in the pleural bath[31,45,46]. This pattern of Cl^- transport is similar to that described above for fetal alveolar buds. Passive hydrophilic solute flow was dominated by restricted diffusion through paths with equivalent pores of less than 0.1 nm radius[47,48]. Other amphibian lungs, such as those of *Necturus*, absorb Na^+ and water[49] but the columnar epithelial cells that line the organ are more typical of the gastrointestinal tract than the lung. The lung epithelium of *Xenopus* resembles mammalian alveolar epithelium more closely. This preparation generates an amiloride-sensitive short circuit current, which implies the presence of a path for Na^+ absorption[50].

Recently, we tried to separate the alveolar and terminal airway regions from larger airways of the vascular-perfused excised left lobe of the rat lung[51-53]. Three ml of Ringer solution with blue dextran were instilled into the airspace and withdrawn. Then, residual aqueous solution (0.5–0.7 ml) was pushed into distal structures by inflation of the lobe with an immiscible O_2-carrying fluorocarbon (Figure 11.8). When perfused with Ringer solutions, these lobes changed weight in response to the transalveolar colloid osmotic gradient, but were not affected by inhibitors or stimulators of ion transport (Table 11.2). When rat serum or fetal bovine serum was mixed with the Ringer solution (40% v/v), colloid osmotic gradients still induced weight changes, but weight was lost when the osmotic driving forces across the airspace–vascular barrier were balanced. The source of the lost weight

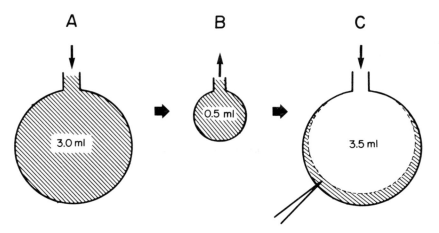

Figure 11.8 Simplified representation of the protocol for blocking, with fluorocarbon (clear), small volumes of aqueous instillate (cross-hatched) in the alveolar region of a vascular-perfused left lung lobe. A = airspace (lumen) inflated through the bronchus with aqueous solution; B = most of the aqueous solution withdrawn; C = aqueous solution pushed to distal regions by inflation of lobe to full capacity with fluorocarbon

appeared to be the airspace lumen because lobes did not lose weight when the entire airspace was filled with fluorocarbon (no Ringer solution instilled). Replacement of Na^+ in the airspace liquid by choline, or inclusion of amiloride, ethyl isopropylamiloride (an inhibitor of Na^+–H^+ exchange), or phloridzin in the instillate reduced the rate of weight loss. Treatment of lobes with CN^- and dinitrophenolate induced a reproducible weight gain. We conclude that volume absorption is probably driven by Na^+ which enters the cell coupled with glucose and by Na^+–H^+ exchange and then is extruded by Na^+ pumps in the basolateral membrane. The role of a Na^+ channel in the apical membrane in the absorptive path has not been defined. Metabolic inhibitors probably damage the vascular–airspace barrier and induce pulmonary oedema.

When aqueous solution in the alveoli was sampled by micropuncture and the concentration of blue dextran was measured in a nanolitre colorimeter, neither basal conditions, nor an osmotic gradient that induced a greater weight loss, concentrated blue dextran in subpleural alveoli (Table 11.3)[52-54]. In contrast, exposure to metabolic inhibitors induced a weight gain that was easily detected as dilution of blue dextran. We were forced to infer that basal volume flow and flow induced by colloid gradients occurred in a region different from subpleural alveoli (e.g. unblocked small airways).

In spite of the lack of evidence for major volume flow across the alveolar epithelium, there was a hint that Na^+ transport was present in this region. When subpleural alveoli of fluorocarbon-blocked lobes were impaled with microelectrodes and the effects of different instillate and perfusion compositions were tested, only amiloride decreased the transepithelial PD of 4 mV (lumen negative) to a value near zero (Figure 11.9)[52,55,56]. Conditions which

245

Table 11.2 Effect of agents or colloid on the rate of change in weight of excised, vascular-perfused, fluorocarbon-blocked left lobes of the adult rat lung

Perfusion	Aqueous instillate	Weight change (mg/min) (mean±SE)
KRB – 6% colloid	None	+2.3±1.2[†]
KRB – 6% colloid, 40% serum	None	+1.2±1.2[†]
KRB – 6% colloid	KRB	−5.8±0.8*
KRB – 6% colloid	KRB – 6% colloid	−0.2±0.6[†]
KRB – 6% colloid, 40% serum	KRB – 6% colloid	−3.2±0.8*
KRB – 6% colloid, 40% serum	KRB – 6% colloid, terbutaline	−5.4±1.2*
KRB – 6% colloid, 40% serum	Na-free KRB – 6% colloid	+1.3±0.7[†]
KRB – 6% colloid, 40% serum	KRB – 6% colloid, amiloride	−1.1±0.5
KRB – 1% colloid, 40% serum	KRB – 6% colloid	−0.6±1.0[†]
KRB – 6% colloid, 40% serum	Glucose-free KRB – 6% colloid, phloridzin	+0.1±1.3[†]
KRB – 6% colloid, 40% serum, CN⁻, DNP	KRB – 6% colloid	+8.5±1.2*[†]

0.5–0.7 ml of aqueous instillate was blocked in the alveoli with fluorocarbon. Values represent means for three or more lobes. KRB = Krebs Ringer bicarbonate solution; terbutaline = 10^{-5} mol L^{-1}; amiloride = 10^{-4} mol L^{-1}; EIPA = ethylisopropyl amiloride 10^{-6} mol L^{-1}; phloridzin = 10^{-3} mol L^{-1}; CN = NaCN $^{-3}$ mol L^{-1}; DNP = dinitrophenol 10^{-4} mol L^{-1}; *Significantly different from zero ($p < 0.05$). [†]Significantly different from KRB–colloid–serum perfused, KRB instillate ($p > 0.05$)

were likely to limit glucose entry from airspace liquid were marginally inhibitory, and agents that affect Cl⁻ transport were without effect.

When the entire lobe was filled with Ringer's solution, amiloride or bumetanide induced, at most, a partial inhibition of alveolar transepithelial PD, whereas replacement of instillate Cl⁻ by gluconate hyperpolarized the PD. These two responses of the liquid-filled lobe resemble the pattern that has been described for large airways. Inhibitors of Na⁺ transport only partially inhibit PD because electrogenic Cl⁻ secretion is increased or unmasked[57]. Replacement of lumenal Cl⁻ results in a bi-ionic PD across Cl⁻ permselective channels in the apical membrane[58].

There was no evidence for Cl⁻ permselectivity in fluorcarbon-blocked lobes even though the backflux of Cl⁻ into the small volume trapped in the alveoli was sufficiently slow (2 mEq L^{-1} min^{-1}) to maintain a substantial Cl⁻ gradient across the epithelium[56]. Consequently, airway epithelia appear to make a major contribution to the voltage across the entire pulmonary epithelium. This contribution obscures PD responses of the alveolar

Table 11.3 Comparison of volume flow from (weight change of) excised, vascular-perfused fluorocarbon-blocked left lung lobes with changes in the concentration of a volume marker in subpleural alveoli

Treatment	Alveolar Blue Dextran Concentration (%) (mean±SE)		
	Instillate	After weight Δ	Projected from weight Δ
Alveolar instillate with 6% colloid	3.2±0.3	2.9±0.3	3.6±0.1
Colloid-free alveolar instillate	3.2±0.2	3.5±0.6	5.3±0.6
Alveolar instillate* with 6% colloid, vascular perfusion with CN⁻ and DNP	3.7	2.4	2.5

The rate of weight loss was monitored for 40 min, then liquid from subpleural alveoli was sampled by micropuncture or *blue dextran concentration was monitored for 40 min by random micropuncture and the concentration from weight change was calculated from the data in Table 11.2

epithelium. Moreover, Cl⁻ permselectivity of the alveolar epithelium appears to be substantially smaller than that of entire pulmonary epithelium and, by difference, the airway epithelium.

We conclude that the subpleural alveolar epithelium has the capacity to transport Na⁺ (actively), but that this transport does not make a major contribution to liquid absorption from the fluorocarbon-blocked lobe. Since the method detects the formation of oedema in, but not liquid absorption from, the subpleural alveolar lumen, we are left with the notion that liquid absorption in fluorocarbon-blocked lobes is mediated by other regions of the lobe that are not occluded by fluorocarbon, e.g. small airways.

Alveolar epithelial cells in culture

The purification of type II cells from freshly disaggregated lung was made easier by relatively selective markers, lamellar bodies and surfactant production. When preparations enriched with type II cells were cultured on impervious surfaces, confluent cell layers were characterized by the appearance and disappearance of liquid-filled hemi-cysts or 'domes'[7,59]. The rate of dome formation was affected by agents that inhibit (amiloride, ouabain) or stimulate (terbutaline) Na⁺ absorption by the pulmonary epithelium, but not by loop diuretics that block airway epithelial Cl⁻ secretion. Cl⁻ secretion was awkward to study because loop diuretic accessibility to cyst liquid between the support and the cells was not known and the composition of cyst liquid was not measured or controlled.

The successful culture of type II pneumocytes on permeable matrices and the development of confluent monolayer barriers that could be mounted as a septum between small flux chambers was an important advance in

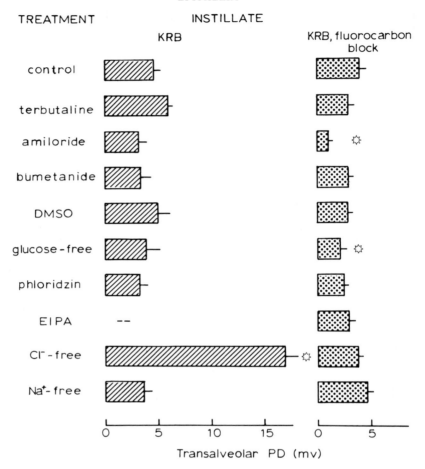

Figure 11.9 Transalveolar PD of vascular-perfused left lobes with airspace (lumen) filled entirely with KRB (cross-hatched) or with KRB blocked in the alveolar region with fluorocarbon (stippled). KRB on the pleural surface was the electrical reference. All PDs were measured with a microelectrode in the lumen of subpleural alveoli and were lumen negative. All drugs or changes in composition were in the aqueous instillate except for DMSO (bumetanide vehicle) and 10^{-4} mol L^{-1} bumetanide which were added to the perfusion. Other drug concentrations were: terbutaline = 10^{-5} mol L^{-1}; amiloride = 10^{-4} mol L^{-1}; phloridzin = 10^{-3} mol L^{-1}; EIPA = 10^{-6} mol L^{-1}

quantitation and experimental control. These preparations were characterized by small transepithelial PD's (<5 mV) and resistances[8,60] that ranged from several hundred to more than 1000 Ω cm^2. The equivalent short-circuit current of type II cell barriers was stimulated by cylic-AMP analogues and agents that are known to raise the intracellular concentration of cyclic-AMP. Exposure of the surface of attachment to ouabain or the opposite surface (lumenal?) to amiloride or Na$^+$-free solution inhibited

Table 11.4 Liquid absorption by cultured rat alveolar epithelial cell monolayers and the pulmonary epithelium of liquid-filled rat lungs

Rat tissue	Short-circuit current		$J_{Absorption}$	Reference
	$(\mu A/cm^2)$	$(\mu Eq\ cm^2\ h^{-1})$	$(\mu l\ cm^2\ h^{-1})$	
Cultured				
Type II cells	4.4	0.16	1.1	8
Type II cells	2.9	0.11	0.8	60
Type I cells?	4.1	0.15	1.0	61
Excised				
Liquid-filled lung	—	—	0.05*	42
Vascular perfused	—	—	0.10†	42

*amiloride-sensitive
†lumenal Na^+-dependent

short-circuit current. These observations are similar to the pattern noted for Na^+ absorption from liquid-filled lung (see above). Recently, a type II preparation that forms electrically tight barriers but resembles, after some time in culture, the morphology of type I cells has been described[61]. Resting equivalent short-circuit current of this preparation is similar to that of conventional type II cell preparations (Table 11.4) and the pattern of drug response is comparable. When the short-circuit current is assumed to equal net Na^+ flux that drives isosmotic volume flow, an 'absorption' of around 1 $\mu l\ cm^{-2}\ h^{-1}$ is estimated for type II cell barriers (Table 11.4). This flow is at least 10 times the volume absorption per unit area for the entire pulmonary epithelium of the isolated vascular-perfused rat lung. Volume absorption per unit surface area of the entire pulmonary epithelium probably over-estimates the contribution of alveolar epithelium because large airways (trachea) excised from the rat are characterized by a ouabain-sensitive short-circuit current[62] of more than 50 $\mu A/cm^2$. Consequently, barriers composed of cultured type II cells probably absorb salt and water much faster than the native alveolar epithelium. Of course, this activity may be damped by dilution of type II cells in the native barrier by type I cells. Unfortunately, the paucity of information about ion transport and permeability of type I cell barriers precludes this conclusion.

SUMMARY, CONCLUSIONS AND FUTURE DIRECTIONS

Unlike studies of large airways, evaluation of solute and water balance across the alveolar epithelium is hampered by complicated architecture which limits the number of direct approaches to characterization of liquid and solute flow across this region. Consequently, there is little information about the function of the intact native epithelium with which cultured cell preparations can be compared. The development of a method to trap aqueous solution in adult alveoli and block much of the airway region with fluorocarbon is a crude attempt to separate airway from native alveolar function. These studies support the inference from experiments with whole lung lavage that a path for Na^+ transport dominates active ion translocation in this region. Moreover, pulmonary oedema induced by metabolic inhibitors appears to flow through the alveolar barrier. However, in contrast to projections from studies of whole lung, there is no direct evidence for liquid absorption from alveoli that lie just beneath the pleura.

Submersion culture of explants from different regions of the fetal lung provides an opportunity to explore maturation of the integrated alveolar epithelium. Secretion of a Cl^- rich solution at a rate similar to that estimated from liquid production by the entire fetal lung hints that functions of these preparations are representative of the native epithelium. But the fetal pattern of ion transport persists in the face of a maturing cell morphology. Whereas this behavior makes explants an obvious choice to test hormones that may convert a Cl^- secreting to a Na^+ absorbing barrier, the dichotomy of structure and function raises questions about commonly accepted indices of maturation and differentiation.

It is clear that these attempts to separate the alveolar epithelium from the airways and to study integrated epithelial alveolar function are a step in meeting the goals put forward earlier in this chapter. There are limitations to these approaches and important gaps remain in our knowledge of solution transport across the intact alveolar barrier *in vitro*. Until these deficiencies are eliminated and appropriate comparisons of intact and cultured preparations are made, it is premature to equate type II cell (or type I) behavior in culture to the function of the epithelium in the living lung.

Acknowledgements

Preparation of this article was supported, in part, by NIH grant HL34322.

References

1. Thurlbeck, W. M. and Wong, N. S. (1974). The structure of the lungs. In Widdicombe, J. G. (ed.) *Respiratory Physiology*, pp. 1–30. (London: Butterworths)
2. Schultz, S. G. and Curran, P. F. (1968). Intestinal absorption of sodium chloride and water. In Code, C. F. (ed.) *Handbook of Physiology, Section 6: Alimentary Canal. Volume III: Intestinal Absorption*, pp. 1245–1275. (Washington: American Physiological Society)
3. Giebisch, G. and Windhager, E. E. (1973). Electrolyte transport across renal tubular membranes. In Orloff, J. and Berliner, R. W. (eds.) *Handbook of Physiology, Section 8: Renal Physiology*, pp. 315–376. (Washington: American Physiological Society)
4. Stratton, C. J. (1984). Morphology of surfactant producing cells and of the alveolar lining layer. In Robertson, B., Van Golde, L. M. G. and Batenburg, J. J. (eds.) *Pulmonary Surfactant*, pp. 67–118. (Amsterdam: Elsevier)
5. Evans, M. J., Stephens, R. J. and Freeman, G. (1971). Effects of nitrogen dioxide on cell renewal in the rat lung. *Arch. Intern. Med.*, **128**, 57–60
6. Schneeberger, E. E. (1976). Ultrastructural basis for alveolar–capillary permeability to protein. In *Lung Liquids*, Ciba Found. Symp. No. 38, pp. 3–21. (Amsterdam: Elsevier)
7. Goodman, B. E. and Crandall, E. D. (1983). Dome formation in primary cultured monolayers of alveolar epithelial cells. *Am. J. Physiol.*, **243**, C96–C100
8. Mason, R. J., Williams, M. C., Widdicombe, J. H., Sanders, M. J., Misfeldt, D. S. and

Berry, L. C. (1982). Transepithelial transport by pulmonary alveolar type II cells in primary culture. *Proc. Natl. Acad. Sci. USA*, **79**, 6033-6037

9. Guyton, A. C., Moffatt, D. S. and Adair, T. H. (1984). Role of alveolar surface tension in transepithelial movement of fluid. In Robertson, B., Van Golde, L. M. G. and Batenburg, J. J. (eds.) *Pulmonary Surfactant*, pp. 171-185. (Amsterdam: Elsevier)

10. Jost, A. and Policard, A. (1948). Contribution experimental a l'etude de developement prenatal du pouman chez le lapin. *Arch. Anat. Microsc.*, **37**, 323-332

11. Normand, I. C. S., Olver, R. E., Reynolds, E. O. R., Strang, L. B. and Welch, K. (1971). Permeability of lung capillaries and alveoli to non-electrolytes in the foetal lamb. *J. Physiol. London*, **219**, 303-330

12. Bland, R., McMillan, D., Bressack, M. and Dong, L. (1979). Clearance of liquid from lungs of newborn rabbits. *J. Appl. Physiol.*, **47**, 397-403

13. Olver, R. E. and Strang, L. B. (1974). Ion fluxes across pulmonary epithelium and the secretion of lung liquid in the foetal lamb. *J. Physiol. London*, **241**, 327-357

14. Brown, M. J., Olver, R. E., Ramsden, C. A., Strang, L. B. and Walters, D. V. (1983). Effects of adrenaline and spontaneous labour on the secretion of lung liquid in the foetal lamb. *J. Physiol. London*, **344**, 137-152

15. Lawson, E. E., Brown, E. R., Torday, J. S., Madansky, D. L. and Taeusch, H. W. (1978). The effect of epinephrine on tracheal fluid flow and surfactant efflux in fetal sheep. *Am. Rev. Respir. Dis.*, **118**, 1023-1026

16. Adamson, T. M., Boyd, R. D., Platt, H. S. and Strang, L. B. (1969). Composition of alveolar liquid in the fetal lamb. *J. Physiol. London*, **204**, 159-168

17. Cotton, C. U., Boucher, R. C. and Gatzy, J. T. (1988). Bioelectric properties and ion transport across excised canine fetal and neonatal airways. *J. Appl. Physiol.*, **65**, 2367-2375

18. Strang, L. B. (1974). Fetal and newborn lung. In Widdicombe, J. G. (ed.) *Respiratory Physiology*, pp. 31-65. (London: Butterworths)

19. Walters, D. V. and Olver, R. E. (1978). The role of catecholamines in lung liquid absorption at birth. *Pediatr. Res.*, **12**, 239-242

20. Barker, P. M., Brown, M. J., Ramsden, C. A., Strang, L. B. and Walters, D. V. (1988). The effect of thyroidectomy in the fetal sheep on lung liquid reabsorption induced by adrenaline or cyclic AMP. *J. Physiol. London*, **407**, 373-383

21. Bland, R. D. and Boyd, C. A. R. (1986). Cation transport in lung epithelial cells derived from fetal, newborn, and adult rabbits. *J. Appl. Physiol.*, **61**, 507-515

22. Olver, R. E., Schneeberger, E. E. and Walters, D. V. (1981). Epithelial solute permeability, ion transport and tight junction morphology in the developing lung of the fetal lamb. *J. Physiol. London*, **315**, 395-412

23. Cotton, C. U., Lawson, E. E., Boucher, R. C. and Gatzy, J. T. (1983). Bioelectric properties and ion transport of airways excised from adult and fetal sheep. *J. Appl. Physiol.*, **55**, 1524-1549

24. Cotton, C. U., Boucher, R. C. and Gatzy, J. T. (1988). Paths of ion transport across canine fetal tracheal epithelium. *J. Appl. Physiol.*, **65**, 2376-2382

25. Phipps, R. J., Torrealba, P. J. and Wanner, A. (1986). Development of ion transport and glycoprotein secretion in sheep trachea. *Fed. Proc.*, **45**, 1012

26. Olver, R. E. and Robinson, E. J. (1986). Sodium and chloride transport by the tracheal epithelium of fetal, new-born and adult sheep. *J. Physiol. London*, **375**, 377-390

27. Gatzy, J. T., Cotton, C. U., Boucher, R. C., Knowles, M. R. and Gowen, C. W. Jr. (1987). Development of epithelial transport in fetal and neonatal airways. In Walters. D. V., Strang, L. B. and Geubelle, F. (eds.) *Physiology of Fetal and Neonatal Lung*, pp. 77-87. (Lancaster: MTP)

28. McAteer, J. A., Cavanagh, T. J. and Evans, A. P. (1983). Submersion culture of the intact and fetal lung. *In Vitro*, **19**, 210-218

29. Krochmal, E. M., Ballard, S. T., Yankaskas, J. R., Boucher, R. C. and Gatzy, J. T. (1989). Volume and ion transport by fetal rat alveolar and tracheal epithelia in submersion culture. *Am. J. Physiol.*, **256**, F397-F407

30. Zeitlin, P. L., Loughlin, G. M. and Guggino, W. B. (1988). Ion transport in cultured fetal and adult rabbit tracheal epithelia. *Am. J. Physiol.*, **254**, C691-698

31. Gatzy, J. T. (1983). Mode of chloride secretion by lung epithelia. *Am. Rev. Respir. Dis.*,

127, S14–S16
32. Courtice, F. and Phipps, P. (1946). The absorption of fluids from the lungs. *J. Physiol. London,* **105**, 186–190
33. Kylstra, J. A. (1958). Lavage of the lung. *Acta Physiol Pharmacol. Nederl.,* **1**, 163–221
34. Taylor, A. E., Guyton, A. C. and Bishop, V. S. (1965). Permeability of the alveolar membrane to solutes. *Circ. Res.,* **16**, 353–362
35. Theodore, J., Robin, E., Gaudio, R. and Acevedo, J. (1975). Transalveolar transport of large polar solutes (sucrose, inulin, dextran). *Am. J. Physiol.,* **229**, 989–996
36. Taylor, A. E. and Gaar, K. A. Jr. (1970). Estimation of equivalent pore radii of pulmonary capillary and alveolar membranes. *Am. J. Physiol.,* **281** 1133–1140
37. Matthay, M. A., Landolt, C. C. and Staub, N. C. (1982). Differential liquid and protein clearance from alveoli of anesthetized sheep. *J. Appl. Physiol.,* **53**, 96–104
38. Berthiaume, Y., Staub, N. C. and Matthay, M. A. (1987). Beta adrenergic agonists increase lung liquid clearance in anesthetized sheep. *J. Clin. Invest.,* **79**, 335–343
39. Goodman, B. E., Kim, K. J. and Crandall, E. D. (1987). Evidence for active sodium transport across alveolar epithelium of isolated rat lung. *J. Appl. Physiol.,* **62**, 2460–2466
40. Effros, R. M., Mason, G. R., Sietsema, K., Silverman, P. and Hukkanen, J. (1987). Fluid reabsorption and glucose uptake in edematous rat lungs. *Circ. Res.,* **60**, 708–719
41. Basset, G., Crone, C. and Saumon, G. (1987). Significance of active ion transport in transalveolar water absorption: a study on isolated rat lung. *J. Physiol. London,* **384**, 311–324
42. Basset, G., Crone, C. and Saumon, G. (1987). Fluid absorption by rat lung *in situ*: pathways for sodium entry in the luminal membrane of alveolar epithelium. *J. Physiol. London,* **384**, 325–345
43. Basset, G., Bouchonnet, F., Crone, C. and Saumon, G. (1988). Potassium transport across rat alveolar epithelium: evidence for an apical Na^+-K^+ pump. *J. Physiol. London,* **400**, 529–543
44. Gatzy, J. T., Reuss, L. and Finn, A. L. (1979). Amphotericin B and K^+ transport across excised toad urinary bladder. *Am. J. Physiol.,* **237**, F145–F156
45. Gatzy, J. T. (1975). Ion transport across the excised bullfrog lung. *Am. J. Physiol.,* **228**, 1162–1171
46. Gatzy, J. T. (1983). Mode of Cl⁻ secretion by bullfrog lung. *Federation Proc.,* **42**, 1282
47. Gatzy, J. T. (1982). Paths of hydrophilic solute flow across excised bullfrog lung. *Lung Res.,* **3**, 147–161
48. Crandall, E. D. and Kim, K. J. (1981). Transport of water and solutes across bullfrog alveolar epithelium. *J. Appl. Physiol.,* **47**, 846–850
49. Ward, M. R. and Boyd, C. A. R. (1987). Analysis of ion and fluid transport across a vertebrate pulmonary epithelium studied *in vitro*. In Walters, D. V., Strang, L. B. and Geubell, F. (eds.) *Physiology of the Fetal and Neonatal Lung,* pp. 91–103. (Lancaster: MTP)
50. Kim, K. J. (1989). Active Na^+ transport across *Xenopus* lung alveolar epithelium. *FASEB J.,* **3**, A562
51. Ballard, S. T., Boucher, R. C. and Gatzy, J. T. (1986). Liquid clearance from alveoli of the perfused rat lung. *Federation Proc.,* **45**, 1013
52. Ballard, S. T. (1989). Solute and water transport by the pulmonary alveolar epithelium of the rat. *Dissertation*, University of North Carolina at Chapel Hill
53. Ballard, S. T. and Gatzy, J. T. (1990). Volume flow across the alveolar epithelium of the rat. *J. Appl. Physiol.,* (provisionally accepted)
54. Ballard, S. T. and Gatzy, J. T. (1987). Measurement of volume changes in liquid-filled alveoli of isolated perfused rat lung. *Federation Proc.,* **46**, 814
55. Ballard, S. T. and Gatzy, J. T. (1988). Microelectrode analysis of the transepithelial PD of liquid-filled alveoli of rat lung. *FASEB J.,* **2**, A706
56. Ballard, S. T. and Gatzy, J. T. (1990). Alveolar transepithelial PD and ion transport in adult rat lung. *J. Appl. Physiol.* (In press)
57. Knowles, M., Murray, G., Shallal, J., Askin, F., Ranga, V., Gatzy, J. and Boucher, R. (1983). Bioelectric properties and ion flow across excised human bronchi. *J. Appl. Physiol.,* **55**, 1542–1549
58. Knowles, M., Gatzy, J. and Boucher, R. (1983). Relative ion permeability of normal and

cystic fibrosis nasal epithelium. *J. Clin. Invest.*, **71**, 1410–1417
59. Goodman, B. E., Fleischer, R. S. and Crandall, E. D. (1983). Evidence for active Na⁺ transport by cultured monolayers of pulmonary alveolar epithelial cells. *Am. J. Physiol.*, **245**, C78–C83
60. Cott, G. R., Sugahara, K. and Mason, R. J. (1986). Stimulation of net active ion transport across alveolar type II cells monolayers. *Am. J. Physiol.*, **250**, C222–C227
61. Cheek, J. M., Kim, K. J. and Crandall, E. D. (1988). Bioelectric properties of tight monolayers of alveolar epithelial cells. *FASEB J.*, **2**, A707
62. Stutts, M. J. and Gatzy, J. T. (1979). Solute permeability of excised rat trachea. *Pharmacologist*, **21**, 155

12
Regulation of Ion Channels in Cultured Airway Epithelial Cells

J. J. SMITH

INTRODUCTION

The airways are lined with pseudostratified columnar epithelium comprised predominantly of goblet, ciliated and basal cells. The goblet cells secrete mucus into the airway lumen (apical surface) contributing, in part, to the protective layer of mucus that traps infectious, particulate and chemical hazards of inspired air. The more abundant ciliated cells are covered with apical membrane projections (cilia) which propel mucus toward the oropharynx (mucociliary clearance)[1]. Basal cells anchor goblet and ciliated cells to the epithelial extracellular matrix[2]; they may also function as stem cells replenishing goblet and ciliated cells sloughed from the epithelium[3]. Adjacent epithelial cells are attached along their apical–basolateral borders by an elaborate complex of adherence proteins, including tight junctions (zonula occludens)[4]. These junctions limit the diffusion of water, ions and neutral molecules between the cells (paracellular transport). This cellular architecture allows the epithelium to function as a barrier between the airway lumen and the submucosal fluid compartment.

A thin layer of fluid lies between airway mucus and the epithelium. This fluid allows just the tips of cilia to project into mucus, facilitating its transport when cilia beat. Fluid transport across the airway epithelium should influence the depth and composition of this periciliary fluid, and thus may modulate the efficiency of mucociliary clearance. In this way, transepithelial ion transport may be an essential component of this pulmonary defence mechanism.

Fluid moves across the airway in response to the transport of sodium (Na^+) and chloride (Cl^-) ions, i.e. water passively accompanies these ions. Na^+ absorption and Cl^- secretion (transcellular transport) involve the passage of Na^+ and Cl^- through ion channels present in the membrane of airway epithelial cells. This chapter focuses on the role of these channels in fluid transport across the respiratory epithelium.

THE TRANSPORT OF IONS ACROSS CELL MEMBRANES

Ions have a relatively high surface charge and thus a strong attraction to water (polar). This state of hydration accounts for a relative lack of ion permeability of phospholipid membranes, i.e. ions do not readily cross cell membranes. Several integral membrane proteins transfer ions across cell membranes; these include pumps, carriers and channels.

The primary pump used for ion transport across the respiratory epithelium is basolateral Na^+-K^+-ATPase which exchanges Na^+ and K^+ ions against their chemical gradients using ATP as the source of energy. This pump maintains the high K^+ and low Na^+ concentrations in the cell; these transmembrane chemical gradients act as driving forces for ion transport by carriers and channels.

In contrast to active transport by Na^+-K^+-ATPase, carriers and channels passively transfer ions across membranes. The mechanisms of ion translocation by carriers and channels are not precisely known. Carriers function as though binding sites alternate between each side of the phospholipid membrane, allowing ions to be passively carried across the membrane. Channels, on the other hand, are pore-forming macromolecules which span the membrane, allowing ions to diffuse freely through the open pore from either surface of the membrane. Regardless of the mode of action, transport is passive in both cases, i.e. ions are transported down an electrical or chemical gradient.

Carriers simultaneously transfer several ions or molecules across a cell membrane (coupled transport); they may be transfered in the same or opposite direction. Driven by the steep transmembrane Na^+ gradient, a basolateral membrane co-transporter transfers Na^+ and Cl^- into airway epithelial cells. This activity does not result in the transfer of charge across the membrane, i.e. transport is electrically neutral[5].

In contrast to this co-transporter, ion channels transfer charge across the membrane (electrically conductive); thus the rate of transport is driven by both electrical and chemical gradients. Furthermore, ion channels may be open (allowing transport) or closed (blocking transport). Factors that increase the amount of time channels are open, e.g. intracellular second messengers, may increase ion flux through the channel and thus may act as regulators of transport.

Pumps, carriers and channels are distributed asymmetrically between apical and basolateral membranes of epithelial cells. This polar distribution is essential for the vectorial movement of ions across the epithelium. For example, simultaneous ion influx at the basolateral membrane and efflux at the apical membrane leads to secretion into the airway lumen. Identical transport activity at both apical and basolateral membranes should modify intracellular osmolarity or cell volume rather than contribute to transepithelial flux. Yet, in order to maintain transepithelial absorption (or secretion) and simultaneously preserve intracellular osmolarity and membrane voltage, the transport activity at both membranes must be co-ordinated. While the mechanisms used to co-ordinate transport at both membranes have not been established, regulatory factors for individual

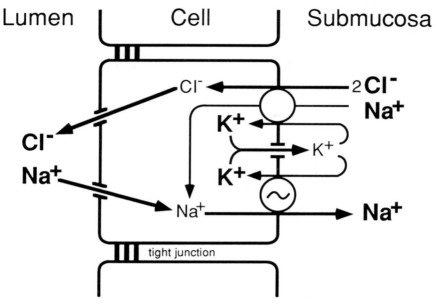

Figure 12.1 Model of ion transport across the airway epithelium (see text)

transporters have been identified. A model[6] of these transport pathways in the airway epithelium is depicted in Figure 12.1.

A TRANSPORT MODEL FOR AIRWAY EPITHELIUM

K^+ transport

Basolateral Na^+-K^+-ATPase maintains intracellular $[K^+]$ above its electrochemical equilibrium[7-9]. This gradient drives K^+ out of the cell when basolateral K^+ channels open. Potassium is thereby recycled across the basolateral membrane by the activity of Na^+-K^+-ATPase and K^+ channels.

Na^+ absorption

Na^+-K^+-ATPase also keeps intracellular $[Na^+]$ below its electrochemical equilibrium by pumping Na^+ into the submucusal fluid compartment. This transmembrane chemical gradient drives Na^+ into the cell when apical membrane Na^+ channels open. Thus, transport via apical Na^+ channels and basolateral Na^+-K^+-ATPase leads to absorption of Na^+ from the airway lumen.

Cl⁻ secretion

Cl⁻ enters airway epithelial cells at the basolateral membrane via furo-semide-sensitive co-transport. This couples the influx of Na^+, and probably K^+, to Cl⁻ influx (Na-K-2Cl transport)[6,8-13]. Driven by the steep trans-membrane Na^+ gradient, Cl⁻ accumulates above its electrochemical equili-brium*. Cl⁻ exits the cell at the apical membrane when Cl⁻ channels open. Thus, the activity of basolateral Na^+-coupled co-transport and apical Cl⁻ channels leads to Cl⁻ secretion into the airway lumen**.

ELECTRICAL PROFILE OF THE AIRWAY EPITHELIUM

The accumulation of intracellular $[K^+]$ above electrochemical equilibrium drives K^+ out of the cell when basolateral K^+ channels open. This conductive efflux causes a build up of negative charge along the inner surface of the cell

*For Na-K-2Cl co-transport (electrically neutral), the net electrochemical force ($\Delta\tilde{\mu}_{Na-K-2Cl}$) is the sum of the transmembrane chemical (μ) gradients for each ion:

$$\Delta\tilde{\mu}_{Na-K-2Cl} = \mu_{Na} + \mu_K + 2\,\mu_{Cl}$$
$$= RT\cdot\ln\frac{[Na]_{out}}{[Na]_{in}} + RT\cdot\ln\frac{[K]_{out}}{[K]_{in}} + 2\,RT\cdot\ln\frac{[Cl]_{out}}{[Cl]_{in}}$$
$$= RT\cdot\ln\left[\frac{[Na]_{out}\cdot[K]_{out}\cdot[Cl]^2_{out}}{[Na]_{in}\cdot[K]_{in}\cdot[Cl]^2_{in}}\right]$$

Net ion flux is zero when the electrochemical potential is zero. Because log 1 = zero, $\Delta\tilde{\mu}$ is zero when:

$$1 = \frac{[Na]_{out}\cdot[K]_{out}\cdot[Cl]^2_{out}}{[Na]_{in}\cdot[K]_{in}\cdot[Cl]^2_{in}}$$

$$[Cl]^2_{in} = \frac{[Na]_{out}\,[K]_{out}\,[Cl]^2_{out}}{[Na]_{in}\,[K]_{in}}$$

With extracellular concentrations of Na^+, K^+, and Cl⁻ at 135, 5, and 140 mmol/L⁻¹, respectively, and cytosolic concentrations for Na^+ and K^+ of 20 and 150 mmol/L⁻¹ respectively[15], the predicted $[Cl]_{in}$ is 66 mmol/L⁻¹ (corresponding to Cl⁻ activity near 50 mmol/L⁻¹).
**When Cl⁻ channels open, diffusion through the channels is determined by the sum of the electrical (Ψ) and chemical (μ) gradients across the cell membrane:

$$\Delta\mu_{Cl} = \Psi_{Cl} + \mu_{Cl}$$

Net Cl⁻ flux is zero when the electrochemical potential ($\Delta\mu_{Cl}$) is zero. This occurs when the electrical potential (Ψ) driving Cl⁻ out of the cell is equal, and opposite in direction, to the chemical potential (μ) driving Cl⁻ into the cell:

$$-\Psi_{Cl} = \mu_{Cl}$$

$$-zF\cdot E_m = RT\cdot\ln\frac{[Cl]_{out}}{[Cl]_{in}}$$

With extracellular [Cl⁻] at 140 mmol/L⁻¹ and apical membrane potential (V_a) of −40 mV[15], predicted $[Cl]_{in}$ is 29 mmol/L⁻¹ (Cl⁻ activity near 22 mmol/L⁻¹). Thus, when $[Cl]_{in}$ is greater than 29 mmol/L⁻¹ (above the electrochemical potential), Cl⁻ efflux would be favoured through the apical membrane channels.

membrane (electrical potential difference). At electrochemical equilibrium, the chemical gradient driving K^+ out of the cell is equal in magnitude to the electrical gradient keeping K^+ in the cell. This is the reversal potential (E_{rev}) for K^+ and is represented by the Nernst equilibrium potential:

$$E_{rev} = \frac{RT}{zF} \cdot \ln \frac{[K^+]_{out}}{[K^+]_{in}} \tag{1.1}$$

where in and out refer to the $[K^+]$ inside and outside the cell[14].

Potassium channels are the predominant conductive pathway for the basolateral membrane. Increasing submucosal $[K^+]$ depolarizes the basolateral membrane (V_b), whereas changes in submucosal $[Na^+]$ or $[Cl^-]$ do not change V_b[15]. Furthermore, submucosal addition of a K^+ channel blocker, Ba^{2+}, also depolarizes V_b indicating that K^+ channel conductance is essential for maintaining V_b. The magnitude of V_b is determined by the relative conductive permeability of the basolateral membrane to K^+, Na^+ and Cl^- ions. This can be represented by the Goldman–Hodgkin–Katz flux equation:

$$E = \frac{RT}{zF} \cdot \ln \frac{P_K[K]_{out} + P_{Na}[Na]_{out} + P_{cl}[cl]_{out}}{P_K[K]_{in} + P_{Na}[Na]_{in} + P_{cl}[Cl]_{in}} \tag{1.2}$$

where P is the permeability coefficient for each ion[14]. Because of the K^+-selective conductance of the basolateral membrane (permeability coefficient is low for Na^+ and Cl^- conductance), V_b approaches the equilibrium reversal potential (E_{rev}) for potassium.

Apical membrane voltage (V_a) is also related to its relative permeability to these ions. When Na^+ and Cl^- channels are closed, the apical membrane resistance (R_a) is very high and there is minimal absorption or secretion[16]. Under these conditions, R_a is much greater than basolateral resistance (R_b) and the apical membrane potential (V_a) is influenced by basolateral membrane K^+ conductance and a minimal degree of apical Na^+ conductance. When apical channels open, R_a decreases (conductive permeability increases) and ions are driven through the channels in response to Na^+ and Cl^- electrochemical gradients. This depolarizes the apical membrane and shifts V_a toward the equilibrium reversal potential for Na^+ (absorption) or Cl^- (secretion). The decrease in R_a may be accompanied by a decrease in R_b (increased K^+ conductance); thus the electrical driving force for Na^+ absorption and Cl^- secretion may be maintained[16]. A model of the electrical circuit for the airway epithelium is depicted in Figure 12.2.

PARACELLULAR TRANSPORT

Water, ions and neutral molecules also cross the epithelium by passive diffusion between adjacent epithelial cells (paracellular transport). Transport through this pathway is restricted by the tight junctions between epithelial cells.

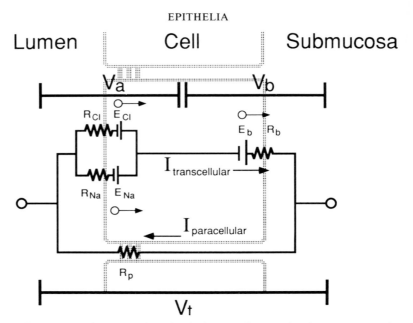

Figure 12.2 Model of the electrical circuit for the airway epithelium. Parallel resistors represent apical membrane Na⁺ and Cl⁻ channels; parallel batteries (electromotive forces) represent the electrochemical gradients for these ions. A compound resistor and electromotive force represent the conductive pathways across the baso-lateral membrane. The electromotive forces direct the flow of current from the lumen to the submucosal compartment (transcellular transport); paracellular current flows in the opposite direction (counter ion transport)

The relative permeabilities of the basolateral and apical membranes usually result in a more negative V_b compared with V_a; thus a transepithelial potential (V_t) exists with the lumen negative relative to the submucosal fluid compartment (Figure 12.3A). This electrical gradient (V_t) leads to the passive diffusion of Na⁺ ions toward the lumen and Cl⁻ ions toward the submucosal compartment through the paracellular pathway (counter ion transport); this pathway runs parallel to transcellular transport. Resistance of the paracellular pathway does not change appreciably with stimulation of secretion[10,16,18]. Thus, when absorption and secretion are minimal, paracellular resistance is less than transcellular resistance; this relationship in resistance is inverted when secretion is stimulated[16]. While paracellular resistance does not change with stimulation of secretion, an increase in transepithelial potential difference may occur. With Cl⁻ secretion, the increase in Cl⁻ conductive permeability depolarizes the apical membrane (shifts the membrane potential toward E_{rev} for Cl⁻). A concomitant increase in basolateral K⁺ conductive permeability hyperpolarizes the basolateral membrane[16]. This results in a greater difference between V_a and V_b, i.e. an increase in the transepithelial electrical gradient, V_t. By increasing V_t, the transport of Na⁺ from the submucosal to mucosal compartments through the paracellular pathway may be enhanced leading to net NaCl secretion.

A. Normal

B. Cystic Fibrosis

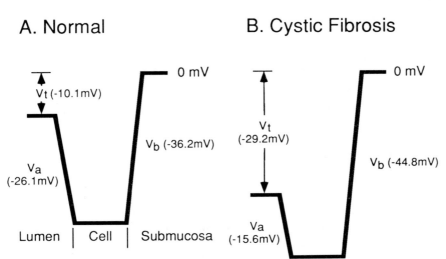

Figure 12.3 The electrical profile across **A**. normal and **B**. cystic fibrosis airway epithelium. V_a, V_b and V_t represent the apical, basolateral and transepithelial membrane voltages, respectively. V_t is greater in CF epithelia (see text). Values from reference 96

STRUCTURAL FEATURES OF ION CHANNELS

Ion channels are glycoprotein macromolecules that function as trans-membrane pores[19-22]. While the amino acid sequence of some channels from excitable tissue have been identified[22-26], the structural characterization of epithelial channels has lagged behind the progress in excitable tissue because of a lack of high-affinity ligands for epithelial channels. Therefore, differences in the structure of ion channels from these two tissue sources remain unknown. Certainly the channel properties, response to antagonists, and physiological functions differ between channels of epithelial and excitable tissue.

Na$^+$ channels have been isolated from amphibian and bovine renal epithelial cells using the binding properties of methylbromoamiloride[27]. Liposomes containing the partially purified channel demonstrate Na$^+$ uptake capability[28]. This channel is nearly 700 kDa in size with disulphide bonds joining several polypeptide subunits ranging in size from 55 to 315 kDa. The epithelial Na$^+$ channel blocker, amiloride, binds exclusively to the 150 kDa subunit[29]. The amino acid sequences of the channel subunits have not been determined; however, the development of antibodies to the renal Na$^+$ channel may hasten further characterization of the channel. Antibody cross-reactivity between bovine and amphibian epithelial Na$^+$ channels[30] indicates a degree of homology between channels from different species, and may facilitate identification of Na$^+$ channels from other epithelial sources.

Two anion channels were recently partially purified from trachea membranes[31]. Based on the structure of a high-affinity ligand for chloride channels, an affinity resin column was developed to purify proteins from solubilized tracheal membrane preparations. The proteins eluted from the affinity column were reconstituted into liposomes and fused with planar lipid bilayers. Two channels were identified in the membrane preparations, but they have not been sufficiently purified to identify their structural features. The conductive properties of these channels will be discussed later.

CHANNEL PROPERTIES

The properties of ion channels in cultured cells are frequently investigated with the patch-clamp technique[32]. With this technique, a small glass pipette (~1 µm in diameter) is placed adjacent to the cell membrane where slight negative pressure applied through the pipette seals a patch of membrane to the pipette (Figure 12.4). The formation of a high-resistance seal (≥ 10 GΩ) between the membrane and the glass pipette isolates the membrane patch such that ion flux (current) in or out of the pipette passes through the patch of membrane. The conductive property of ion channels in the membrane patch allow conductive ion flux to be detected. The rapid transitions between open and closed states of the channels are distinguished by unitary (quantum) changes in current amplitude (Figure 12.5).

Several patch-clamp configurations are used to investigate the properties of ion channels. Single channel events are studied in cell-attached membrane patches as well as cell-free patches where the membrane patch has been excised from the rest of the cell (Figure 12.4). In addition, large populations of channels within a single cell can be investigated by removing the membrane patch within the pipette (transient negative pressure) while maintaining the high-resistance seal between the membrane and pipette (Figure 12.4). The cummulative conductive effect of many channels produces the 'whole cell' current (net ion flux in and out of the cell).

With the patch-clamp technique, several methods are used to identify ion channels. First, ion substitutions in the pipette or bathing solutions allow identification of the ions responsible for the current and the relative ion selectivity of channels. Second, by varying transmembrane voltage the channel conductance may be determined (the slope of the current vs voltage plot). Third, the effect of transmembrane voltage on the probability of channels being open (P_o) can be determined (voltage-dependence). Finally, channels may be identified by their response to specific agonists or antagonists.

As previously noted, ion channels can be either open or closed. Inactivated channels remain in the closed state; activated channels alternate between open and closed states. Some channels are activated by transmembrane voltage, i.e. channel activation is determined by the voltage potential across the cell membrane. Ion channels in membrane patches excised from epithelial cells can be activated by depolarization (voltage activation)[33-35]. However, this property may be physiologically more

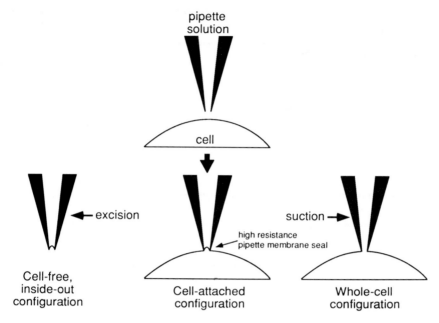

pipette
solution

cell

← excision

suction →

high resistance
pipette membrane seal

Cell-free,
inside-out
configuration

Cell-attached
configuration

Whole-cell
configuration

Figure 12.4 Recording configurations for the patch-clamp technique. With the pipette adjacent to a cell, gentle suction allows the formation of a high-resistance (GΩ) seal between the pipette and cell membrane. Single channel activity is recorded with cell-attached and cell-free configurations. The membrane patch may be disrupted with transient negative pressure so that the 'whole cell' currents (the cumulative activity of many channels within a cell) can be recorded

relevant in excitable tissue where shifts in membrane potential provide unique cellular functions, e.g. propagation of a nerve impulse. Channels may also be ligand-activated, i.e. specific ligands or second messengers activate the channel. For example, Cl⁻ channel activation in airway epithelial cells is regulated by the intracellular concentration of cAMP. This second messenger presumably regulates phosphorylation of the channel or a membrane-associated regulatory protein[33,34]. Phosphorylation appears to be a common mechanism of regulating channels in many cells, including excitable cells[36,37].

With the patch-clamp technique, ion channels have been identified in virtually all cells including cultured airway epithelial cells (Table 12.1). This technique is generally limited to the investigation of cultured cells because high-resistance seals are more easily obtained with these cells. Theoretically, the expression or properties of ion channels may be modified by culture conditions, such as the soluble factors in culture media, the type of matrix support, the degree of cell confluence and differentiation, cell age, and the cell source (primary isolate, explant or transformed cell line). However, the transport properties of cultured airway cells are consistent with the properties of native tissue. Cultured respiratory epithelial monolayers and

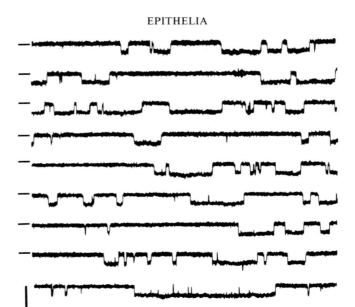

Figure 12.5 Single channel activity from a cell-free, inside-out patch excised from a cultured CF airway epithelial cell. Pipette and bathing solutions are identical (in mmol L⁻¹): Na gluconate 140, $CaCl_2$ 2, $MgCl_2$ 2, HEPES 10, pH=7.3; membrane potential is −40 mV in reference to the pipette solution (near physiological apical membrane voltage). The closed channel current level is marked at the beginning of each tracing. Channel openings are represented by downward deflections in (inward) current (the flow of cations toward the cytosolic surface of the membrane or anions in the opposite direction). Calibration marks are 1 pA and 1 s; filtered at 500 Hz

isolated cells have absorptive and secretory properties similar to the native tissue; the responses to numerous secretagogues and antagonists are qualitatively comparable. In addition, ion transport alterations found in cystic fibrosis tissue are also expressed in cultured CF cells[13,38–42]. This suggests that, while culture conditions may influence epithelial transport properties, heritable differences in transport appear to be maintained in cultured airway cells.

ION CHANNELS IN CULTURED AIRWAY CELLS

Cation channels

Potassium channels have been identified in cultured canine[43] and human[39] airway epithelial cells. The channel conductance is near 20 pS (picosiemens) at 0 mV with an inwardly rectifying current–voltage relationship (conductance is non-linear and greater with inward currents). This channel is highly selective[44] for K^+ over Na^+, but equally conductive to K^+ and Rb^+. The channel is not voltage-activated, and its activity is blocked by Ba^{2+} and

264

charybdotoxin but not by tetraethylammonium[44]. Potassium channels with very high conductance have also been identified. In inside-out membrane patches from human airway cells, Kunzelmann and co-workers[45] observed K^+ channels with conductance near 300 pS while bathed in symmetrical KCl solutions (Table 12.1).

Single-channel recordings of Na^+ channels in human airway epithelial cells have been reported only recently. Two such cation channels have been identified[46,47]. One channel has a linear conductance of 7–8 pS. This channel is highly selective for Na^+ over K^+ and Cl^- ions, and its activation is inhibited by amiloride. A larger channel (with a linear conductance of 18–20 pS) is non-selective for cations and activation is not inhibited by amiloride. Both channels appear to be inhibited by choline and caesium[47].

Anion channels

Outwardly rectifying Cl^- channels have been identified in cultured canine and human airway cells (conductance is greater with outward currents). Conductance ranges from 25–50 pS at 0 mV in human and canine cells (Table 12.1). This channel is moderate to highly selective for Cl^- ions, and channel activity increases in response to agonists that increase intracellular levels of cAMP[33,34,39,40,48]. Stilbene, diphenylamine 2-carboxylate and related derivatives inhibit channel activation[39,48,49].

Cl^- channels with a linear conductance of 20 pS have also been reported in human airway cells[40,50]. In addition, channels with large conductance (250–400 pS) have been observed in human[50] and canine[51] airway cells and rat alveolar (type II) epithelial cells[52].

As previously noted, two anion channels have been purified from trachea membrane preparations using an affinity resin column[31]. The proteins isolated with this technique were reconstituted into liposomes and fused with planar lipid bilayers. One channel had a linear conductance near 100 pS with a high degree of selectivity for Cl^- over K^+. The other channel was less selective for Cl^- and had a linear single-channel conductance of 400 pS with at least one subconductance state. Both channels were not inhibited by stilbene or anthranilic acid analogues[31].

Valdivia and colleagues[53] purified apical-membrane vesicles from bovine tracheal epithelium and identified several anion channels using planar bilayer techniques. One channel was calcium independent, voltage dependent, and highly selective for Cl^- over Na^+. The conductance was 71 pS. This channel was sensitive to stilbene derivatives. In addition, incubation with ATP and the catalytic subunit of cAMP-dependent protein kinase (PKA) increased channel activity without changing the current–voltage curve[53].

There are some differences between the properties of channels isolated from membrane preparations and channels identified in patch-clamp studies. It is possible that the techniques used to isolate and purify channel proteins separate the conductive portion of the channel from other cellular components that influence channel properties. In addition, the composition of the planar bilayer may differ sufficiently from native membrane so that channel properties become modified. On the other hand, the nature of the

Table 12.1 Ion channels in cultured respiratory epithelial cells

Source	Selectivity	Conductance (pS)		Activation (Inhibitors)	Reference
		Cation channels			
Canine and human tracheal	K^+	20	R	α_1-Adren Ca^{2+}	39,44
Human nasal	K^+	300	L	Ca^{2+}	45
	Na^+	7	L	(Ba^{2+}, TEA, quinidine, lidocaine) (amiloride, choline, caesium)	46,47,50
	Na^+	8	L	(amiloride)	47,51
	cation	18	L	(choline, caesium)	48,51
		Anion channels			
Canine tracheal	Cl^-	30–50	R	β-Adrenergic (9-AC and DPC)	48,51
Human nasal and tracheal	Cl^-	20	L	β-Adren/cAMP	40,50
	Cl^-	25–50	R	β-Adren/cAMP (9-AC and DPC)	33,34,39 40,49,50
Transformed human nasal	Cl^-	30	R	β-Adren/cAMP	41

R=rectifying conductance
L=linear conductance

patch-clamp procedure may also distort the channel environment and, thereby, modulate channel properties. While both techniques provide unique approaches to the characterization of ion channels, comparison with native tissue remains essential.

General mechanisms of channel regulation

Many physiological and pharmacological agonists modulate ion transport across the airway epithelium, e.g. isoprenaline (isoproterenol) and bradykinin. Their actions are mediated by receptors which increase the concentration of intracellular second messengers, such as cyclic nucleotides, Ca^{2+}, or phospholipid derivatives[54,55]. These second messengers directly or indirectly regulate the activity of the pumps, carriers, and channels transporting ions across both apical and basolateral membranes.

This general mechanism of receptor-mediated transport regulation enables extracellular signals to modulate transport across the epithelium. Agonist–receptor interactions at one membrane increase transport activity at both membrane surfaces. For example, numerous secretagogues increase the intracellular concentration of cAMP leading to activation of apical Cl^- channels. The electrochemical gradient across the apical membrane drives Cl^- toward the airway lumen, i.e. Cl^- is secreted (Figure 12.1). However, to sustain this secretory response an increase in the activity of the basolateral Na-K-2Cl co-transporter, K^+ channels and Na^+-K^+-ATPase may also be necessary. Increasing Na^+-coupled co-transport maintains intracellular $[Cl^-]$ above its electrochemical equilibrium; activation of K^+ channels maintains the electrical driving force for Cl^- efflux[16]. Furthermore, Na^+-K^+-ATPase activity maintains the low intracellular $[Na^+]$ since Cl^- influx at the basolateral membrane is coupled to Na^+ entry[56]. The requirement for these transporters to sustain Cl^- secretion is supported by the inhibitory effects of bumetanide, Ba^{2+}, and ouabain which block co-transport, K^+ channel and Na^+-K^+-ATPase activities, respectively.

While the regulation of these transporters may be coupled at least during sustained secretion, the nature of this linkage and the messengers involved are unknown. Difficulties in isolating and reliably measuring the activity of individual transporters limits our understanding of their regulation. The conductive properties of channels, especially the rapid transition between conductive (open) and non-conductive (closed) states, make the patch-clamp technique useful for investigating channel regulation. The mechanisms of regulating ion channel activation in airway cells will now be considered in more detail.

CHANNEL REGULATION IN AIRWAY CELLS

K+ channels

Welsh and McCann[43] used the patch-clamp technique to investigate activation of K^+ channels in cultured canine airway cells. The addition of adrenaline and selective α_1-adrenergic agonists increased both cytosolic [Ca^{2+}] (measured by quin-2 fluorescence) and, in cell-attached patches, the probability of K^+ channels being open. The calcium ionophore, A23187, also increased cytosolic [Ca^{2+}] and the opentime probability of these K^+ channels. Furthermore, exposing the cytosolic surface of cell-free membrane patches to Ca^{2+} increased K^+ channel activation. Finally, this K^+ channel activity in airway cells was inhibited by the extracellular addition of charybdotoxin[44] (an antagonist of Ca^{2+}-activated K^+ channels)[57]. This indicates that cytosolic Ca^{2+} is a second messenger for K^+ channel activation in airway cells.

There is also evidence that basolateral K^+ conductance in airway cells may be regulated by a Ca^{2+}-independent mechanism. This has been investigated using monolayers of cultured airway cells mounted in Ussing chambers. In these chambers, the epithelial layer separates the luminal and submucosal bathing solutions so that net transepithelial transport of Na^+ and Cl^- ions is measured as short-circuit current (I_{sc}). In the presence of amiloride to block Na^+ conductive transport, isoprenaline[58] stimulates sustained Cl^- secretion ($>$ 20 minutes). As previously noted, submucosal addition of a K^+ channel blocker, Ba^{2+}, blocks this secretory response[58], indicating that K^+ channels are required for sustained Cl^- secretion.

Several findings suggest that the prolonged increase in K^+ conductance is Ca^{2+}-independent:

(1) Isoprenaline stimulates a transient ($<$100 s) increase in cytosolic [Ca^{2+}] while Cl^- secretion and the increase in basolateral K^+ conductance are sustained[5,9,58].

(2) Charybdotoxin blocks only the initial ($<$100 s) component of secretion (Figure 12.6), while Ba^{2+} blocks all of the secretory response[58].

(3) In cells depleted of intracellular Ca^{2+} (measured by fura-2 fluorescence during continuous exposure to A23187)[60], the transient Cl^- secretory response is blocked (Ca^{2+}-activated K^+ channels), yet the sustained secretory response is maintained[58].

Thus, sustained activation of K^+ channels by isoprenaline does not require a prolonged increase in cytosolic [Ca^{2+}]. In another study, Clancy and co-workers[61] used I^- efflux from cultured airway epithelial cells to assay Cl^- channel activity. (I^- is not transported by the Na-K-2Cl co-transporter[62] yet Cl^- channels are I^- conductive.) They found that by varying [K^+] in the bathing solution in the presence of valinomycin to clamp the membrane voltage, cAMP-mediated I^- efflux (isoprenaline) was voltage dependent. That is, cAMP-mediated efflux was enhanced by hyperpolarization as would be expected with K^+ channel activation in intact cells. While isoprenaline-stimulated I^- efflux was enhanced by K^+ channel activation, the maximum effect of isoprenaline on I^- efflux was not lower in Ca^{2+}-depleted cells[61]. These findings suggest that at least some K^+ channel activation by isoprenaline is independent of cytosolic [Ca^{2+}]. These studies are consistent with two mechanisms of activating K^+ channels: one Ca^{2+}-dependent and another Ca^{2+}-independent.

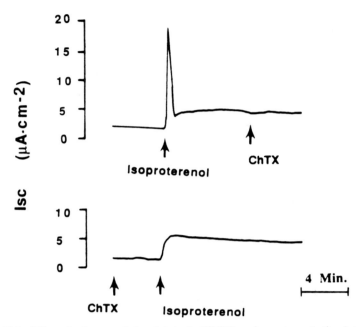

Figure 12.6 Effect of submucosal charybdotoxin (ChTX) on isoproterenol-stimulated I_{sc} (Cl secretion). In the top tracing, isoproterenol-stimulated I_{sc} is not decreased by subsequent addition of ChTX. In the bottom tracing, prior addition of ChTX blocks the transient peak in isoproterenol-stimulated I_{sc}. (Figure from reference 58)

Na⁺ channels

The regulation of Na^+ absorption by canine airway cells has been investigated by Cullen and Welsh[63] using cultured monolayers mounted in Ussing chambers. With monolayers bathed in Cl⁻-free solution, Na^+ absorption was measured as I_{sc}. The baseline rate of Na^+ absorption decreased in the presence of indomethacin, suggesting a role for prostaglandins in the regulation of Na^+ absorption. Furthermore, absorption increased in response to prostaglandin E_2, isoprenaline, and 2-chloroadenosine (agonists that increase intracellular levels of cAMP). Finally, Na^+ absorption increased in response to membrane-permeable cAMP analogues indicating that cAMP may act as a second messenger for Na^+ absorption under Cl⁻-free conditions[63]. Direct evidence for Na^+ channel activation by cAMP in airway cells has not been reported. Cyclic-AMP can increase Na^+ permeability in other epithelia[64-66], but its mechanism of action has not been characterized. How cAMP-mediated agonists stimulate Cl⁻ secretion under most physiological conditions and Na^+ absorption under Cl⁻-free conditions is unknown.

On a more chronic level, Na^+ absorption in cultured airway cells can be stimulated by the mineralocorticoid, aldosterone. Two days after monolayers were exposed to aldosterone for a 24-hour period, amiloride-sensitive I_{sc} was 50% greater than monolayers not exposed to aldosterone[63]. This

Table 12.2 Airway epithelial cell Cl⁻ secretagogues

Agonist	Secondary messenger(s)	Tissue/cells	Reference
Epinephrine	cAMP, Ca^{2+}	Canine and human	54,58,75
Isoproterenol	cAMP, Ca^{2+}	Canine and human	54,75
PGE_2	cAMP	Canine	17
Neurokinin A and B	cAMP	Canine	90
Bradykinin	PGE_2, cAMP, Ca^{2+}	Canine	55,91
Substance P	cAMP, Ca^{2+}	Canine	92
Leukotrienes C_4 and D_4	PGs	Canine	93
A23187	Ca^{2+} (PGE_2)	Canine and human	70,71,76
$PGF_{2\alpha}$	Unknown	Canine	17
Adenosine	cAMP	Canine	94
Eosinophil MBP	PGE_2	Canine	95

effect should be mediated by mineralocorticoid receptors, but the specific mechanism of aldosterone-stimulated absorption, and its physiological significance in the airway, is unclear.

cAMP-mediated Cl⁻ channel regulation

Many agonists stimulate an increase in intracellular [cAMP] in airway epithelial cells along with an increase in Cl⁻ secretion: β-adrenergic agonists, prostaglandin E_2, bradykinin, sustance P, and neurokinins A and B (Table 12.2). In addition, membrane-permeable cAMP analogues mimic this secretory effect[39,40]. These studies indicate that cAMP acts as an intracellular messenger for receptor-mediated Cl⁻ secretion.

The production of prostaglandin E_2 by airway cells appears to regulate the baseline rate of Cl⁻ secretion. First, exogenous PGE_2 increases intracellular cAMP levels and stimulates secretion. Second, endogenous PGE_2 production correlates directly with intracellular cAMP levels in airway epithelia. Third, indomethacin decreases PGE_2 production, intracellular [cAMP] and baseline Cl⁻ secretion in canine airway cells[17,54]. Furthermore, while indomethacin reduces baseline Cl⁻ secretion, it does not reduce the secretory response[17] to exogenous PGE_2. These findings indicate that endogenous PGE_2 may regulate, at least partially, intracellular cAMP concentration and thus the baseline rate of Cl⁻ secretion in airway epithelia.

The role of cAMP in channel regulation has been investigated more directly in cultured airway cells using the patch-clamp technique. In cell-free patches, cAMP-dependent protein kinase (PKA) activates Cl⁻ channels in the presence of cAMP and ATP[33,34]. The mechanism of action is depicted in Figure 12.7. PKA exists predominantly in inactive form with two regulatory and two catalytic subunits combined. Binding of four cAMP molecules to the regulatory subunits causes the release of the catalytic subunits which have phosphorylating activity. In airway cells, the catalytic subunits presumably phosphorylate the channel itself, or a membrane-associated protein coupled to channel activation. This mechanism of activation

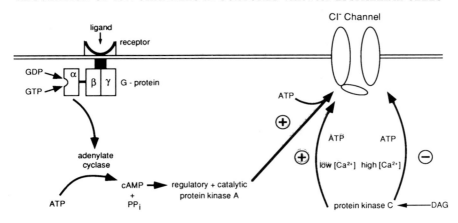

Figure 12.7 Model of cAMP-mediated Cl⁻ channel activation. Agonist binding at the extracellular receptor site induces a conformational change in the guanine nucleotide binding regulatory protein (composed of α, β, and γ subunits) such that GTP replaces GDP binding to the α-subunit. The GTP-α complex dissociates from the β and γ subunits and stimulates adenylate cyclase activity. The concentration of cAMP increases with subsequent binding to the regulatory subunit of protein kinase A. This releases catalytic subunit which activates the channel by phosphorylation[33,34]. Intrinsic GTPase activity of the α-subunit hydrolyses GTP to GDP, whereby GDP-α complex recombines with β and γ G-protein subunits. Alternatively, in the presence of low cytosolic [Ca²⁺], diacylglycerol binding at a regulatory site of protein kinase C may also activate Cl⁻ channels[69,78]. PKC may inactivate Cl⁻ channels in the presence of elevated cytosolic [Ca²⁺][69]. To date, the coupling of PKC to receptor-mediated Cl⁻-secretion has not been reported in airway epithelium

suggests that the channel, or regulatory protein, exists in phosphorylated and dephosphorylated states corresponding to activated and inactivated states of the channel, respectively. The probability of Cl⁻ channels being open may depend on a balance of competing kinase and phosphatase activities[36,67,68]. Receptor-mediated increases in [cAMP] may shift this balance toward the phosphorylated state, thereby stimulating Cl⁻ secretion.

Other kinases may also regulate Cl⁻ channel activation. Li and co-workers[69] investigated the effects of protein kinase C (PKC) on channel activation. Cell-free membrane patches excised from cultured canine and human airway cells were exposed to PKC in the presence of ATP and diacylglycerol. When [Ca²⁺] was <10 nmol L⁻¹, PKC activated Cl⁻ channels in the majority of cell-free patches. In control patches (no PKC), Cl⁻ channels were activated in none of the patches. The presence of Cl⁻ channels was confirmed in electrically silent patches using membrane depolarization to activate the channels. Neither PKC nor ATP alone activated the channels. These results indicate that PKC can activate apical Cl⁻ channels in the presence of low [Ca²⁺].

The effect of PKC in the presence of higher Ca²⁺ concentrations differs. With [Ca²⁺] ≥1 µmol L⁻, PKC inactivated Cl⁻ channels. Furthermore, after exposure to PKC at high [Ca²⁺], the channels could not be activated by depolarization. Yet, in control patches exposed to high [Ca²⁺] (without PKC), the channels were activated by depolarization. This indicates that high $[Ca^{2+}]$ alone does not interfere with activation by depolarization, while high $[Ca^{2+}]$ in the presence of PKC inhibits channel activation.

PKC also inactivates Cl^- channels activated by prior depolarization[69]. At high $[Ca^{2+}]$, PKC inactivated Cl^- channels in most patches, while at low $[Ca^{2+}]$ the channels were not inactivated by PKC. As mentioned above, high $[Ca^{2+}]$ alone (no PKC) did not inactivate Cl^- channels. These findings indicate that PKC may have both activating and inactivating effects on Cl^- channels in airway cells depending on the local Ca^{2+} concentration. Channel activation by PKC requires low $[Ca^{2+}]$; inactivation by PKC requires a higher $[Ca^{2+}]$.

This divergent effect of PKC is consistent with previous reports that phorbol esters (potent PKC activators) initially stimulate Cl^- secretion in airway cells, and later reduce secretion[70]. Phorbols also inhibit subsequent secretory responses to cAMP-mediated secretagogues suggesting that channel inactivation by PKC is not reversed by PKA[70,71].

While physiological relevance in airway cells has not been established, these reports suggest that PKC may have a dual role in Cl^- channel regulation via two phosphorylation sites on the channel or regulatory protein. Phosphorylation at one site activates Cl^- channels; phosphorylation at a different site inactivates the channels. However, while receptor-mediated modulation of Cl^- secretion by PKC may be physiologically relevant in the intestinal epithelium[72], receptor-mediated Cl^- secretion in airway cells has not been linked to PKC activity.

Calcium-mediated Cl^- channel regulation

Cl^- secretion from airway cells can be stimulated by the calcium ionophore A23187[39,40,48,49,61,73]. Willumsen and Boucher[73] used transcellular and intracellular electrodes to investigate this effect in cultured human airway cells. In the presence of amiloride, A23187 depolarized the apical membrane, decreased the fractional resistance of the apical membrane, and increased equivalent short-circuit current. Bumetanide, an inhibitor of Cl^- entry, attenuated this secretory response. Furthermore, in Cl^--free solution, intracellular Cl^- activity decreases in response to A23187[73]. These observations indicate that A23187 regulates apical membrane Cl^- conductance in human airway cells. Using I^- efflux as an assay of Cl^- activity in cultured airway epithelial cells, Clancy and colleagues[61] found that both isoprenaline and A23187 stimulated I^- efflux; these effects were additive. Isoprenaline-stimulated I^- efflux was Ca^{2+} independent; A23187-stimulated I^- efflux was Ca^{2+} dependent[61]. This suggests that the Ca^{2+} and cAMP-dependent mechanisms of Cl^- channel activation are independent pathways. Frizzell and colleagues[40] identified Cl^- channel activation by A23187 in cell-attached patches of airway epithelial cells. While this suggests that an increase in intracellular $[Ca^{2+}]$ activates Cl^- channels, the mechanism of action is not through direct activation of the channel by Ca^{2+}. First, channel activation in excised patches is not consistently observed over a wide range of Ca^{2+} concentrations[39,69]. Second, the onset of channel activation in excised patches exposed to Ca^{2+} may be delayed for several minutes[74]. Third, Cl^- channel activation is not observed with some receptor-mediated increases in cytosolic $[Ca^{2+}]$, such as α_1-adrenergic agonists[75].

How increases in cytosolic $[Ca^{2+}]$ lead to Cl^- secretion is still unclear. Cytosolic $[Ca^{2+}]$ may stimulate phospholipase A_2 activity which releases arachidonate and may lead to prostaglandin-mediated Cl^- channel activation. This is consistent with the stimulatory effect of A23187 on PGE_2 production[76] and cytosolic $[cAMP]$[71] in cultured canine and human tracheal cells. However, indomethacin's inhibitory effect on A23187-induced Cl^- secretion in airway cells is partial[17,54,70,77], indicating that the Cl^- secretory effect of A23187 is mediated, only in part, by PGE_2 production. Furthermore, A23187 stimulates Cl^- secretion in airway cells from patients with cystic fibrosis, a disease characterized by the lack of response to cAMP-mediated secretagogues.

Cytosolic $[Ca^{2+}]$ may stimulate PKC activity and thereby activate Cl^- channels[69,78]. Li and colleagues[69] reported that PKC can activate Cl^- channels but may also inactivate channels in cell-free patches when $[Ca^{2+}]$ exceeds 1 $\mu mol \ L^{-1}$. Several reports[70,71] indicate that phorbol esters (potent stimulators of PKC) transiently stimulate Cl^- secretion followed by a persistent decrease in secretion that is refractory to isoprenaline stimulation. These findings suggest that while PKC may mediate both activation and inactivation in airway epithelial cells, activation by PKC appears to have a shorter response time than inactivation. Thus, an increase in cytosolic $[Ca^{2+}]$ may initiate PKC-mediated activation prior to the effects of PKC-mediated inactivation. However, secretory responses in cystic fibrosis epithelia, to be discussed later, suggest that A23187-induced secretion is not mediated by PKC.

Increases in cytosolic $[Ca^{2+}]$ may also enhance Cl^- secretion by activation of basolateral K^+ channels[55,58]. This would increase the electrical driving force for Cl^- efflux via already activated Cl^- channels but would not account for apical membrane depolarization or the decrease in apical membrane fractional resistance induced by A23187[73].

Thus, the mechanism by which A23187, or an increase in cytosolic $[Ca^{2+}]$, induces Cl^- secretion has not been established. Whether calcium-dependent kinase or phosphatase activity plays a role in Cl^- channel activation is also unknown.

Non-physiological mechanisms of Cl^- channel activation

There are several additional methods used to activate Cl^- channels in cell-free membrane patches. Because these methods are effective only in cell-free patches, they are considered to be non-physiological mechanisms of channel activation.

Activation by depolarization

Depolarizing voltages applied to cell-free membrane patches activate Cl^- channels[33-35]. Activation is proportional to the duration and magnitude of the depolarizing voltage[35]. In cell-attached patches, membrane depolarization, does not activate Cl⁻ channels, but following excision, depolarization of equivalent magnitude activates the channels[35]. This indicates that while a strong electrical field influences the open state of these channels in cell-free patches, a similar electrical field applied to cell-attached patches is insufficient for activation of the same channels. Activation in cell-free patches by depolarization is readily reversed when hyperpolarizing voltages are applied to the patch; channels may be repeatedly switched back and forth between activated and inactivated states by changes in membrane voltage[35]. This reversibility suggests that activation does not result from the removal of an inhibitory factor from the channel environment.

Activation by increasing the ambient temperature

After membrane patch excision from airway cells at room temperature (21–23°C), Cl⁻ channels may be activated by increasing the ambient temperature of the excised patch to 37°C[35]. Because channel activation is not induced during cell-attached recordings by increasing ambient temperature to 37°C, this method of activation is also considered non-physiological. Following activation by increasing the temperature of cell-free patches, the channels remain activated when the temperature is returned to 21–23°C; at room temperature these channels may then be inactivated by membrane hyperpolarization. However, hyperpolarization does not inactivate channels when the excised patch remains at 37°C[35].

The reversibility of this method of activation is further evidence that activation does not result from the loss of an inhibitory factor from the channel environment. Speculating on a molecular model, Cl⁻ channels appear to exist in at least two relatively stable states corresponding to activated and inactivated channels. An energy barrier may hinder channel transition between these two states. Perhaps both depolarization and the increase in ambient temperature provide sufficient energy to overcome this barrier and lead to channel activation.

Trypsin-induced activation in excised patches

Channels may also be activated by exposing the cytosolic surface of cell-free patches to trypsin[35]. Activation appears to be due to the enzymatic action of trypsin because channel activation does not occur when heat-treated trypsin is used, or when trypsin inhibitors are present. The conductive properties of the channel are not altered by the trypsin incubation. This indicates that trypsin removes or alters a component involved in channel inactivation rather than affecting the conductive portion of the channel.

Activation with trypsin incubation differs from the other non-phsyiological methods in that channel activation is not reversible. Once activated by trypsin, the channels do not return to an inactivated state even when membrane hyperpolarization and phosphorylation with PKC are used[35].

DEFECTIVE CHANNEL REGULATION IN CYSTIC FIBROSIS RESPIRATORY EPITHELIUM

Cystic fibrosis (CF) is a common genetic disease characterized clinically by recurrent pulmonary infections, pancreatic insufficiency and an elevated Cl^- concentration in eccrine sweat gland secretions. Patients with this disease have altered ion transport in airway, intestinal and sweat gland duct epithelia.

Enhanced Na^+ absorption

Knowles and colleagues[79] first reported that patients with CF have an increased potential difference across the nasal epithelium. This higher potential difference was almost completely abolished with the mucosal application of amiloride, a Na^+ channel blocker. This higher transepithelial voltage indicates a greater difference between apical and basolateral membrane voltages in the nasal epithelia of CF patients (Figure 12.3B). The observation that amiloride abolished most of the transepithelial voltage suggests that an increase in apical Na^+ conductance may account for a relatively depolarized apical membrane, and thus the greater difference between apical and basolateral membrane voltages.

Using radiolabelled Na^+, Cl^- and mannitol, Boucher and colleagues[42] investigated transcellular and paracellular transport across nasal epithelia excised from CF and non-CF subjects. They found that the baseline rate of Na^+ absorption across CF tissue was 2–3 times greater than the rate across non-CF tissue. Furthermore, while isoprenaline stimulated cAMP-mediated Cl^- secretion in non-CF tissue, it did not enhance Cl^- secretion in CF tissue, but instead further increased Na^+ absorption. Paracellular transport was normal in the CF nasal epithelia[42]. These ion transport alterations in CF do not appear to be secondary to chronic disease because, prior to the development of clinical disease, infants with CF have transport alterations similar to those identified in adult CF patients[80]. Furthermore, ion transport is altered in excised CF respiratory tissue[42], cultured CF epithelial cells[38–40], and in transformed CF epithelial cells[41].

How Na^+ absorption is regulated in the airway has not yet been determined. On a chronic basis, airway monolayers cultured in the presence of aldosterone have an increased rate of Na^+ absorption[63]. However, the higher rate of absorption in CF airway epithelium does not appear to be secondary to increased aldosterone activity because aldosterone excretion is not increased in CF patients. Furthermore, blocking aldosterone activity does not reduce the nasal transepithelial potential in either CF or non-CF patients[81]. Finally, if increased Na^+ absorption of CF epithelia was due to a circulating factor, absorption by CF and non-CF should be similar when cells from both sources are grown under identical culture conditions.

On a more acute basis, cultured CF monolayers respond to cAMP-mediated Cl^- secretagogues with a paradoxical increase in Na^+ absorption[42,82]. Non-CF cells bathed in Cl^--free solution also increase amiloride-sensitive absorption and net transepithelial Na^+ flux when exposed to these secretagogues[63,82]. Furthermore, membrane-permeable cAMP analogues mimic this absorptive effect[63] suggesting that cAMP may act as a second messenger for Na^+ channel activation. However, the increase in absorption in CF airway cells is not due to increased cAMP levels because baseline cAMP levels and the levels measured in response to cAMP-mediated agonists are comparable in normal and CF cells[39,42].

On the single channel level, two apical Na^+ channels have been identified in CF and non-CF airway cells, although there are no details about their activation characteristics. Two channels have been reported: one channel is amiloride-sensitive with a linear conductance of 7–8 pS and a high degree of Na^+ selectivity[46,47]. The other is a non-selective cation channel not inhibited by amiloride (18 pS)[47]. Whether these channels differ in abundance or open-time probabilities between CF and non-CF cells remains to be established. Thus, the mechanism leading to increased Na^+ absorption in CF respiratory epithelium is still unknown.

Cl⁻ impermeability

Quinton[83] first reported that CF sweat duct epithelia have reduced Cl⁻ permeability. Knowles and colleagues[84] reported that Cl⁻ permeability is also decreased in the respiratory epithelium of CF tissue. This reduced Cl⁻ permeability was subsequently localized to the apical membrane in cultured CF airway cells[85] and more recently to defective regulation of apical membrane Cl⁻ channels[39,40].

Cell-attached patch-clamp recordings of cultured normal airway cells show occasional Cl⁻ channel openings under baseline conditions. Channel opening increases markedly when the cells are exposed to Cl⁻ secretagogues or membrane-permeable analogues of cAMP. However, the cell-attached patches of CF cells demonstrate no Cl⁻ channel openings under baseline conditions and during exposure to cAMP-mediated secretagogues[39,40]. The channels are present in CF cell membranes because they can be activated in cell-free patches with depolarization. This defective response is distal to cAMP production because increases in [cAMP] in response to Cl⁻ secretagogues are normal in CF cells[39]. Furthermore, Cl⁻ channel activation is not observed in CF cells exposed to membrane-permeable analogues of cAMP[39,40].

Cl⁻ channel activation by cAMP-dependent protein kinase (PKA) has been examined in cell-free membrane patches. In the presence of ATP, PKA activates channels in cell-free patches excised from normal cells, but not in patches from CF cells[33,34]. Whether phosphorylation is blocked, or phosphorylation occurs without channel activation in CF is unknown. However, these studies indicate that Cl⁻ channel activation in CF is defective at, or distal to, the level of phosphorylation of the channel or a membrane-associated regulatory protein.

Channel activation in patches from CF cells with protein kinase C (PKC) has also been investigated[69,78]. At low Ca^{2+} concentration (10 nmol L^{-1}),

PKC activated Cl^- channels in cell-free patches excised from normal cells, but not in patches from CF cells. This lack of channel activation with PKC in CF is similar to the findings with PKA[69,78]. However, channel inactivation by PKC at high Ca^{2+} concentration (10 μmol L^{-1}) was effective in CF. In patches excised from both normal and CF cells, Cl^- channels initially activated by depolarization were subsequently inactivated by the addition of PKC[69]. Furthermore, patches excised from normal and CF cells exposed to PKC and $[Ca^{2+}]$ could not be activated by depolarization[69].

The failure of both PKA and PKC to activate channels in CF indicates that defective channel activation is not linked to a single regulatory pathway. However, phosphorylation by both PKA and PKC may have a common regulatory component or pathway. Indeed, both kinases may phosphorylate the same site on the Cl^- channel or membrane regulatory protein. In addition, the presumed phosphorylation site for activation should differ from the site for inactivation because PKC-mediated channel inactivation was similar in CF and normal patches.

The lack of Cl^- channel activity in CF cells is unlikely to result from an endogenous increase in PKC-mediated inactivation. First, total PKC activity is similar in CF and non-CF airway epithelium, and greater than 80% of this activity resides in the cytosolic compartment of CF and non-CF tissue[86]. Second, an equivalent degree of PKC phosphorylation of endogenous substrates is observed in CF and NL tissue, i.e. no evidence of increased PKC responsiveness in CF tissue[86]. Third, if membrane-associated PKC activity was increased in CF, Cl^- channel activation by depolarization should be absent in CF excised patches exposed to high $[Ca^{2+}]$. However, channels in both CF and normal patches could be activated by depolarization in the presence of high $[Ca^{2+}]$ but not in the presence of exogenous PKC and high $[Ca^{2+}]$[69].

Ca^{2+} activation of Cl^- channels in CF

In native CF airway epithelium and primary cultures of CF airway cells, A23187 stimulates Cl^- secretion[40,70,73,77]. Boucher and colleagues[86] found that both A23187 and bradykinin stimulated Cl^- secretion in CF airway epithelium, and the secretory responses were quantitatively similar in CF and non-CF tissues. Furthermore, phorbol esters stimulate Cl^- secretion in normal epithelia but this response is absent in CF epithelia. These findings indicate that, in addition to the cAMP-mediated pathway, Cl^- secretion by the airway epithelium may be regulated by receptor-mediated increases in cytosolic $[Ca^{2+}]$. This also suggests that Ca^{2+}-mediated Cl^- secretion by A23187 and bradykinin is probably not mediated by a PKC-dependent pathway.

Thus, in CF cells, A23187 and bradykinin activate Cl^- channels. Whether these agonists activate Cl^- channels that are distinct from those activated by cAMP-mediated agonists remains to be determined. Quantitative differences in the responses to these agonists in CF and non-CF cells have not been identified. The secretory responses to A23187 and bradykinin in CF cells appear to be mediated by an increase in cytosolic $[Ca^{2+}]$, but not via a PKC pathway. Whether phosphorylation plays a role in Cl^- channel activation by A23187 or bradykinin is unclear.

Non-physiological Cl^- channel activation in CF

As previously noted, several non-physiological methods have been used to activate Cl^- channels in cell-free patches. With these methods, the channels in CF patches appear to respond similarly to those in non-CF patches[35]. Activation by membrane depolarization, by increasing the ambient temperature, and by exposing the cytosolic membrane surface to trypsin are effective in CF cell-free patches[35]. While these studies do not indicate the nature of the Cl^- channel defect in CF, they are consistent with the concept that Cl^- channel conductive properties are preserved in CF but physiological regulation of the channel is defective.

Currently, defective regulation of Cl^- channels in CF airway cells appears to be limited to channel activation by phosphorylation. This pathway is defective for phosphorylation mediated by both PKA and PKC. However, channel activation is at least qualitatively normal in intact CF cells following exposure to bradykinin and the calcium ionophore, A23187. Furthermore, Cl^- channels in both CF and normal cell-free patches can be activated by the non-physiological methods noted above.

Whether the defective regulatory site is located on the Cl^- channel itself or a membrane-associated regulatory protein is unknown. Whether the increase in Na^+ absorption found in CF cells is related to defective apical Cl^- conductance *per se* or whether both Cl^- and Na^+ conductances are altered by a shared transport regulatory defect is unclear. A specific defect in regulation of CF Na^+ channels has not been established to date. The recent identification of the CF gene defect[87–89] should help define the molecular basis for the ion transport alterations identified in CF epithelia.

The cystic fibrosis gene

The product of the putative CF gene has been called cystic fibrosis transmembrane conductance regulator (CFTR). Data from RNA blot hybridization indicate that the gene is expressed in many epithelial tissues: nasal polyps, pancreas, lung, colon, sweat gland epithelial cells, liver, parotid gland and kidney. Furthermore, a signal for gene expression is absent in brain, adrenal gland, or skin fibroblast and lymphoblast cell lines[89]. Analysis of the DNA sequence predicts a gene product approximately 168 kDa in size. Twelve hydrophobic transmembrane domains are predicted indicating that CFTR is a membrane-bound rather than secretory protein. A large polar domain contains multiple sites for potential phosphorylation by PKA and PKC. There are also two large hydrophilic regions with sequences resembling nucleotide (ATP)-binding sites. The most common gene mutation found in 70% of patients with CF[89] results in the loss of a single amino acid residue, phenylalanine, from one of these nucleotide-binding regions. Additional variants of CFTR may be identified in the near future.

This discovery should help identify the regulatory pathways of ion transport in epithelia. There are many questions that need to be addressed. For example, does gene expression vary within subpopulations of airway epithelial cells, e.g. ciliated vs goblet cells, or cells within airway submucous glands? What is the intracellular distribution of CFTR? Is its distribution limited to the apical membrane, and can CFTR be localized to any subcellular compartments? What is the turnover rate of the gene product? Does the gene product have conductive properties? Does occupancy at the nucleotide binding folds modulate Cl⁻ channel activation? Can CFTR be phosphorylated by PKA or PKC and does phosphorylation differ between the CF and non-CF gene products? Are additional mutant gene variants altered structurally at, or near, the site of phenylalanine deletion? Does expression of normal CFTR in CF epithelial cells correct the ion transport abnormalities? Does CFTR link conductive transport to other secretory functions, such as exocytosis? Does the gene product participate in cellular activities unrelated to ion transport? Hopefully, answers to these and many other questions will provide additional insight about the regulation of ion transport in the airway epithelia and how it is altered in cystic fibrosis.

References

1. Sleigh, M. A., Blake, J. R. and Liron, N. (1988). The propulsion of mucus by cilia. *Am. Rev. Respir. Dis.,* **137**, 726–741
2. Evans, M. J. and Plopper, C. G. (1988). The role of basal cells in adhesion of columnar epithelium to airway basement membrane. *Am. Rev. Respir. Dis.,* **138**, 481–483
3. Inayama, Y., Hook, G. E. R, Brody, A. R. *et al.* (1989). In vitro and in vivo growth and differentiation of clones of tracheal basal cells. *Am. J. Pathol.,* **134**, 539–549
4. Gumbiner, B. (1987). Structure, biochemistry, and assembly of epithelial tight junctions. *Am. J. Physiol.,* **253** (*Cell Physiol.,* 22), C749–C758
5. Welsh, M. J., Smith, P. L. and Frizzell, R. A. (1982). Chloride secretion by canine tracheal epithelium. II. The cellular electrical potential profile. *J. Membr. Biol.,* **70**, 227–238
6. Welsh, M. J. (1987). Electrolyte transport by airway epithelia. *Physiol. Rev.,* **67**, 1143–1184
7. Widdicombe, J. H., Basbaum, C. B. and Highland, E. (1985). Sodium-pump density of cells from dog tracheal mucosa. *Am. J. Physiol.,* **248**, (*Cell Physiol.,* 17), C389–C398
8. Widdicombe, J. H., Ueki, I. F., Bruderman, I. and Nadel, J. A. (1979). The effects of sodium substitution and ouabain on ion transport by dog tracheal epithelium. *Am. Rev. Respir. Dis.,* **120**, 385–392
9. Welsh, M. J. (1983). Evidence for basolateral membrane potassium conductance in canine tracheal epithelium. *Am. J. Physiol.,* **244** (*Cell Physiol.,* 13), C377–C384
10. Welsh, M. J. (1983). Inhibition of chloride secretion by furosemide in canine tracheal epithelium. *J. Membr. Biol.,* **71**, 219–226
11. Widdicome, J. H., Nathanson, I. T. and Highland, E. (1983). Effects of "loop" diuretics on ion transport by dog tracheal epithelium. *Am. J. Physiol.,* **245** (*Cell Physiol.,* 14), C388–C396
12. Welsh, M. J. (1984). Energetics of chloride secretion in canine tracheal epithelium: comparison of the metabolic cost of chloride transport with the metabolic cost of sodium transport. *J. Clin. Invest.,* **74**, 262–268
13. Widdicombe, J. H. (1990). Use of cultured airway epithelial cells in studies of ion transport. *Am. J. Physiol.,* **258** (*Lung Cell. Mol. Physiol.,* 2), L13–L18

14. Gögelein, H. (1988). Chloride channels in epithelia. *Biochim. Biophys. Acta*, **947**, 521–547
15. Widdicombe, J. J., Basebaum, C. B. and Highland, E. (1981). Ion contents and other properties of isolated cells from dog tracheal epithelium. *Am. J. Physiol.*, **241** (*Cell Physiol.*), C184–C192
16. Welsh, M. J., Smith, P. L. and Frizzell, R. A. (1983). Chloride secretion by canine tracheal epithelium. III. Membrane resistance and electromotive forces. *J. Membr. Biol.*, **71**, 209–218
17. Al-Bazzaz, F., Yadava, V. P. and Westenfelder, C. (1981). Modification of Na and Cl transport in canine tracheal mucosa by prostaglandins. *Am. J. Physiol.* (*Renal Fluid Electrolyte Physiol.*), **240**, F101–F105
18. Welsh, M. J. and Widdicombe, J. H. (1980). Pathways of ion movement in the canine tracheal epithelium. *Am. J. Physiol.*, **239** (*Renal Fluid Electrolyte Physiol.*), F215–F221
19. Hille, B. (1984). *Ionic Channels of Excitable Membranes*. (Sunderland, MA: Sinauer Assoc.)
20. Lear, J. D., Wasserman, Z. R. and DeGrado, W. F. (1988). Synthetic amphiphilic peptide models for protein ion channels. *Science*, **240**, 1177–1181
21. Jan. L. Y. and Jan, Y. N. (1989). Voltage-sensitive ion channels. *Cell*, **56**, 13–25
22. Catterall, W. A. (1988). Structure and function of voltage-sensitive ion channels. *Science*, **242**, 50–61
23. Noda, M., Shimizu, S., Tanabe, T., Takai, T., Kayano, T., Ikeda, T., Takahashi, H., Nakayama, H., Kanaoka, Y., Minamino, N., Kangawa, K., Matsu, H., Raftery, M. A., Hirose, T., Inayama, S., Hayashida, H., Miyata, T. and Numa, S. (1984). Primary structure of *Electrophorus electricus* sodium channel deduced from cDNA sequence. *Nature (London)*, **312**, 121–127
24. Salkoff, L., Butler, A., Wei, A., Scavarda, N., Giffen, C., Goodman, R. and Mandel, G. (1987). Genomic organization and deduced amino acid sequence of a putative sodium channel gene in *Drosophila*. *Science*, **237**, 744–749
25. Papazian, D. M., Schwarz, T. L., Tempel, L., Jan, Y. N. and Jan, L. Y. (1987). Cloning of genomic and complementary DNA from *Shaker*, a putative potassium channel gene from *Drosophila*. *Science*, **237**, 749–753
26. Noda, M., Ikeda, T., Suzaki, H., Takeshima, H., Takahashi, T., Kuno, M. and Numa, S. (1986). Expression of functional sodium channels from cloned cDNA. *Nature (London)*, **322**, 826–828
27. Benos, D. J., Saccomani, G., Brenner, B. M. and Sariban-Sohraby, S. (1986). Purification and characterization of the amiloride-sensitive sodium channel from A6 cultured cells and bovine renal papilla. *Proc. Natl. Acad. Sci. USA*, **83**, 8525–8529
28. Sariban-Sohraby, S. and Benos, D. J. (1986). Detergent solubilization, functional reconstitution, and partial purification of epithelial amiloride-binding protein. *Biochemistry*, **25**, 4639–4646
29. Benos, D. J., Saccomani, G. and Sariban-Sohraby, S. (1987). The epithelial sodium channel. *J. Biol. Chem.*, **262**, 10613–10618
30. Sorscher, E. J., Accavitti, M. A., Keeton, D., Steadman, E., Frizell, R. A. and Benos, D. J. (1988). Antibodies against purified epithial sodium channel from bovine renal papilla. *Am. J. Physiol.*, **255**, C835–C843
31. Landry, D. W., Akabas, M. H., Redhead, C., Edelman, A., Cragoe, E. J. and Al-Awqati, Q. (1989). Purification and reconstitution of chloride channels from kidney and trachea. *Science*, **244**, 1469–1472
32. Hamill, O., Marty, A., Neher, E., Sakmann, B. and Sigworth, F. J. (1981). Improved patch-clamp techniques for high-resolution current recording from cells and cell-free membrane patches. *Pflügers Arch.*, **391**, 85–100
33. Li, M., McCann, J. D., Liedtke, C. M., Nairn, A. C., Greengard, P. and Welsh, M. J. (1988). Cyclic AMP-dependent protein kinase opens chloride channels in normal but not cystic fibrosis airway epithelium. *Nature (London)*, **331**, 358–360
34. Schoumacher, R. A., Shoemaker, R. L., Halm, D. R., Tallent, E. A., Wallace, R. W. and Frizzell, R. A. (1987). Phosphorylation fails to activate chloride channels from cystic fibrosis airway cells. *Nature (London)*, **330**, 752–754
35. Welsh, M. J., Li, M. and McCann, J. D. (1989). Activation of normal and cystic fibrosis Cl⁻ channels by voltage, temperature, and trypsin. *J. Clin. Invest.*, **84**, 2002–2007

36. Levitan, I. (1985). Phosphorylation of ion channels. *J. Membr. Biol.*, **87**, 177–190
37. Ewald and Levitan, I. (1985). Modulation of single Ca^{2+} dependent K channel activity by protein phosphorylation. *Nature (London)*, **315**, 503–506
38. Boucher, R. C., Cotton, C. U., Gatzy, J. T., Knowles, M. R. and Yankaskas, J. R. (1988). Evidence for reduced Cl and increased Na$^+$ permeability in cystic fibrosis human primary cell cultures. *J. Physiol.*, **405**, 77–103
39. Welsh, M. J. and Liedtke, C. M. (1986). Chloride and potassium channels in cystic fibrosis airway epithelia. *Nature (London)*, **322**, 467–470
40. Frizzell, R. A., Rechkemmer, G. and Shoemaker, R. L. (1986). Altered regulation of airway epithelial cell chloride channels in cystic fibrosis. *Science*, **233**, 558–560
41. Jetten, A. M., Yankaskas, J., Jackson, M. J., Willumsen, N. J. and Boucher, R. C. (1989). Persistence of abnormal chloride conductance regulation in transformed cystic fibrosis epithelia. *Science*, **244**, 1472–1475
42. Boucher, R. C., Stutts, M. J., Knowles, M. R., Cantley, L. and Gatzy, J. T. (1986). Na$^+$ transport in cystic fibrosis respiratory epithelia. *J. Clin. Invest.*, **78**, 1245–1252
43. Welsh, M. J. and McCann, J. D. (1985). Intracellular calcium regulates basolateral potassium channels in a chloride-secreting epithelium. *Proc. Natl. Acad. Sci. USA*, **82**, 8823–8826
44. McCann, J. D., Matsuda, J., Garcia, M., Kaczorowski, G. and Welsh, M. J. (1990). Basolateral K$^+$ channels in airway epithelia: I. Regulation by Ca^{2+} and block by charybdotoxin. *Am. J. Physiol. (Lung Cell Mol. Physiol.* 2), **258**, L334–L342
45. Kunzelmann, K., Pavenstadt, H. and Greger, R. (1989). Properties and regulation of chloride channels in cystic fibrosis and normal airway cells. *Pflügers Arch.*, **415**, 172–182
46. Disser, J. and Frömter, E. (1989). Properties of Na$^+$ channels of respiratory epithelium from CF and non-CF patients. *Pediatr. Pulm.*, **7** (Suppl. 4), 115a
47. Man, S. F. P., Duszyk, M. and French, A. S. (1989). Sodium channels in the apical membrane of human airway epithelial cells. *Am. Rev. Respir. Dis.*, **139**, A477
48. Welsh, M. J. (1986). Single apical membrane anion channels in primary cultures of canine tracheal epithelium. *Pflügers Archiv.*, **407** (Suppl. 2), S116–S122
49. Welsh, M. J. (1986). An apical-membrane chloride channel in human tracheal epithelium. *Science*, **232**, 1648–1650
50. Duszyk, M., French, A. S. and Man, S. F. P. (1989). Cystic fibrosis affects chloride and sodium channels in human airway epithelia. *Can. J. Physiol. Pharmacol.*, **67**, 1362–1365
51. Shoemaker, R. L., Frizzell, R. A., Dwyer, T. M. and Farley, J. M. (1986). Single chloride channel currents from canine tracheal epithelial cells. *Biochim. Biophys. Acta*, **858**, 235–242
52. Schneider, G. T., Cook, D. I., Gage, P. W. and Young, J. A. (1985). Voltage sensitive, high-conductance chloride channels in the luminal membrane of culture pulmonary alveolar (type II) cells. *Pflügers Arch.*, **404**, 354–357
53. Valdivia, H. H., Dubinsky, W. P. and Coronado, R. (1988). Reconstitution and phosphorylation of chloride channels from airway epithelium membranes. *Science*, **242**, 1441–1444
54. Smith, P. L., Welsh, M. J., Stoff, J. S. and Frizzell, R. A. (1982). Chloride secretion by canine tracheal epithelium: I. Role of intracellular cAMP levels. *J. Memb. Biol.*, **70**, 217–226
55. Smith, J. J., McCann, J. D. and Welsh, M. J. (1990). Bradykinin stimulates airway epithelial Cl$^-$ secretion via two second messenger pathways. *Am. J. Physiol. (Lung Cell. Mol. Physiol.* 2), **258**, L369–L377
56. Shorofsky, S. R., Field, M. and Fozzard, H. A. (1986). Changes in intracellular sodium with chloride secretion in dog tracheal epithelium. *Am. J. Physiol.*, **250**, C646–C650
57. Miller, C., Moczydlowski, E., Latorre, R. and Phillips, M. (1985). Charybdotoxin, a protein inhibitor of single Ca^{2+}-activated K$^+$ channels from skeletal muscle. *Nature (London)*, **313**, 316–318
58. McCann, J. D. and Welsh, M. J. (1990). Basolateral K$^+$ channels in airway epithelia: II. Role in Cl$^-$ secretion and evidence for two types of K$^+$ channel. *Am. J. Physiol. (Lung Cell. Mol. Physiol.* 2), **258**, L343–L348
59. Welsh, M. J. (1983). Barium inhibition of basolateral membrane potassium conductance in

tracheal epithelium. *Am. J. Physiol.,* **244** (*Renal Fluid Electrolyte Physiol.,* 13), F639–F645

60. McCann, J. D., Bhalla, R. C. and Welsh, M. J. (1989). Release of intracellular calcium by two different second messengers in airway epithelium. *Am. J. Physiol.,* **257** (*Lung Cell. Mol. Physiol.,* 1), L116–L124

61. Clancy, J. P., McCann, J. D., Li, M. and Welsh, M. J. (1990). Calcium-dependent regulation of airway epithlial chloride channels. *Am. J. Physiol.,* **258** (*Lung Cell. Mol. Physiol.,* 2), L25–L32

62. Widdicombe, J. H. and Welsh, M. J. (1980). Anion selectivity of the chloride transport process in dog tracheal epithelium. *Am. J. Physiol.,* **239** (*Cell Physiol.,* 8), C112–C117

63. Cullen, J. J. and Welsh, M. J. (1987). Regulation of sodium absorption by canine tracheal epithelium. *J. Clin. Invest.,* **79**, 73–79

64. Van Driessche, W. and Zeiske, W. (1985). Ionic channels in epithelial cell membranes. *Physiol. Rev.,* **65**, 833–903

65. Helman, S. I., Cox, T. C. and Van Driessche, W. (1983). Hormonal control of apical membrane Na transport in epithelia. *J. Gen. Physiol.,* **82**, 201–220

66. Garty, H. and Edelman, I. S. (1983). Amiloride-sensitive trypanization of apical sodium channels. *J. Gen. Physiol.,* **81**, 785–803

67. Brautigan, D. L.. (1988). Molecular defects in ion channel regulation in cystic fibrosis predicted from analysis of protein phosphorylation/dephosphorylation. *Int. J. Biochem.,* **20**, 745–752

68. Kume, H., Takai, A., Tokuno, H. and Tomita, T. (1989). Regulation of Ca^{2+}-dependent K^+-channel activity in tracheal myocytes by phosphorylation. *Nature (London),* **341**, 152–154

69. Li, M., McCann, J. D., Anderson, M. P., Clancy, J. P., Liedtke, C. M., Nairn, A. C., Greengard, P. and Welsh, M. J. (1989). Regulation of chloride channels by protein kinase C in normal and cystic fibrosis airway epithelia. *Science,* **244**, 1353–1356

70. Welsh, M. J. (1987). Effect of phorbol ester and calcium inophore on chloride secretion in canine tracheal epithelium. *Am. J. Physiol.,* **253** (*Cell Physiol.*), C828–C834

71. Barthelson, R. A., Jacoby, D. B. and Widdicombe, J. H. (1987). Regulation of chloride secretion in dog tracheal epithelium by protein kinase C. *Am. J. Physiol.,* **253** (*Cell Physiol.,* 22), C802–C808

72. Cohn, J. A. (1990). Protein disease C mediates cholinergically regulated protein phosphorylation in a Cl^--secreting epithelium. *Am. J. Physiol.,* **258** (*Cell Physiol.,* 27), C227–C233

73. Willumsen, N. J. and Boucher, R. C. (1989). Activation of an apical Cl^- conductance by Ca^{2+} ionophores in cystic fibrosis airway epithelia. *Am. J. Physiol.,* **256** (*Cell Physiol.,* 25), C226–C233

74. Frizzell, R. A. (1987). Cystic fibrosis: A disease of ion channels? *TINS,* **10**, 190–193

75. Welsh, M. J. (1986). Adrenergic regulation of ion transport by primary cultures of canine tracheal epithelium: Cellular electrophysiology. *J. Membr. Biol.,* **91**, 121–128

76. Widdicombe, J. H., Ueki, I. F., Emergy, D., Margolskee, D., Yergey, J. and Nadel, J. A. (1989). Release of cyclooxygenase products from primary cultures of tracheal epithelia of dog and human. *Am. J. Physiol.,* **257** (*Lung Cell. Mol. Physiol.,* 1), L361–L365

77. Widdicombe, J. H. (1986). Cystic fibrosis and beta-adrenergic response of airway epithelial cell cultures. *Am. J. Physiol.,* **251** (*Regulatory Integrative Comp. Physiol.,* 20), R818–822

78. Hwang, T-C., Lu, L., Zeitlin, P. L., Gruenert, D. C., Huganir, R. and Guggino, W. B. (1989). Cl^- channels in CF: lack of activation by protein kinase C and cAMP-dependent protein kinase. *Science,* **244**, 1351–1353

79. Knowles, M. R., Gatzy, J. T. and Boucher, R. C. (1981). Increased bioelectric potential difference across respiratory epithelia in cystic fibrosis. *N. Engl. J. Med.,* **305**, 1489–1495

80. Gowen, C. W., Lawson, E. E., Gingras-Leatherman, J., Gatzy, J. T., Boucher, R. C. and Knowles, M. R. (1986). Increased nasal potential difference and amiloride sensitivity in neonates with cystic fibrosis. *J. Pediatr.,* **108**, 517–521

81. Knowles, R. R., Gatzy, J. T. and Boucher, R. C. (1985). Aldosterone metabolism and transepithelial potential difference in normal and cystic fibrosis subjects. *Pediatr. Res.,* **19**, 676–679

82. Pedersen, P. S., Brandt, N. J. and Larsen, E. H. (1986). Qualitatively abnormal beta-adrenergic response in cystic fibrosis sweat duct cell culture. *IRCS Med. Sci.,* **14**, 701–702

83. Quinton, P. M. (1983). Chloride permeability in cystic fibrosis. *Nature (London)*, **301**, 421–422

84. Knowles, M. R., Stutts, M. J., Spock, A., Fischer, N. L., Gatzy, J. T. and Boucher, R. C. (1983). Abnormal ion permeation through cystic fibrosis respiratory epithelium. *Science*, **221**, 1067–1070

85. Widdicombe, J. H., Welsh, M. J. and Finkbeiner, W. E. (1985). Cystic fibrosis decreases the apical membrane chloride permeability of monolayers cultured from cells of tracheal epithelium. *Proc. Natl. Acad. Sci. USA*, **82**, 6167–6171

86. Boucher, R. C., Cheng, E. H. C., Paradiso, A. M., Stutts, M. I., Knowles, M. R. and Earp, H. S. (1989). Chloride secretory response of cystic fibrosis human airway epithelia. *J. Clin. Invest.*, **84**, 1424–1431

87. Rommens, J. M., Iannuzzi, M. C., Kerem, B., Drumm, M. L., Melmer, G., Dean, M., Rozmahel, R., Cole, J. L., Kennedy, D., Hidaka, N., Zsiga, M., Buchwald, M., Riordan, J. R., Tsui, L-C and Collins, F. S. (1989). Identification of the cystic fibrosis gene: chromosome walking and jumping. *Science*, **245**, 1059–1065

88. Riordan, J. R., Rommens, J. M., Kerem, B., Alon, N., Rozmahel, R., Grezelczak, Z., Zielenski, J., Lok, S., Plavsic, N., Chou, J-L., Drumm, M. L., Iannuzzi, M. C., Collins, F. S. and Tsui, L-C. (1989). Identification of the cystic fibrosis gene: cloning and characterization of complementary DNA. *Science*, **245**, 1066–1073

89. Kerem, B., Rommens, J. M., Buchanan, J. A., Markiewicz, D., Cox, T. K., Chakravarti, A., Buchwald, M. and Tsui, L-C. (1989). Identification of the cystic fibrosis gene: genetic analysis. *Science*, **245**, 1073–1080

90. Tamaoki, J., Ueki, I. F., Widdicombe, J. H. and Nadel, J. A. (1988). Stimulation of Cl secretion by Neurokinin A and Neurokinin B in canine tracheal epithelium. *Am. Rev. Respir. Dis.*, **137**, 899–902

91. Leikauf, G. D., Ueki, I. F., Nadel, J. A. and Widdicombe, J. H. (1985). Bradykinin stimulates Cl secretion and prostaglandin E_2 release by canine tracheal epithelium. *Am. J. Physiol.*, **248** (*Renal Fluid Electrolyte Physiol.*, 17), F48–F55

92. Al-Bazzaz, F. J., Kelsey, J. G. and Kaage, W. D. (1985). Substance P stimulation of chloride secretion by canine tracheal mucosa. *Am. Rev. Respir. Dis.*, **131**, 86–89

93. Leikauf, G. D., Ueki, I. F., Widdicombe, J. H. and Nadel, J. A. (1986). Alteration of chloride secretion across canine tracheal epithelium by lipoxygenase products of arachidonic acid. *Am. J. Physiol.*, **250** (*Renal Fluid Electrolyte Physiol.*, 19), F47–F53

94. Pratt, A. D., Clancy, G. and Welsh, M. J. (1986). Mucosal adenosine stimulates chloride secretion in canine tracheal epithelium. *Am. J. Physiol.*, **251** (*Cell Physiol.*, 20), C167–C174

95. Jacoby, D. B., Ueki, I. F., Widdicombe, J. H., Loegering, D. A., Gleich, G. J. and Nadel, J. A. (1988). Effect of human eosinophil major basic protein on ion transport in dog tracheal epithelium. *Am. Rev. Respir. Dis.*, **137**, 13–16

96. Williamsen, N. J., Davis, C. W. and Boucher, R. C. (1989). Celluar Cl transport in cultured cystic fibrosis airway epithelium. *Am. J. Physiol.*, **256** (*Cell Physiol.*, 25), C1045–C1053

Section V
SKIN AND SKIN GLANDS

Introduction

In this section of the book, there are three chapters, the first of which, by Terence Kealey, describes the isolation and properties of human skin glands and hair follicles, and considers their application to the study of specific human epithelial diseases. By repeatedly snipping a skin biopsy with scissors until a porridge-like consistency is obtained, apocrine and eccrine sweat glands, sebaceous glands, hair follicles and intact pilosebaceous units can be picked out under a dissecting microscope. The viability of human eccrine sweat glands isolated in this way has been confirmed by metabolic, ultrastructural and electrophysiological studies, on both freshly-isolated and maintained organs, and by the establishment of viable cells in culture. Extensive studies of lipid synthesis in isolated apocrine, sebaceous and pilosebaceous units, whether freshly-isolated or maintained briefly in organ culture, have confirmed that they retain biological activity. Isolated skin glands are especially suited to *in vitro* studies in the maintained state because their small size facilitates gaseous and nutrient diffusion. Dr Kealey considers the role of androgens in the function of sebaceous glands and in the aetiology of acne.

In the following chapter by Christopher Jones, the electrophysiological and morphological properties of cells in secretory and reabsorptive segments of the human eccrine sweat gland are compared with those of cells in primary cultures established from these regions. The recent identification and cloning of the gene for cystic fibrosis has placed renewed pressure on the need to develop satisfactory *in vitro* systems with which to investigate the expression of the CF gene and to test possible therepautic regimens. This chapter considers in detail some of the properties of the native sweat gland secretory and reabsorptive epithelia and assesses potential signatures which could be used to investigate the veracity of the cells grown from these regions in primary culture. The evidence presented suggests that only cells in the secretory tubule and in primary cultures obtained from this region respond to acetylcholine. Sensitivity to amiloride is present in a proportion of cultured cells obtained from secretory and ductal regions. The responsiveness to isoprenaline (isoproterenol) is investigated and found to be bicarbonate-dependent in the perifused reabsorptive duct. The dye-coupling status of different cell types in the secretory and reabsorptive segments of the gland is described.

The chapter by Catherine Lee opens with a general introduction comprising a brief history of cell culture, an explanation of the culture system and a discussion on the need for culture models in the study of epithelial physiology. Methods of primary epithelial cell culture with reference to human epidermal cells, sweat and sebaceous glands are then considered in more detail. The criteria which are used to determine whether the cultured cells are indeed epithelial are discussed, the most precise of which is the identification of cytokeratins by immunocytochemistry. The benefits and drawbacks of primary cell cultures are compared with those of cell lines and the techniques used to obtain immortalization are described, together with the properties of the cell lines obtained.

Note added in proof

Since the identification and cloning of the gene sequence mutated in cystic fibrosis, and referred to by several contributors to this book, it has been shown, using cultured CF airway[1] and pancreatic epithelial cells[2], that the defective transport of chloride, characteristic of the disease, can be corrected in a complementation experiment by introducing normal copies of the CFTR gene into the defective cells.

1. Rich, D. P., Anderson, M. P., Gregory, R. J., Cheng, S. H., Paul, S., Jefferson, D. M., McCann, J. D., Klinger, K. W., Smith, A. E. and Welsh, M. J. (1990). Expression of cystic fibrosis transmembrane conductance regulator corrects defective chloride channel regulation in cystic fibrosis airway epithelial cells. *Nature*, **347**, 358–363
2. Drumm, M. L., Pope, H. A., Cliff, W. H., Rommens, J. M., Marvin, S. A., Tsui, L-C., Collins, F. S., Frizzell, R. A. and Wilson, J. M. (1990). Correction of the cystic fibrosis defect *in vitro* by retrovirus-mediated gene transfer. *Cell*, **62**, 1227–1233

13
Isolated Human Skin Glands and Appendages: Models for Cystic Fibrosis, Acne Vulgaris, Alopecia and Hidradenitis Suppurativa

T. KEALEY

INTRODUCTION

There are a number of diseases of human epithelia whose aetiology is unknown and whose treatment is unsatisfactory. These include cystic fibrosis, acne vulgaris, alopecia (baldness) and hidradenitis suppurativa. These particular diseases are either restricted to skin glands or they express themselves in an important part through abnormal skin gland activity. Thus, cystic fibrosis is characterized by abnormalities of the lungs and pancreas[1], but its pathognomonic sign is a raised eccrine sweat NaCl[1]. Acne vulgaris is a disease of pilosebaceous units[2], alopeica is a condition of hair follicles[3] and hidradenitis suppurativa is a disease of apocrine sweat glands[4]. These diseases are all specific to humans, and the lack of animal models has restricted their research. Study of these diseases has, in the past, been further restricted by the difficulties involved in isolating the human skin glands and hair follicles. This is because the glands are small, delicate, and sometimes translucent, and therefore, difficult to visualize against the dermal collagen of a skin biopsy. Microdissection can, therefore, be slow, difficult and of low yield.

In the past, *in vitro* experiments were made on small biopsies of human skin, within which glands or hair follicles were still embedded[5-7]; but these preparations have obvious limitations, of which the greatest are the restrictions on gas exchange and nutrient diffusion. To overcome these drawbacks, and despite the difficulties with the technique, experimenters have microdissected human skin glands and hair follicles[8-12]. Such preparations have often yielded enormous information – in particular Quinton[8] and Sato and Sato[9] discovered the chloride impermeability and β-adrenergic resistance of cystic fibrosis, respectively, while working with micro-

289

dissected human eccrine sweat glands – but research would be facilitated by alternative methods for gland isolation.

To improve the yield of glands and hair follicles, and to develop simpler techniques, digestive enzymes have been employed. Kealey[13], in 1983, isolated viable human eccrine sweat glands by the collagenase digestion of skin biopsies. These biopsies were obtained as slivers of skin removed from the incision of operations on patients who had no systemic disease (appendicitis, for example, or coronary artery bypass grafts). Ethical Committee permission was granted for this work.

Okada *et al.*[14], also in 1983, isolated viable human eccrine sweat glands by dispase digestion. Glands isolated by digestive enzymes retain their viability as judged by a number of criteria: these include light and electron microscopy, ATP contents, cyclic AMP and cyclic GMP responses to secretagogues, and the responses of glycolysis and glucose oxidation to secretagogues. Other groups have since used collagenase digestion to isolate human eccrine sweat glands[15], but we discovered, in 1984, that shearing[16], which had been a preparatory step to collagenase digestion, was itself sufficient to release, not only eccrine sweat glands[16], but also sebaceous glands[17], apocrine sweat glands[17] and hair follicles[18].

The yields are high – up to 100 eccrine and 40 sebaceous glands from a sliver (150 mm long, 3 mm wide and as deep as the subcutaneous fat) of chest skin removed from the incision during cardiac surgery, and up to 40 apocrine glands from an axillary sliver (25 mm long, 3 mm wide) removed during lymph node biopsy for carcinoma of the breast. Up to 40 ceruminous glands can be isolated from an external auditory meatus incision (Kealey, unpublished observations).

Shearing is very simple. All that is required is the repeated snipping of a skin biopsy with sharp scissors in medium until a porridge-like consistency is achieved. The glands and appendages are then picked out under a binocular microscope. Figure 13.1 illustrates an isolated human eccrine sweat gland and Figure 13.2 illustrates isolated human sebaceous glands, hair follicles and an entire human pilosebaceous unit.

The technique works because of the periglandular capsule of fibroblasts. As Figure 13.3 illustrates, skin glands *in situ* are surrounded by very thin fibroblastic cytosolic projections, separated by collagen fibres, and these layers shear over each other under the pressure of scissors. Soft skin, such as that of the neonatal rat, requires the use of blunt scissors because the blades of sharp scissors tend to cut through the skin cleanly, rather than shear it. Human skin is sufficiently tough to necessitate the use of sharp scissors even to shear it. Because the shearing occurs through the capsule, the underlying gland pops out intact, and with no apparent electron-micrographic evidence of damage.

Many indices show that the glands and hair follicles are viable on isolation. Light and electronmicrographs show no evidence of tissue damage and we have used the energy charge, as defined by Atkinson[19] ((ATP) + 0.5 (ADP)/(ATP) + (ADP) + (AMP)) to show that the glands and hair follicles are viable biochemically on isolation (Table 13.1).

Using glands isolated by shearing, we have undertaken a number of

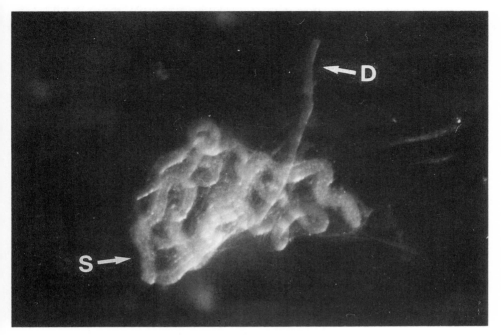

Figure 13.1 An isolated human eccrine sweat gland. The collecting duct (D) can be seen leading up to the skin surface. S = secretory tubule. Magnification 50×. Reproduced by permission from the *Biochemical Journal, 212*, 143–148(c), 1983, The Biochemical Society, London

studies, including those on fuel metabolism, lipid synthesis, second messenger turnover, electrophysiology, organ maintenance and cell culture.

METABOLIC STUDIES

The skin is a tissue which engages in aerobic glycolysis, as defined by Krebs[22]. That is to say, it preferentially metabolizes glucose to lactate despite the presence of oxygen (see the discussion by Newsholme and Leech[23]). With the partial exception of the sweat gland, the reason for the skin's unusual metabolism is not known[23]. The lactate content of the sweat is high (15–60 mmol/L^{-1})[24] whereas that of the blood is only 1 mmol/L^{-1}. The pH of the sweat, however, is low at approximately 5, and this may be caused in part by the lactic acid. This, in turn, may be bacteriostatic. In 1973, Sato and Dobson[25] microdissected monkey eccrine sweat glands and studied their rates of glucose oxidation by monitoring the conversion of [U-^{14}C]glucose to $^{14}CO_2$, and they studied their rates of aerobic glycolysis by measuring the appearance of lactate in the bathing medium. We have extended these observations to the human gland[13,26].

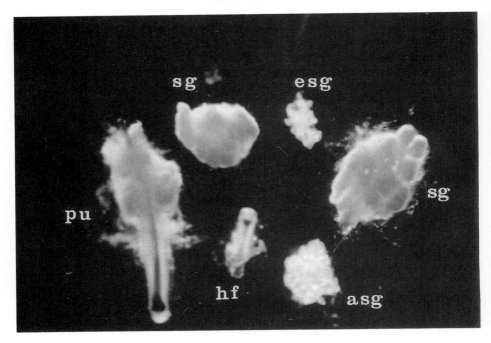

Figure 13.2   Isolated human skin glands and appendages. One pilosebaceous unit (pu), two sebaceous glands (sg), one hair follicle (hf), an apocrine sweat gland (asg) and an eccrine sweat gland (esg) can be seen. Magnification 20×. Reproduced by permission from *Acne and Related Disorders*, (eds.) Marks, R. and Plewig, G., 1989, Martin Dunitz Publisher, London

Table 13.2 shows that glucose is largely metabolized to lactate, not CO_2. It can be calculated that the flux of glucose-derived pyruvate through lactate dehydrogenase is 16 times that through pyruvate dehydrogenase[13]. This ratio of 16:1 contrasts with that of 0.5:1 seen, for example, in working heart muscle[29]. We have calculated that the sweat gland synthesizes sufficient lactate *in vitro* to account for the lactate content of sweat *in vivo* and that it is not necessary to postulate that the gland might sequester lactate from the blood[13]. We have also calculated[13] that the metabolism of glucose seen *in vitro* is sufficient to account for the fluid secretion seen *in vivo*[13] if: (1) it is assumed that fluid secretion is mediated by the $(Na^+ + K^+)$-stimulated ATPase; (2) that the consumption of 1 mol ATP transports 3 mol Na^+; (3) that each mole of lactate produced from glucose yields 1 mol ATP; and (4) that 36 mol ATP are produced for each mole of glucose that is oxidized to 6 CO_2.

Pyruvate dehydrogenase does not appear to be regulated in the sweat gland. Dichloroacetate will not simulate $[1-^{14}C]$pyruvate oxidation, nor does the secretagogue activation of metabolism alter the fate of pyruvate, i.e. 10^{-5} mol L^{-1} acetylcholine stimulates both glucose oxidation and lactate production 2.3 fold, while 10^{-5} mol L^{-1} isoprenaline, a weak sudorific agonist, stimulates them both 1.5 fold[26]. The β-adrenergic stimulation

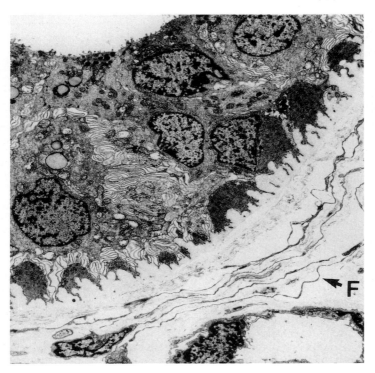

Figure 13.3 An electronmicrograph of the secretory coil of a human eccrine sweat gland *in situ*. Outside the basal lamina, three or four connective rings of fibroblasts (F) can be seen. Shearing occurs between them. Magnification 2800×. Electronmicrograph made by Dr C. J. Jones

Table 13.1 Adenine neuclopide content of human skin glands isolated by shearing (mean ± SEM)

	Eccrine[16]	Apocrine[17]	Sebaceous[17]	Hair follicle[a]
ATP (pmol/gland)	81.0 ± 12.7	310 ± 34.1	148.8 ± 30.3	9.19 ± 0.78
ADP (pmol/gland)	13.8 ± 3.3	90.4 ± 16.3	30.6 ± 4.7	6.26 ± 0.53
AMP (pmol/gland)	3.8 ± 1.0	40.1 ± 11.8	14.9 ± 4.7	1.21 ± 0.08
Energy charge[19]	0.90	0.81	0.84	0.72

Adenine nucleotides were measured by luciferin luciferase[20] as described by Spielman *et al.*[21]
[a]Rat, not human: Philpott, M. P., Kealey, T. (unpublished data)

of eccrine sweat gland glycolysis indicates that, in this tissue, phospho-fructokinase 2 is not inhibited by cyclic AMP[26].

We have shown that glucose is the major sweat gland fuel: the glandular oxidation of ^{14}C-labelled fatty acids and ketone bodies only yields one fifth as much ATP as does the metabolism of glucose[26]. The lack of sweat gland oxidative capacity towards fats and ketone bodies probably accounts for an

Table 13.2   Rate of glucose oxidation and glycolysis in human eccrine sweat glands

	Amount of substrate oxidized or hydrolysed (nmol gland^{-1} h^{-1} (mean±SEM)	
	5 mmol L^{-1} [U-^{14}C]glucose	5 mmol L^{-1} [2-3H]glucose
Control	0.062 ± 0.009	0.63 ± 0.12
10^{-5} mol L^{-1} isoprenaline	0.097 ± 0.23	0.86 ± 0.17
10^{-5} mol L^{-1} acetylcholine	0.147 ± 0.038*	1.41 ± 0.21**

Five pairs of glands, each from 3 different subjects, were maintained in bicarbonate-buffered medium[27] and the rates of glucose oxidation determined by monitoring the production of $^{14}CO_2$ and the rates of glycolysis calculated from the rate of [2-^{3}H]glycolysis into ^3H_2O as described by Hammerstedt[28].
*$p < 0.05$, **$p < 0.01$ compared with the control value as determined by Student's t-test

interesting anomaly: the lack of a glucose/fatty acid cycle. The inhibition of glucose metabolism effected by fat and ketone body oxidation in most tissues is not seen in the sweat gland[26].

These data suggest that the most likely model for the secretagogue-stimulation of metabolism is that glucose flux through the cell membrane and phosphofructokinase is increased allosterically by changes in the concentration of ATP, ADP and AMP caused by the energetic requirement of secretion. Because pyruvate dehydrogenase is unregulated, the increased production of pyruvate causes similar increases in flux through both pyruvate dehydrogenase and lactate dehydrogenase. The acetylcholine-induced depletion of glycogen, however, that is seen both *in vivo* and *in vitro*, might be further stimulated by a calcium-activation of phosphorylase[26].

LIPID SYNTHESIS

Apocrine sweat gland

The function of the sebaceous and apocrine sweat glands is to synthesize lipid. (It is also the function of the ceruminous glands, but we have not yet studied them systematically.) The roles of the lipid produced by the different glands are distinct. The apocrine lipid is odiferogenic. Shelley *et al.*[30] showed, in 1953, that the primary apocrine secretion was both odourless and sterile, but that skin surface bacteria broke down the apocrine secretion into volatile odours, probably volatile fatty acids. Leyden *et al.*[31] showed, by *in vivo* sampling, that the apocrine sweat lipids comprised: 76.1% cholesterol, 19.2% triglycerides and free fatty acids, 3.6% wax esters, 0.9% cholesterol esters and 0.2% squalene. We have now shown, by exposing human apocrine glands *in vitro* to 2 mmol L^{-1} [U-^{14}C]acetate[32], that uptake into chloroform/methanol extractable material is linear over 6 h. The major lipids formed were: 12.3% cholesterol, 44.4% glycerides and fatty acids, 38.8% phospholipids, and no cholesterol esters, wax esters or squalene. These *in vivo* and *in vitro* figures are more compatible than they appear at first sight. The large phospholipid synthesis *in vitro* will be a consequence of the normal cells' membrane lipid turnover, and it would not be expected to appear on the surface of the skin; while the *in vivo* appearance of cholesterol esters, wax esters and squalene can be attributed to contamination from sebum, because the apocrine sweat gland empties into the pilosebaceous duct.

In vitro, therefore, the isolated human apocrine sweat gland maintains its normal lipogenic function, which is to synthesize lipids, predominantly triglycerides, as a substrate for breakdown by skin surface bacteria into volatile fatty acids. The nature of these latter remains to be determined.

Sebaceous glands

These secrete sebum, which is almost pure lipid. The composition of sebum *in vivo*[33] is: 57.5% glycerides and fatty acids, 26% wax esters, 12% squalene, 3% cholesterol esters and 1.5% cholesterol. When human sebaceous glands are isolated by shearing and maintained in either 2 mmol L^{-1} [U-^{14}C]glucose or 2 mmol L^{-1} [U-^{14}C]acetate, uptake into chloroform/methanol extractable material is linear over 6 h[17,34]. It can be calculated from the known density of sebaceous glands and from the known *in vivo* rates of sebum secretion that the rate of sebum secretion/gland *in vitro* matches that seen *in vivo*[17,34]. The pattern of lipogenesis *in vitro*[34] is: 57.8% triglycerides and fatty acids, 20.2% squalene, 12.8% phospholipids, 7.1% wax monoesters and cholesterol esters and 2.1% cholesterol. This approximates very well to the *in vivo* situation, except that the 12.8% phospholipid uptake *in vitro* reflects a membrane turnover that would not be seen *in vivo*. Sebaceous glands isolated by shearing retain, therefore, their biological action.

Sebaceous glands secrete sebum into the follicular duct of hair follicles. From these, sebum spreads along the hair shaft. The function of sebum over the terminal hairs (the large visible hairs that are found on the scalp, over the pubic areas, over the beard area and chest in post-pubertal men and over the legs in adults of both sexes) is presumably the same as the function of sebum in all other mammals' terminal hairs: namely waterproofing. The function of the sebum synthesized by the sebaceous glands that empty their sebum into sebaceous follicular ducts is less clear.

Sebaceous follicles are found in the areas where acne develops: the face and the torso. They contain vellus hairs, but these are vestigial. The sebaceous glands of sebaceous follicles, however, are very large, and may reach 1 mg in wet weight[17]. Their growth is androgen-dependent, and the onset of their sebum secretion coincides with puberty[35]. Acne, which also coincides with puberty, is a consequence of the inflammation of sebaceous follicles[2]. I have suggested that acne might be caused by the loss of vitamin A from the sebaceous follicular duct cells into the sebum[36,37].

Despite many studies, no role can be found for sebaceous follicular sebum[2]. It appears to offer no protection against sunlight, against dehydration or against bacteria. I suggest that the role of sebaceous follicular sebum is, in fact, to cause acne. The evidence is as follows.

Firstly, acne is an androgen-mediated phenomenon[35] with marked psychological consequences[2]. Male balding[3], and the development of body odour[30], are also androgen-dependent phenomena with marked socio-biological effects.

Secondly, human sebum is unique amongst mammals for containing triglycerides (57.5%). *Propionibacterium acnes (P. acnes)* is a normal sebaceous duct commensal, and it appears to hydrolyse triglycerides to free fatty acids[38]. These free fatty acids are highly inflammatory[38], and may account for much of the inflammation of acne.

The development of acne, therefore, may be a physiological exploitation of the loss of terminal hair in humans, to create a further sociobiological signal of puberty.

ECCRINE SWEAT GLAND SECOND MESSENGERS

A number of studies have been made on second messengers in human eccrine sweat glands because of their potential importance in cystic fibrosis.

Cholinergic second messengers

The major sweat gland secretagogue is acetylcholine[39], which is delivered by the post-ganglionic neurone of the anomalous sympathetic system[40]. The acetylcholine stimulation of sweat secretion is blocked by atropine[41], and is calcium dependent[5], which confirms that this site is muscarinic. The immediate second messengers of muscarinic acetylcholine are inositol 1,4,5-triphosphate (IP_3) and diacylglycerol, which are produced by the hydrolysis of phosphatidyl inositol 4,5-biphosphate[42]. We have studied the cholinergic stimulation of IP_3 appearance in human eccrine sweat glands.

Methods

Eccrine sweat glands (50 per experiment) were incubated for 18 h in Williams E medium supplemented with 20 μmol L^{-1}-myo-[2-^{3}H]inositol (S.A. 9.3 Ci/mmol). Control experiments showed the uptake was linear up to 24 h at 37.2 ± 17.0 fmol gland^{-1} h^{-1} (mean$\pm$SEM). Glands were made up to 10 μmol L^{-1} acetylcholine for variable lengths of time; further reaction was stopped with trichloroacetic acid and the inositol phosphates were separated by use of ion-exchange chromatography as described by Berridge *et al.* (1983)[43]. Table 13.3 shows that there was a significant rise in glandular inositol phosphate contents following incubation with pilocarpine. Control experiments showed: (1) that the rise in IP_3 content was blocked by atropine, and (2) that the exposure of glands to 1 mmol L^{-1} lithium chloride, an inhibitor of inositol 1-phosphatase[43], raised the inositol mono-phosphate content three-fold.

Table 13.3 Inositol phosphate ester content of sweat glands following stimulation with 10^{-5} mmol L^{-1} pilocarpine

Time of stimulation (min) with 10^{-5} mol L^{-1} pilocarpine	Inositol-2-monophosphate (fmol/gland) (mean±SEM)	Inositol 1,4-biphosphate (fmol/gland) (mean±SEM)	Inositol 1,4,5-triphosphate (mean±SEM)
0(n=6) (control)	3.34 ± 0.43	1.31 ± 0.22	0.30 ± 0.10
0.5(n=5)	4.24 ± 0.26	1.56 ± 0.35	0.40 ± 0.20
1(n=4)	* 6.25 ± 0.85	*4.45 ± 1.27	1.05 ± 0.54
5(n=4)	***17.8 ± 2.0	***7.60 ± 1.27	***1.18 ± 0.12

Statistical significance of difference from control levels was assessed using Student's t-test $*p < 0.05$, $***p < 0.001$. Groups of glands were incubated with myo[2^{-3}H]inositol as described in the text. After 0.5, 1 or 5 min, stimulation was stopped by addition of TCA (15% w/v). Water-soluble extracts were subjected to anion exchange chromatography on a Dowex 1×8 column, and inositol phospate esters' contents were determined as described in the text

Table 13.4 Cyclic nucleotide metabolism in human eccrine sweat glands

Conditions of incubation	Cyclic AMP (fmol/gland) (mean±SEM)	Cyclic GMP (fmol/gland) (mean±SEM)
Unstimulated	28.8 ± 3.1	5.71 ± 1.69
10^{-5} mol L^{-1} acetylcholine		
2 min	191.1 ± 75.8	162.1 ± 48.1*
5 min	408.4 ± 245.0	112.9 ± 25.0*
10 min	178.4 ± 93.2	68.3 ± 13.7*
10^{-5} mol L^{-1} isoprenaline		
2 min	162.7 ± 31.6*	8.91 ± 3.50
5 min	200.2 ± 16.6***	8.90 ± 2.5
10 min	202.5 ± 14.8***	7.60 ± 1.09

Each observation was made in duplicate on 5 glands, each from 3 subjects[16]. Glands were incubated in bicarbonate-buffered medium[27], exposed to secretagogues, and their cyclic nucleotides were measured by radioimmunoassay[45]. Statistical significance of difference from control levels was assessed by Student's t-test *$p < 0.05$, ***$p < 0.001$

Cyclic GMP

Muscarinic cholinergic stimulation of tissues causes a rise in cyclic GMP through the diacylglycerol activation of cyclo-oxygenase[44]. Table 13.4 shows that sheared glands show a marked rise in cyclic GMP content following 2 min exposure to 10^{-5} mol L^{-1} acetylcholine, which then progressively declines. The findings of Sato and Sato[46] on microdissected monkey eccrine sweat glands are similar.

Cyclic AMP

β-Adrenergic agonists stimulate a small amount of sweat secretion[41]. These agents activate adenyl cyclase and raise cyclic AMP levels in human eccrine sweat glands 10-fold[13,17] (Table 13.4). Sato and Sato[9] have reported similar findings for microdissected human eccrine sweat glands. Curiously, acetylcholine also causes a transient, but very marked, rise in cyclic AMP contents (Table 13.4).

The collecting duct

The sweat gland consists of two parts: a fluid-secreting coil and a collecting or reabsorptive duct that directs the sweat to the surface of the skin and which actively reabsorbs NaCl from the primary secretion. One of the major defects in cystic fibrosis, the raised NaCl content of sweat, is caused by a chloride impermeability of the collecting duct[8]. This might be constitutive (i.e. the chloride channels never function) or regulatory (i.e. the chloride channels are normal but their activation is abnormal). We investigated, therefore, the response of second messengers in the collecting duct of the sweat gland to agonists.

Table 13.5 Cyclic nucleotide metabolism of collecting ducts and whole glands

Conditions of incubation	Cyclic nucleotide metabolism (fmol/duct)		Cyclic GMP metabolism (fmol/whole gland) ($n=5$)
	Cyclic GMP ($n=5$)	Cyclic AMP ($n=3$)	
Unstimulated	1.62 ± 0.75	19.96 ± 0.68	12.34 ± 4.15
10^{-5} mol L^{-1} acetyl-choline			
0.5 min	—	—	48.04 ± 10.23
1 min	1.27 ± 0.68	—	59.05 ± 3.17
2 min	2.72 ± 1.48	—	79.22 ± 19.36*
5 min	1.80 ± 0.92	—	58.10 ± 17.89*
10^{-5} mol L^{-1} isopre-naline			
2 min	—	35.38 ± 3.36	—
5 min	—	57.00 ± 5.08*	—

All results are mean±SEM. *$p <0.005$ cf. unstimulated. Duplicate assays were performed on batches either of two whole glands or three ducts isolated by microdissection of sheared glands

Table 13.5[47] shows that the collecting duct shows a cyclic AMP response to isoprenaline. This is, however, found in almost all tissues and may not be significant. There is no ductal cyclic GMP response to acetylcholine. This latter biochemical evidence indicates that there may not be a direct ductal regulation by acetylcholine.

This receives further biochemical corroboration from the findings of Sato and Dobson that microdissected sweat gland collecting ducts, in contra-distinction to microdissected sweat gland secretory coils, show no acetyl-choline stimulation of glucose oxidation or of lactate release[25]. But one proviso needs to be noted. Electronmicrography of isolated human eccrine sweat glands shows that secretagogues stimulate fluid secretion[13]: this can be seen by the dilatation of intercellular canaliculi. Isolated collecting ducts, however, may not be able to respond to acetylcholine by reabsorbing NaCl and so stimulating metabolism, because their lumens may be collapsed and empty of fluid and ions. A definitive catabolic experiment would have to be made on lumen-perfused ducts.

Electrophysiology of human eccrine sweat glands[47,48]: This will be discussed elsewhere in this volume (Chapter 14 by C. J. Jones).

ORGAN MAINTENANCE AND CELL CULTURE

These two techniques are very important in experimental dermatology. Organ maintenance is the long-term maintenance of intact organs: that is to say they are preserved in a state which retains their native architecture, cell–cell interaction, cell–connective tissue interaction and cellular differentiation as much as possible. Cell culture, in contrast, is the promotion of cell growth, often in monolayers, and this is generally achieved at the expense of native architecture and of some differentiated functions (see review[49]).

Organ maintenance and cell culture are important in experimental dermatology because many of the diseases of skin are a consequence of abnormal rates of cell division or of abnormal cell differentiation. Acne, for example, reflects hyperproliferation and hyperkeratinization of the cells of the pilosebaceous follicular duct[2]. Alopecia involves the conversion of terminal scalp hairs into vellus ones[3], which is a gradual process spread over many cycles of individual hair growth (each cycle may last for up to two years). Furthermore, culture may offer the best physiological tool for determining the biological actions of pharmacological or endocrinological agents. The retinoids, for example, which are highly effective in acne, act by regulating sebocyte cell division and differentiation[50], while testosterone regulates apocrine sweat gland growth. These phenomena can best be studied by culture *in vitro*.

Because organ maintenance retains normal tissue architecture, it tends to retain normal cell differentiation. It is, therefore, appropriate to studies of diseases of differentiation. Cell culture offers different advantages: we have cultured sweat gland cells to increase tissue availability and to expose the luminal membrane. These have been of importance in cystic fibrosis research (see below).

Organ maintenance

We have now successfully maintained in organ culture the four major human skin glands and appendages: eccrine sweat glands[16], apocrine sweat glands[32], sebaceous glands[50] and hair follicles[51]. These tissues are particularly suitable for organ maintenance because they are small, and so their viability will not be threatened by a lack of gaseous or nutrient diffusion. This means that they can be maintained whole, without being cut through the basement membrane into small masses of cells. We have used the same technique for all the glands and hair follicles: they are maintained on porous polycarbonate filters which float at the air/surface interface of supplemented Williams E medium at 37° C in 5% CO_2:95% air. From these studies, certain general principles can be derived. First, terminally differentiated or Go cells can be successfully maintained and they will continue to express differentiated functions. So, for example, human apocrine sweat glands will demonstrate an unchanged electronmicrographic structure and will synthesize the same amounts and the same classes of lipid from [U-^{14}C]acetate after 10 days maintenance as they did when freshly isolated[32]. Organ-maintained, terminally differentiated or Go cells will even adapt appropriately. So, for example, human eccrine sweat glands will, after 7 days' maintenance, demonstrate a five-times greater rise in cyclic AMP content after stimulation with 10^{-5} mol L^{-1} isoprenaline than freshly isolated glands[16]. This maintenance supersensitivity reflects the denervation supersensitivity of tissues *in situ*[52]. Appropriately, the maintained human eccrine sweat gland demonstrates a maintenance hyposensitivity of cyclic GMP response to acetylcholine[16], and this mimics the known *in vivo* denervation hyposensitivity of human eccrine sweat glands[53], and the described *in vivo* denervation hyposensitivity of other exocrine glands' muscarinic responses[54].

The retention of fully differentiated function by dividing stem cells in organ maintenance is, however, less complete; no general rules can be derived, as is illustrated by maintained sebaceous glands and hair follicles.

Sebaceous glands

Sebum secretion is holocrine. Sebocytes arise from peripheral stem cells, and, as they move centrally, they engorge with lipid droplets until they burst, releasing their contents into the sebaceous duct. Tracer experiments following the *in vivo* injection of [^{3}H]thymidine or [U-^{14}C]leucine, have suggested a sebocyte transit time of 7–28 days[55,56].

In vitro, on organ maintenance, the rates of [^{3}H]thymidine and [U-^{14}C] leucine uptake into TCA precipitable material are unchanged over 14 days[50]. (TCA, or trichloroacetic acid, precipitates biological macromolecules, such as DNA and proteins.) [^{3}H]Thymidine autoradiography also shows a normal peripheral sebocyte nuclear localization[50]. But these new cells do not differentiate properly. They show few, if any, lipid droplets, and over 14 days' maintenance, the glandular rate of lipid synthesis falls progressively[50].

It appears, microscopically, that sebocytes which were already committed to differentiation on isolation continue to differentiate: maintained sebaceous glands lose the gradual centripedal progression of maturing cells but show, instead, a large peripheral zone of undifferentiated cells, which border abruptly on a central core of highly mature cells; these latter presumably represent the final maturation stage of those sebocytes that were already committed to differentiation when the glands were isolated. This seems to confirm the general rule suggested above that differentiated cells (or differentiating terminal cells) retain their differentiated function on organ maintenance, unlike stem cells. Nonetheless, the organ-maintained sebaceous gland has proved a good model for studying the biological effects of retinoids and androgens. 13-*cis*-Retinoic acid causes a 4-fold reduction in DNA synthesis, and testosterone doubles the rates of lipid synthesis[50].

Hair follicles

Hair growth is mediated by the division and keratinization of the epithelial matrix cells of the hair shaft. [^{3}H]Thymidine autoradiography of freshly isolated hair follicles shows intense nuclear uptake by such cells[18], but, on organ maintenance, hairs cease to grow, and they change from anagen (the growth phase) to catagen (the transition to the resting phase) or a catagen-like state[51].

Cell culture

This is discussed by other contibutors to this book. Let it be noted here that we have successfully grown primary epithelial cells from human eccrine sweat glands[57], sebaceous glands and hair follicles (our unpublished observations). While the two latter were no surprise, cell turnover being a normal *in vivo* feature of such tissues, the success with eccrine sweat glands has shed light on their cell biology. Mitotic figures within the secretory coil of sweat glands are not seen *in vivo*[58], and it has been suggested that they are terminally differentiated, like neurones or muscle cells. But we have shown that new secretory coil cells will start to grow within 36 h of explant[57]. This establishes that sweat gland secretory coil cells *in vivo* are in Go and not terminally differentiated. It has long been known, of course, that distal collecting duct cells divide *in vivo*[58], to maintain their structure within the epidermis which turns over rapidly, so the primary culture *in vitro*[57] of collecting cells was less of a surprise.

CONCLUSION

We have shown that human skin glands and hair follicles can be isolated intact and viable and in large numbers by shearing. The metabolic and cell biological studies discussed here indicate that they may prove useful models for studying dermal biology and pathology.

Acknowledgements

I thank Dr J. H. Barth, Miss J. Gowdy, Dr C. Jones, Dr C. Lee, Mr M. Philpott, Mrs C. Ridden, Mr J. Ridden, Mr M. Tiffen and Miss C. Welsh for their collaboration over the last few years. I thank Professor C. N. Hales, Dr P. Kent, Dr D. Harrison, Miss R. Maxwell, Professor Sir Philip Randle and Dr A. Skillen for their help and advice. This work has been supported by the Cystic Fibrosis Research Trust, the Wellcome Trust, the MRC, the SERC and the British Association of Dermatologists.

Note in proof: We have now successfully maintained growing hair follicles in anagen during organ maintenance[59].

References

1. Talamo, R. C., Rosenstein, B. J. and Berninger, R. W. (1983). In Stanbury, J. B., Wyngarden, J. B., Frederickson, D. S., Goldstein, J. L. and Brown, M. S. (eds.) *Metabolic Basis of Inherited Disease*, pp. 1889–1917. (New York: McGraw-Hill)
2. Kligman, A. M. (1974). An overview of acne. *J. Invest. Dermatol.,* **62**, 268–287
3. Munro, D. D. and Darley, C. R. (1986). In Fitzpatrick, T. B., Eisen, A. Z., Wolff, K., Freedberg, I. M. and Austen, K. F. (eds.) *Dermatology in General Medicine*, pp. 395–418. (New York: McGraw-Hill)
4. Lever, W. F. and Schaumberg-Lever, G. (1983). In *Histopathology of the Skin*, pp. 293–294. (New York: Lippincott)
5. Prompt, C. A. and Quinton, P. M. (1978). Functions of calcium in sweat secretion. *Nature (London),* **272**, 171–172
6. Cooper, M. F., McGrath, H. and Shuster, S. (1976). Sebaceous lipogenesis in human skin. *Br. J. Dermatol.,* **94**, 165–172
7. Hardy, M. H. (1949). The development of mouse hair *in vitro* with some observations on pigmentation. *J. Anat. (London),* **83**, 364–384

8. Quinton, P. M. (1983). Chloride impermeability in cystic fibrosis. *Nature (London)*, **301**, 421–442
9. Sato, K. and Sato, F. (1984). Defective beta adrenergic response of cystic fibrosis sweat glands in vivo and in vitro. *J. Clin. Invest.*, **73**, 1763–1771
10. Sato, K. (1980). Pharmacological responsiveness of the myoepithelium of the isolated human axillary apocrine sweat gland. *Br. J. Dermatol.*, **103**, 235–243
11. Hay, J. B. and Hodgins, M. B. (1978). Distribution of androgen metabolizing enzymes in isolated tissues of human forehead and axillary skin. *J. Endocrinol.*, **79**, 29–39
12. Frater, R. (1983). Inhibition of growth of hair follicles by a lectin-like substance from rat skin. *Austr. J. Biol. Sci.*, **36**, 411–418
13. Kealey, T. (1983). The metabolism and hormonal responses of human eccrine sweat glands isolated by collagenase digestion. *Biochem. J.*, **212**, 143–148
14. Okada, N., Kitano, Y. and Morimoto, T. (1983). The isolation of a viable eccrine sweat gland by dispase. *Arch. Dermatol. Res.*, **275**, 130–133
15. Collie, G., Buchwald, M., Harper, P. and Riordan, J. R. (1985). The culture of sweat gland epithelial cells from normal individuals and patients with cystic fibrosis. *In Vitro Cell. Dev. Biol.*, **21**, 517–602
16. Lee, C. M., Jones, C. J. and Kealey, T. (1984). Biochemical and ultrastructural studies of human eccrine sweat glands isolated by shearing and maintained for seven days. *J. Cell Sci.*, **72**, 259–274
17. Kealey, T., Lee, C. M., Thody, A. J. and Coaker, T. (1986). The isolation of human sebaceous glands and apocrine sweat glands by shearing. *Br. J. Dermatol.*, **114**, 181–188
18. Green, M. R., Clay, C. S., Gibson, W. T., Hughes, T. C., Smith, C. G., Westgate, G. E., White, M. and Kealey, T. (1986). The rapid isolation in large numbers of intact, viable, individual hair follicles from skin: biochemical and ultrastructural characterization. *J. Invest. Dermatol.*, **87**, 768–770
19. Atkinson, D. E. (1968). The energy charge of the adenylate pool as a regulatory parameter. Interaction with feedback modifiers. *Biochemistry*, **7**, 4030–4034
20. Stanley, P. E. and Williams, S. G. (1969). Use of liquid scintillation spectrometer for determining adenosine triphosphate by the luciferase enzyme. *Anal. Biochem.*, **29**, 381–392
21. Spielman, H., Jacob-Muller, B. and Schulz, P. (1981). Simple assay of 0.1–1.0 pmol of ATP, ADP and AMP in single somatic cells using purified luciferin luciferase. *Anal. Biochem.*, **113**, 172–178
22. Krebs, H. A. (1972). The Pasteur effect and the relations between respiration and fermentation. *Essays Biochem.*, **8**, 1–34
23. Newsholme, E. A. and Leech, A. R. (1983). *Biochemistry for the Medical Sciences.* (Chichester and New York: Wiley)
24. Quinton, P. M. (1987). Physiology of sweat secretion. *Kidney Int.*, **32** (Suppl. 21), S102–S108
25. Sato, K. and Dobson, R. L. (1973). Glucose metabolism of the isolated eccrine sweat gland. *J. Clin. Invest.*, **52**, 2166–2174
26. Welch, C., Bryant, A. and Kealey, T. (1989). An NADH-linked luciferase assay for glycogen: the preparation of glycogen-free, viable human eccrine sweat glands. *Anal. Biochem.*, **176** 228–233
27. Krebs, H. A. and Henseleit, K. (1932). Untersuchungen uber die Harnstoffbilding im Tierkorper. *Hoppe-Seyler's Z. Physiol. Chem.*, **210**, 33–36
28. Hammerstedt, R. H. (1973). The use of Dowex-1-borate to separate 3HOH from 2-[^{3}H]-glucose. *Anal. Biochem.*, **56**, 292–293
29. Neely, J. R., Denton, R. M., England, P. J. and Randle, P. J. (1972). The effects of increased heart work on the tricarboxylate and its interactions with glycolysis in the perfused rat heart. *Biochem. J.*, **128**, 147–159
30. Shelley, W. B., Hurley, H. J. and Nicols, A. C. (1953). Axillary odor: experimental study of the role of bacteria, apocrine sweat and deodorants. *Arch. Dermatol.*, **68**, 430–446
31. Leyden, J. J., McGinley, K. J., Holzl, E., Labows, J. N. and Kligman, A. M. (1981). The microbiology of the human axilla and its relationship to axillary odor. *J. Invest. Dermatol.*, **77**, 413–416.
32. Barth, J. H., Ridden, J., Philpott, M. P., Grenall, M. J. and Kealey, T. (1989). Lipogenesis by isolated human apocrine sweat glands: Testosterone has no effect during long-term

organ maintenance. *J. Invest. Dermatol.,* **92** 333–336

33. Greene, R. S., Downing, D. T., Pochi, P. E. and Straus, J. S. (1970). Anatomical variation in the amount and composition of human skin surface lipid. *J. Invest. Dermatol.,* **54**, 240–247

34. Cassidy, D. M., Lee, C. M., Laker, M. F. and Kealey, T. (1986). Lipogenesis in isolated human sebaceous glands. *FEBS Lett.,* **200**, 173–176

35. Pochi, P. E. and Straus, J. S. (1974). Endocrinologic control of the development and activity of the human sebaceous gland. *J. Invest. Dermatol.,* **62**, 191–201

36. Kealey, T. (1988). Hypovitaminosis A of follicular duct as cause of acne vulgaris. *Lancet,* **2**, 449

37. Kealey, T. (1988). Hypovitaminosis A, follicular ducts, and acne. *Lancet,* **2**, 1260–1261

38. Straus, J. S. and Kligman, A. M. (1960). The pathological dynamics of acne vulgaris. *Arch. Dermatol.,* **82**, 779–790

39. Randal, W. C. and Kimura, K. (1955). The pharmacology of sweating. *Pharmacol. Rev.,* 7, 365–397

40. Dale, H. H. and Feldberg, W. (1934). The chemical transmission of secretory impulses to the sweat glands of the cat. *J. Physiol. (London),* **82**, 121–128

41. Sato, K. (1977). The physiology, pharmacology and biochemistry of the eccrine sweat gland. *Rev. Physiol. Biochem. Pharm.,* **79**, 51–131

42. Berridge, M. J. (1984). Inositol triphosphate and diacylglycerol as second messengers. *Biochem. J.,* **22**, 345–360

43. Berridge, M. J., Dawson, R. H., Downes, C. P., Heslop, J. P. and Irvine, R. F. (1983). Changes in levels of inositol phosphates after agonist-dependent hydrolysis of membrane phosphoinositides. *Biochem. J.,* **212**, 473–482

44. Goldberg, N. D. and Maddox, M. K. (1977). Cyclic GMP metabolism and involvement in biological regulation. *Ann. Rev. Biochem.,* **46**, 823–896

45. Harper, J. F. and Brooker, G. (1975). Femtomole sensitive radioimmunoassay after 2'^{o} acetylation by acetic anhydride in aqueous solution. *J. Cyclic Neucleotide Res.,* **1**, 207–218

46. Sato, K. and Sato, F. (1984). Cyclic GMP accumulation during cholinergic stimulation of eccrine sweat glands. *Am. J. Physiol.,* **247**, C234–C239

47. Jones, C. J. and Kealey, T. (1987). Electrophysiological and dye-coupling studies on secretory, myoepithelial and duct cells in human eccrine sweat glands. *J. Physiol.,* **389**, 461–481

48. Jones, C. J., Hyde, D., Lee, C. M. and Kealey, T. (1986). Electrophysiological studies on isolated human eccrine sweat glands. *Q. J. Exp. Physiol.,* **71**, 123–132

49. Freshney, R. I. (1987). *Culture of Animal Cells.* (New York: Alan R. Liss)

50. Ridden, J. and Kealey, T. (1990). Organ maintenance of human sebaceous glands: In vitro effects of 13-*cis* retinoic acid and testosterone. *J. Cell Sci.,* **95**, 125–136

51. Philpott, M., Green, M. R. and Kealey, T. (1989). Studies on the biochemistry and morphology of freshly isolated and maintained rat hair follicles. *J. Cell Sci.,* **93**, 409–418

52. Thesleff, S. and Sellin, L. C. (1980). Denervation supersensitivity. *Trends Neurol. Sci.,* 3, 122–126

53. Silver, A., Versaci, A. and Montagna, W. (1963). Studies of sweating and sensory function in cases of peripheral nerve injuries of the hand. *J. Invest. Dermatol.,* **40**, 243–258

54. Downes, C. P., Dibner, M. D. and Hanley, M. P. (1983). Sympathetic denervation impairs agonist-stimulated phosphatidylinositol metabolism in rat parotid glands. *Biochem. J.,* **214**, 865–870

55. Epstein, E. H. and Epstein, W. C. (1966). New cell formation in sebaceous glands. *J. Invest. Dermatol.,* **46**, 453–458

56. Plewig, G. and Christopher, E. (1974). Renewal rate of human sebaceous glands. *Acta Dermatovener (Stockholm),* **54**, 177–182

57. Lee, C. M., Carpenter, F., Coaker, T. and Kealey, T. (1986). The primary culture of epithelia from the secretory coil and collecting duct of normal human and cystic fibrotic eccrine sweat glands. *J. Cell Sci.,* **83**, 103–118

58. Christopher, E. and Plewig, G. (1973). The formation of the acrosyringium. *Arch. Dermatol.,* **107**, 378–382

59. Philpott, M., Green, M. R. and Kealey, T. (1990). Human hair growth in vitro. *J. Cell Sci..* (in press)

14
Electrophysiological and Morphological Studies on Secretory and Reabsorptive Segments of the Human Eccrine Sweat Gland and on Primary Cell Cultures Established from these Regions

C. J. JONES

INTRODUCTION

Epithelial tissues are routinely dissected from laboratory animals for use in scientific research but investigations on human epithelia have been hampered by the difficulty in obtaining fresh normal samples routinely. Skin represents an obvious exception in view of its accessibility and has provided cells for fibroblast culture (for example references 1 and 2) and glands for direct study (for example references 3–6), organ maintenance[7], primary cell culture[8–14] and established cell lines[15,16]. The glands can either be microdissected from small skin plugs obtained by punch biopsy, or isolated by the shearing procedure from larger skin specimens obtained from individuals undergoing elective surgery. In the first report on shearing[4], the skin specimens were pretreated with collagenase, but this was later found to be unnecessary[3,5,7,17]. The isolation of skin glands without the need to preincubate with collagenase was especially advantageous, particularly if the glands were to be used for direct physiological and biochemical investigations where enzyme pretreatment might prove deleterious. Further information on the bulk isolation of skin glands can be found in Chapter 13 by Terence Kealey.

The present chapter will focus on the human eccrine sweat gland, a cutaneous exocrine gland with thermoregulation as its primary function. It is one of the most accessible human exocrine glands and may serve as a general model for other exocrine glands, particularly with respect to ion and fluid secretion. Normally a single blind-ended tubule, the eccrine sweat gland consists of a coiled portion located in the dermis and an uncoiled

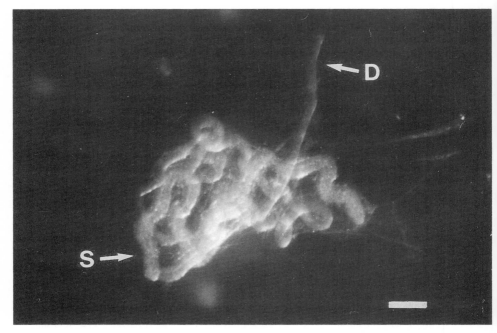

Figure 14.1   Coiled region of a human eccrine sweat gland isolated by shearing. The secretory (S) and reabsorptive ductal (D) components can be distinguished. The reabsorptive duct has a smoother surface appearance and narrower outer diameter than the secretory tubule. Scale bar = 200 μm. Figure reproduced by kind permission of Dr Terence Kealey

portion connecting with the skin surface. The coiled portion comprises a secretory component generating a sodium chloride-rich isotonic fluid[18,19] which then passes into the coiled reabsorptive duct, where some of the sodium and chloride are reabsorbed in excess of water[20] (Figure 14.1). Further reabsorption may occur in the ascending duct which also serves as a conduit, conducting hypotonic sweat to the skin surface.

The eccrine sweat gland, therefore, gives one the opportunity to study secretory and reabsorptive processes in a human exocrine gland. The major problem with such investigations is the small size of the gland; the tubule is approximately 6–8 mm in length and only 30–60 μm in outer diameter. Yet, in spite of this, it has been possible to devise microtechniques to cannulate the secretory tubule and collect the primary secretion[19], to microperfuse the reabsorptive duct lumen[20] and to analyse the elemental content of samples less than 1 nl in volume[21].

The reason why so much attention has been focussed on the eccrine sweat gland, in the face of undoubted practical difficulties, is the involvement of this organ in cystic fibrosis (CF), an inherited disease of ion transport known only in man. The levels of sodium and chloride are significantly elevated in the sweat of patients with CF[22]. The CF sweat gland fails to secrete in response to the β-adrenergic agonist isoprenaline (isoproterenol)[23] and the reabsorptive duct demonstrates a markedly reduced capacity to reabsorb chloride[24,25]. Unlike the exocrine pancreas, the sweat gland does not show ductal inspissation and progressive pathological degeneration during the course of the disease, probably because sweat is relatively serous in nature.

The presence of an abnormal chloride conductance in cystic fibrosis has been confirmed in both *in vivo* and *in vitro* studies with intact nasal tissues, with primary cultures of respiratory epithelial cells and with transformed cell lines established from respiratory epithelia. These findings are discussed in detail in Chapter 10 by Michael Van Scott, James Yankaskas and Richard Boucher. In cultured airway epithelial cells, the patch-clamp technique has been used to focus upon the operation of chloride channels and this is reviewed in Chapter 12 by Jeffrey Smith. It has been shown in cell-attached recordings that cyclic AMP-mediated agonists, including epinephrine, forskolin, isoprenaline and 8-bromo-cyclic AMP, evoke chloride channel activity in normal airway epithelial cells but not in those derived from CF subjects[26,27]. Intracellular cyclic AMP levels increase normally in CF cells in response to β-agonists[23,27,28] and, in view of the fact that patent chloride channels could be demonstrated in excised membrane patches from both normal and CF cells, the present focus is directed towards the regulation of the chloride channel.

To date, it has not been possible to confirm these patch-clamp findings in cell cultures established from normal and CF human eccrine sweat glands and the reasons for this are unclear. Little is known of the properties of individual cell types in the human eccrine sweat gland and this in turn limits the degree to which cultures derived from these cells can be characterized.

Now that the gene for cystic fibrosis has been identified[29], cDNA cloned and characterized[30] and a protein structure predicted[30], the need for stable cell cultures and immortalized cell lines from normal and CF epithelia is redoubled. Also, once agents have been developed to counteract the deleterious effects of the CF mutation, the eccrine sweat gland and stable cultures of affected cells could serve as useful assay systems.

The present chapter will review the known properties of different cell types in the human eccrine sweat gland and will consider the extent to which primary cell cultures established from the secretory and reabsorptive regions of the gland express the properties of their progenitors.

PROPERTIES OF THE HUMAN ECCRINE SWEAT GLAND

The secretory tubule

The wall of the secretory tubule comprises a pseudostratified columnar epithelium of clear (non-granular) and dark (granular) cells, bounded on the basal surface by a discontinuous layer of myoepithelial cells[31,32] between which basal infoldings of the secretory cells can be discerned (Figure 14.2). The clear and dark cells are considered to fulfil different secretory roles. The presence in the human eccrine sweat gland of more than one type of

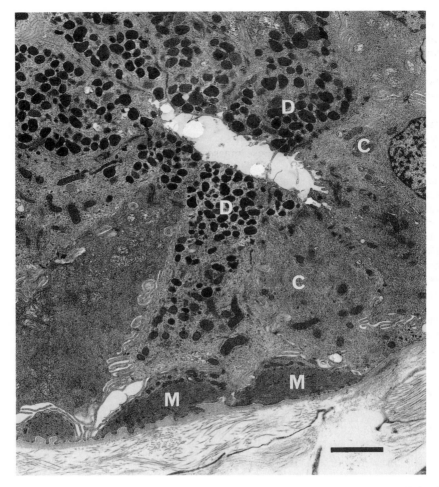

Figure 14.2 Electron micrograph showing a transverse section through the secretory tubule of the human eccrine sweat gland. The clear (C), dark (D) and myoepithelial (M) cells can be distinguished. Scale bar = 2 μm

secretory cell is a feature shared in common with several exocrine glands, including the human submandibular[33] and bronchial submucosal glands[34]. Electrophysiological studies of glandular epithelia[35] have concentrated largely on acini comprising one secretory cell type (homocrine glands), for example the lacrimal, pancreas, parotid and, in the case of the rat and mouse, the submandibular glands[36].

In the human eccrine sweat gland, dark cells, so named because of their cytoplasmic basophilia[37], show in their pseudostratification a structural bias towards the main lumen and possess osmiophilic, acid mucopolysaccharide-rich granules, the contents of which are thought to be secreted into the lumen by exocytosis[31,38]. Clear cells do not contain these granules and are considered from ultrastructural comparisons of resting and actively secreting glands to be primarily responsible for fluid secretion, much of which is thought to pass into intercellular canaliculi. These are branches of the main tubule lumen and serve to amplify the surface area for ion and fluid transport. The apical surface of the clear cells is further amplified by microvilli, which are present in an ordered array (Figure 14.3). Microvilli are also found on the apical surface of the dark cells but here the arrangement is disordered (Figure 14.3) and this would be consistent with an environment exposed to frequent exocytotic events. Myoepithelial cells (Figure 14.4), which contain dense masses of myofilaments, are located on the epithelial side of the basement membrane and normally run parallel to the long axis of the secretory tubule[39]. Qualitative photomicrographic studies made with resting and acetylcholine-stimulated sweat glands obtained from the monkey palm[40], and subsequent isometric measurements made with microdissected lengths of secretory tubule[41] have attributed a contractile function to these cells. Sato et al.[41] concluded that the myoepithelial cells probably provide structural support and enable the secretory epithelium to develop the high luminal hydrostatic pressures that may be necessary to deliver sweat to the skin surface.

Between the secretory tubule and the reabsorptive duct is found a short narrow segment of tubule, the transition zone, consisting of a layer of columnar epithelial cells superimposed on a thin basal cell layer[42]. Because of its small size, very little work has been done on this region and its function is unclear.

The reabsorptive duct

The reabsorptive duct, on the other hand, has been extensively investigated. In the coiled part of the gland, the duct wall comprises initially two layers of essentially cuboidal epithelial cells which show marked structural differences between the inner and outer cell layers (Figure 14.5). Cells of the inner layer possess a luminal surface amplified with stubby microvilli and the cytoplasm contains abundant tonofilaments linking with spot desmosomes liberally distributed along the highly convuluted border with neighbouring apical cells[31] (Figure 14.5). The tonofilaments are considered to provide structural support. In studies where sweat glands were maintained in organ culture for up to 7 days, the inner layer cells were shown to withstand considerable luminal dilatation[7] (Figure 14.6). The cells of the outer layer, which appear less closely linked than their inner layer neighbours, are very rich in mitochondria (Figure 14.5) and some of the ATP generated will fuel the operation of the Na^+/K^+ ATPase which serves to maintain an effective gradient for sodium reabsorption from the duct lumen. The enzyme is known to be located basolaterally from studies of $[^3H]$ouabain binding[43] and from the demonstration that sodium reabsorption is inhibited by basolateral but not luminal ouabain[20].

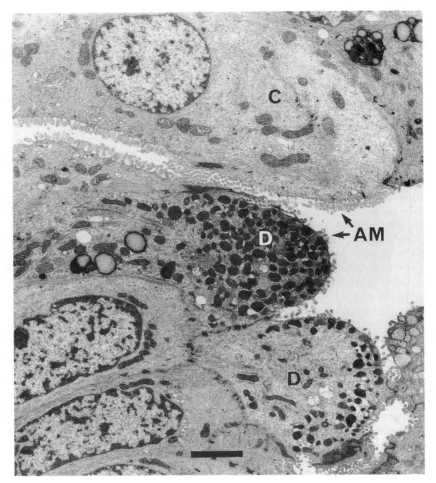

Figure 14.3 Electron micrograph showing the luminal surface of the secretory epithelium in the human eccrine sweat gland. Apical microvilli (AM) of the clear cell (C) are ordered whereas those of the dark cells (D) are disordered. Scale bar = 2 μm

Prolactin-like immunoreactivity

The hormone, prolactin, has been implicated in the regulation of ion and fluid transport across epithelial cell membranes in both mammals and fishes. In mammals, prolactin increases the reabsorption of sodium in mammary cells[44-48] and the reabsorption of sodium, chloride and water in renal[49-51] and intestinal epithelia[52-54]. Because of this and the fact that chloride transport is known to be abnormal in cystic fibrosis, sections through the human eccrine sweat gland have been investigated for prolactin-like immunoreactivity[55]. Clear cells of the secretory coil were found to stain strongly for the presence of prolactin or a closely related substance whereas

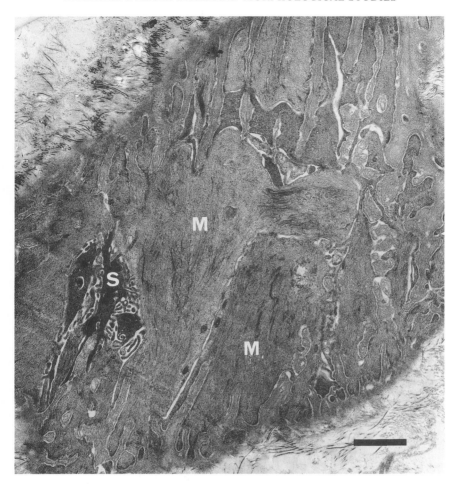

Figure 14.4 Electron micrograph of myoepithelial cells (M) on the basal surface of the secretory tubule. The basal infoldings of secretory cells (S) can be discerned between the myoepithelial cells. Scale bar = 2 μm

virtually no stain was detected in the dark cells. In the reabsorptive duct, patchy staining was observed in cells of the outer layer but none was observed in the inner layer. These results are illustrated in colour in reference 55. The finding of prolactin-like immunoreactivity in certain cell types in the human eccrine sweat gland raises the possibility of a role for this substance in sweat gland function, the nature of which remains to be determined.

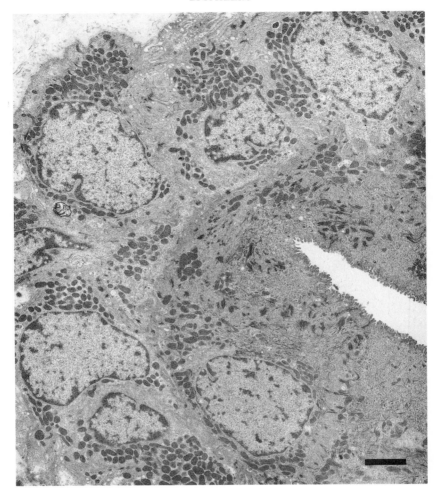

Figure 14.5 Electron micrograph of a transverse section through the reabsorptive duct showing the apical and basal cell layers. Scale bar = 2 μm. (from reference 7, with permission)

ELECTROPHYSIOLOGICAL PROPERTIES OF DIFFERENT CELLULAR COMPARTMENTS: THE CELL SIGNATURE CONCEPT

Because the coiled region of the human eccrine sweat gland contains at least five different cell types, excluding those present in the transition zone, any attempt to associate electrophysiological responsiveness with a particular cell type will require some form of labelling, permitting subsequent identification of the cell type from which intracellular recordings were made. As an added bonus, if the substance used for labelling also passes

312

Figure 14.6 Transverse section through the reabsorptive duct of a sweat gland maintained in organ culture for 7 days. The epithelial cells are withstanding considerable luminal distension. Scale bar = 2 μm. (from reference 7, with permission)

through gap junctions, the dye-coupling status of the cell type can be ascertained. Should a unique response or combination of responses be confirmed for a particular cell type or coupled compartment, then routine labelling would no longer be necessary. The labelling of cells in sweat glands obtained from patients with cystic fibrosis will determine whether the same electrophysiological and dye-coupling properties apply there. Such signatures derived from individual cell types or dye-coupled compartments could then serve as benchmarks with which to assess the properties of cells in culture relative to their progenitors.

Electrophysiological recordings were made initially from cells in the coiled portion of the human eccrine sweat gland and later from cells in short lengths of secretory and reabsorptive tubule microdissected from this region. To ascertain which cell type was impaled with the recording microelectrode, the fluorescent naphthalimide dye, Lucifer Yellow CH (mol.wt. 457)[56,57], was introduced from the microelectrode tip into the cells by hyperpolarizing current once the electrophysiological recordings had been made. A procedure was developed for the embedding of small specimens in glycol methacrylate-based resin for photomicrography and serial sectioning[58]. Treated glands and fragments thereof were embedded and then examined by transmission fluorescence microscopy, first as whole mounts and subsequently as 5 µm serial sections. Sections containing labelled cells were photographed and then stained with Toluidine Blue to complete the identification of the cell types involved.

In addition, microdissected lengths of reabsorptive duct were investigated by luminal microperfusion and intracellular impalement in the absence of bicarbonate and by intracellular impalement in the presence of bicarbonate. By combining luminal microperfusion with intracellular impalement, transepithelial (V_t) and basolateral (V_b) resting potentials could be recorded simultaneously. The acquisition of stable intracellular impalements was greatly aided by immobilizing sweat gland segments between suction micropipettes and by using a piezoelectric microstepper to advance the recording microelectrode.

ELECTROPHYSIOLOGICAL AND DYE-COUPLING PROPERTIES OF CELL TYPES IN NORMAL AND CF ECCRINE SWEAT GLANDS

Two studies have been performed. The first involved the use of sweat gland coils which contained both secretory and reabsorptive regions[59] and the second utilized microdissected fragments of secretory coil and reabsorptive duct[60].

In the former study, forty-five successful Lucifer Yellow microiontophoreses were performed, three of which were shown to be in secretory cells, twelve in myoepithelial cells and thirty in cells of the reabsorptive duct. Secretory cells were therefore identified as the most difficult to impale and reabsorptive duct cells as the easiest in the coiled part of the gland. In the case of sweat gland coils obtained from patients with cystic fibrosis, six microiontophoreses were performed and one secretory cell, two myoepithelial cells and three reabsorptive duct cells were subsequently found to be labelled. All sites of labelling in normal and CF sweat gland coils were confirmed by analysis of 5 µm serial sections.

The second study focussed on microdissected fragments of secretory coil in an effort to investigate further the dye-coupling status and electrophysiological properties of the clear and dark cells.

Secretory cells

The reason for the difficulty in impaling secretory cells is unknown. The presence of an almost complete layer of myoepithelial cells at the basal

314

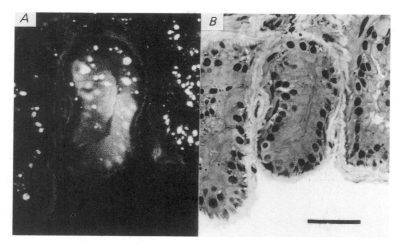

Figure 14.7   Extensive dye-coupling between columnar cells in the secretory epithelium of the human eccrine sweat gland. The cells in this region do not show pseudostratification and so are considered to be in the transition zone.
A 5 μm transverse section observed by transmission fluorescence microscopy. **B** The same section after staining with Toluidine Blue and observed in normal light. Scale bar = 50 μm. (from reference 59, with permission)

surface may be a factor. Alternatively, secretory cells may be more sensitive to impalement and so decouple from neighbouring cells. Unfortunately, secretory cell impalements were relatively short-lived and so there was little opportunity for experiment.

All of the secretory cells labelled in normal glands showed dye-coupling to neighbouring cells. In one case (V_b = −63 mV), Lucifer Yellow spread to two selected neighbours; in another case (V_b = −40 mV), seven secretory cells were involved in labelling; and in the third case (V_b = −52 mV), more than fifty adjacent columnar cells were found to be dye-coupled (Figure 14.7). In the latter case, no pseudostratification was observed in the columnar epithelium, and it seems likely that the region labelled lay in the transition zone rather than the secretory tubule. Only one secretory cell was successfully impaled (V_b = −48 mV) in eccrine sweat glands obtained from two patients with cystic fibrosis and the result showed selective dye-coupling (Figure 14.8). The labelled cells showed a luminal bias in their pseudostratification and this, taken together with their increased basophilia on staining with Toluidine Blue, would argue in favour of selective dark cell–dark cell coupling.

In view of the latter finding and because secretory cells proved so difficult to impale successfully, a further study was undertaken in which micro-dissected fragments of secretory coil were immobilized between two suction micropipettes and intracellular impalements attempted with the help of a piezoelectric microstepper[60]. Even with these modifications, it proved much more difficult to impale cells in fragments of secretory tubule compared with reabsorptive duct. In an effort to secure stable impalements of

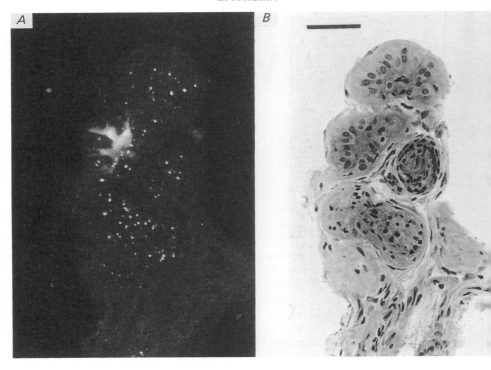

Figure 14.8 **A** 5 μm section of an isolated CF eccrine sweat gland observed by fluorescence microscopy and showing dye-coupling between selected neighbouring secretory cells. **B** After staining the section, the cells involved in dye-coupling are confirmed as secretory and show apical bias and increased Toluidine Blue basophilia, features which are consistent with the properties of dark cells. Scale bar = 50 μm. (from reference 59, with permission)

secretory cells, microelectrode advances of 1–2 μm using the piezoelectric stepper were combined with capacitance overcompensation. Entry potentials of −20 to −60 mV were recorded but these proved difficult to maintain. Lucifer Yellow was introduced if the initial value remained stable for two minutes.

Two basic patterns of dye-labelling were observed. Labelling was either confined to single cells (Figure 14.9) or showed limited spread to selected adjacent cells. In all seven cases where label had spread, serial sectioning of the embedded fragments, followed by staining with Toluidine Blue revealed that only dark cells had been labelled (Figure 14.10). On no occasion was a clear cell shown to be involved in dye-coupling. The fact that dark cells showed preferential labelling was surprising because they are the smaller of the two secretory cell types and their cell bodies are more remote from the impaling microelectrode. The reason why clear cells are so difficult to label is not known. Lucifer Yellow is secreted by certain epithelia, for example malpighian tubule cells of Locusta (J. H. Anstee and D. Hyde, personal

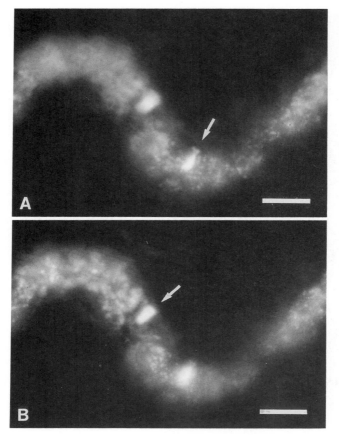

Figure 14.9 Whole mount of a secretory tubule segment containing two sites of Lucifer Yellow labelling, each of which appears to involve a single cell. **A** Smaller wine glass-shaped cell has the morphology of a dark cell. **B** Larger cell with broader basal aspect may be a clear cell. Scale bar = 50 μm. (from reference 60, with permission)

communication) and Rhodnius (S. H. P. Maddrell, personal communication). Perhaps clear cells of the human eccrine sweat gland also secrete Lucifer Yellow. Another possibility is that dye-coupling may not exist in clear cells or that they may decouple very easily under the experimental conditions. They may be more sensitive to impalement damage than dark cells.

Myoepithelial cells

Myoepithelial cells were found to be dye-coupled in both normal and CF eccrine sweat glands. This coupling involved only adjacent myoepithelial cells and not the secretory cells with which they also made contact (Figure

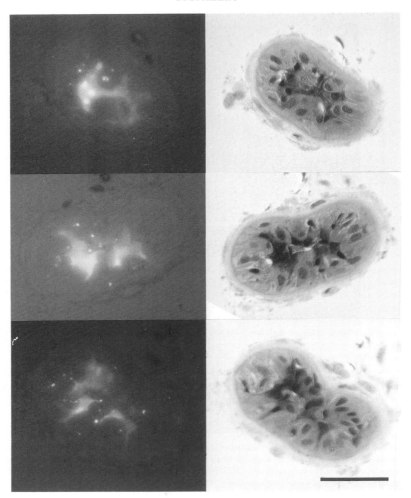

Figure 14.10 Three non-serial 5 μm sections through a group of dye-coupled cells in the secretory tubule. **Left panel**: sections observed with fluorescent optics. **Right panel**: same sections, now stained with Toluidine Blue and observed in normal light. The dye-coupled cells are unequivocally identified as dark cells. Scale bar = 50 μm. (from reference 60, with permission)

14.11). Resting potentials for myoepithelial cells in normal glands were in the range −35 to −65 mV (mean = −52 mV, SEM ± 2.4 mV, n = 12). In eight of these cells, spontaneous depolarizing transients were observed, the amplitude but not frequency of which increased with increasing membrane polarization (Figure 14.12). When glands were maintained for up to thirty hours in organ culture, the frequency and amplitude of the transients fell (Figure 14.13). Administration of acetylcholine (10^{-6}–10^{-7} mol L^{-1}) produced microelectrode dislodgement in cells displaying frequent transients

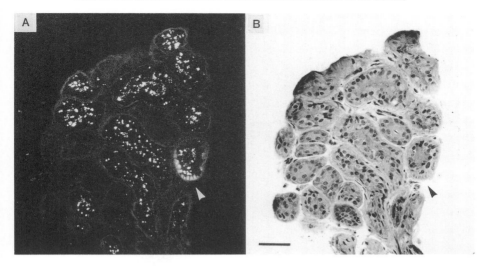

Figure 14.11 5 μm section through the coiled portion of the human eccrine sweat gland showing dye-coupling exclusively between myoepithelial cells (arrowed). **A** Section observed with fluorescent optics. **B** Same section, now stained with Toluidine Blue and observed in normal light. Scale bar = 50 μm. (from references 59, with permission)

and caused membrane potential depolarization in cells showing few or no transients (Figure 14.13). In the latter case, a group of depolarizing transients was sometimes observed during or after the acetylcholine-induced depolarization. Spontaneous depolarizing transients were also observed in recordings from myoepithelial cells in CF eccrine sweat glands.

Reabsorptive duct cells

In experiments conducted on intact sweat gland coils, thirty cells were labelled in the reabsorptive duct, twenty-six of which showed complete dye-coupling to neighbouring cells in both apical and basal cell layers of the duct wall (Figure 14.14). No relationship was observed between dye-coupling status and basal resting potential which lay in the range −40 to −82 mV (mean = −60 mV, SEM =±2.4, n = 30). Input impedances were recorded between 5 and 30 MΩ. The finding that reabsorptive duct cells are capable of complete dye-coupling is in accord with the presence of numerous gap junctions between the cells and the observation that exploratory impalements along ductal segments produced similar readings for basal resting potential and low input impedance. With CF sweat glands, only one of the three duct cells labelled (V_b = −60 mV) showed complete dye-coupling (Figure 14.15), whereas the remaining two cells (V_b = −58 and −62 mV) showed no spread of label. The results, therefore, indicate the capability for complete dye-coupling and for decoupling in CF sweat glands.

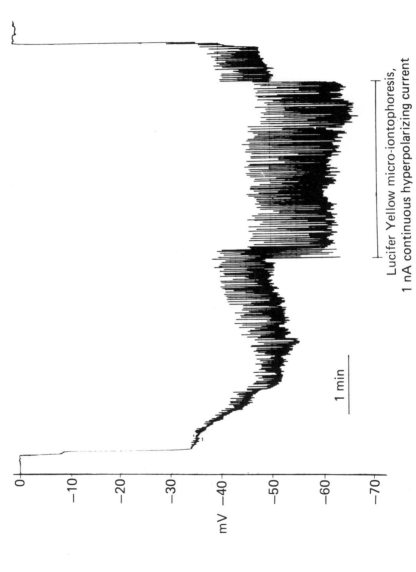

Figure 14.12 Electrical recording made from a myoepithelial cell. Spontaneous depolarizing transients, which increase in amplitude but not frequency with increasing membrane polarization, are superimposed upon the basal membrane potential record. (From reference 59, with permission)

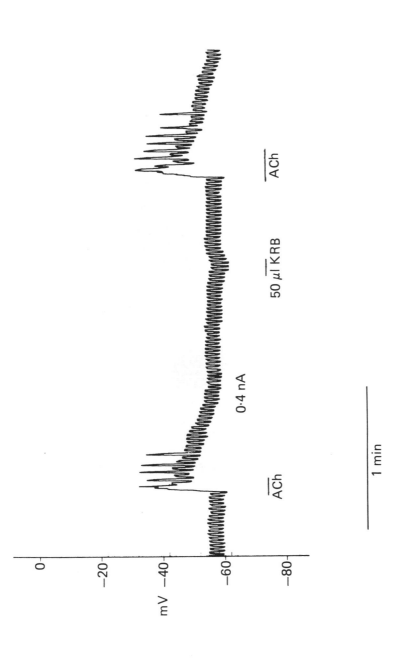

Figure 14.13 Myoepithelial cell electrical recording made from a gland which had been maintained in organ culture. Acetylcholine (10^{-6}–10^{-7} mol L^{-1}) induces membrane potential depolarization and associated depolarizing transients. The oscillations in the recording are explained by 0.4 nA depolarizing current pulses injected into the cell. (From reference 59, with permission)

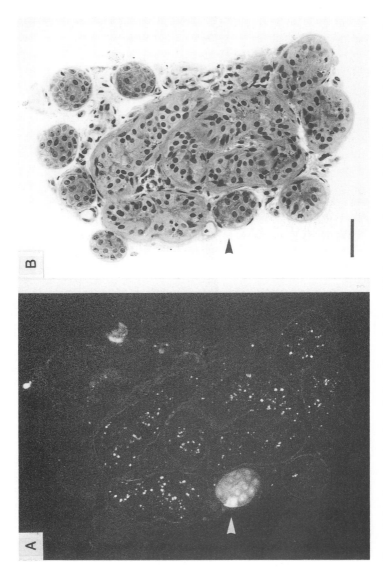

Figure 14.14 5 μm section through the reabsorptive duct of the human eccrine seat gland showing complete dye-coupling between the inner and outer cell layers. **A** Section observed by transmission fluorescence microscopy. **B** The same section after staining with Toluidine Blue and observed in normal light. Scale bar = 50 μm. (From reference 59, with permission)

322

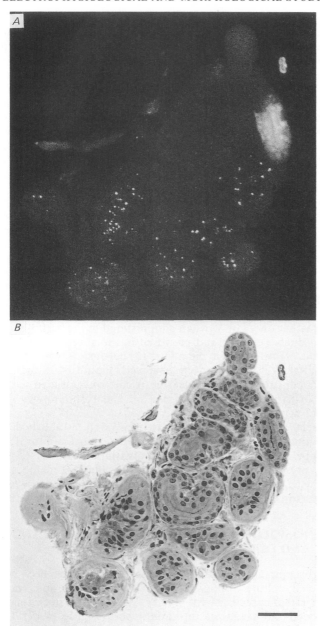

Figure 14.15 Complete dye-coupling between the inner and outer cell layers of the reabsorptive duct in an eccrine sweat gland isolated from a patient with cystic fibrosis. **A** 5 μm section observed by transmission fluorescence microscopy. **B** The same section after staining with Toluidine Blue and observed in normal light. Scale bar = 50 μm. (from reference 59, with permission)

On giving acetylcholine (10^{-6}–10^{-7} mol L^{-1}), ten out of twenty-five cells tested showed no change in V_b, whereas the remainder exhibited depolarizing or hyperpolarizing responses consistent with an apparent reversal potential between −65 and −70 mV. This was in agreement with findings in a previous study by Jones et al.,[3] where the majority of cells impaled would have been duct cells. The finding that some cells did not respond to acetylcholine and others needed several doses to elicit a response might imply an effect secondary to an induction of secretion from the secretory coil. This secretion would subsequently flow into the duct where it would be subject to modification. In experiments with isolated fragments of reabsorptive duct, acetylcholine had no effect on basal resting potential[61] nor did it induce a significant increase in the content of the cholinergic second messenger, cyclic GMP[59]. Moreover, the cholinergic agonist, methacholine (10^{-5}–10^{-6} mol L^{-1}) administered in the bathing solution to isolated microperfused and perifused human sweat ducts in the absence of bicarbonate produced no change in V_b or V_t[61] and the same is true for V_b of ducts perifused in the presence of bicarbonate[62]. These findings would therefore argue against a direct action of cholinergic agonists on the reabsorptive sweat duct.

Cells of the reabsorptive duct have also been exposed to isoproterenol (isoprenaline) in the search for a possible link between the failure of CF sweat glands to secrete in response to isoproterenol[23] and the reduced capacity to reabsorb chloride demonstrated in the duct[24,25]. Isoproterenol had no effect either on V_b or V_t when administered in the bathing solution to normal sweat ducts perfused and perifused in the absence of bicarbonate[61] (Figure 14.16). However, on changing to perifusion with bicarbonate-containing buffer, V_b responses were observed (Figure 14.16), mimicked by forskolin, which took the form of an oscillator, initiated either by a depolarizing or hyperpolarizing response[62,63]. This response was also found to be chloride dependent.

With respect to ion channels, it is now well established that the reabsorptive duct epithelium is equipped with apical amiloride-sensitive sodium channels[20,61,64] and basolateral potassium channels[59,64]. Chloride channels are present in both the apical and basolateral membranes[65] and recent work with chloride-sensitive microelectrodes has confirmed this (M. M. Reddy, personal communication).

DIFFERENT PHYSIOLOGICAL SIGNATURES OF SWEAT GLAND SECRETORY AND DUCT CELLS IN PRIMARY CULTURE

Electronmicroscopic and electrophysiological studies have been performed on primary cell cultures obtained from the secretory and reabsorptive regions of the human eccrine sweat gland[10]. When grown in culture, the cells became flattened and organized into multilayers (Figure 14.17). In spite of these changes in shape, it was still possible to impale the cells with a microelectrode and make intracellular recordings.

The resting potentials for cells cultured from the secretory tubule (−35±2 mV, n = 36) were significantly greater than those recorded for cultured duct cells (−22±1 mV, n = 58, $p \leq 0.01$). On giving the cholinergic

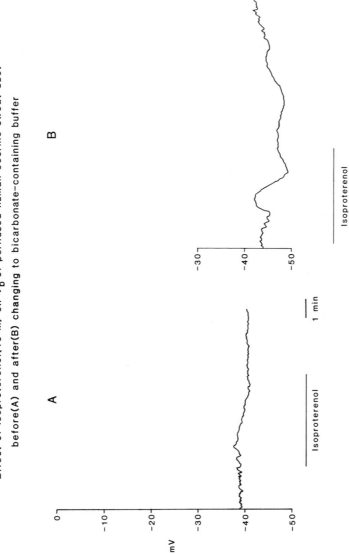

Figure 14.16 Effect of isoproternol (10^{-5} mol L') on the basolateral membrane potential of perifused human eccrine sweat duct before (**A**) amd after (**B**) changing to bicarbonate-containing buffer. (From reference 63, with permission)

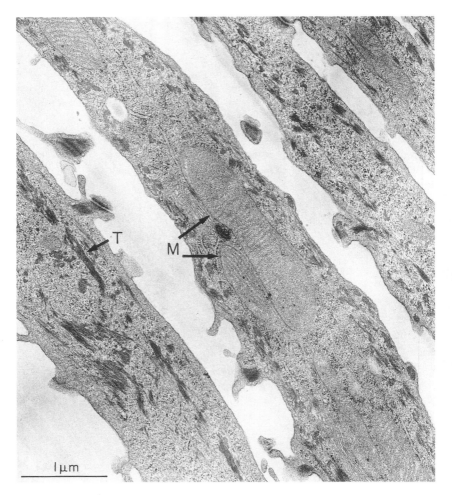

Figure 14.17 Electron micrograph showing layers of cells with flattened morphology in a primary cell culture established from an explanted fragment of secretory tubule. M = mitochondria; T = tonofilaments; Scale bar = 1 μm

agonist, methacholine (10^{-5}–10^{-6} mol L^{-1}), the cultured secretory cells could be distinguished unequivocally by their atropine-sensitive hyperpolarization (-20 ± 2 mV, $n = 43$) (Figure 14.18), whereas no cultured duct cells responded (Figure 14.19). Administration of the sodium conductance antagonist, amiloride (10^{-5} or 10^{-6} mol L^{-1}) caused 44% of cultured secretory cells to respond by hyperpolarization (-8 ± 1 mV, $n = 8$), whereas 87% of the cultured duct cells hyperpolarized (-15 ± 1 mV, $n = 46$) and by a significantly increased margin ($p \leq 0.01$). Replacement of chloride with gluconate in the bathing medium caused membrane potential depolarization in both cultured secretory and duct cells, consistent with the presence of a chloride

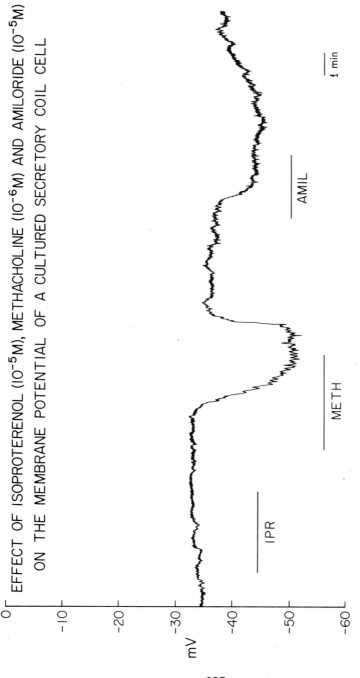

EFFECT OF ISOPROTERENOL (10^{-5}M), METHACHOLINE (10^{-6}M) AND AMILORIDE (10^{-5}M) ON THE MEMBRANE POTENTIAL OF A CULTURED SECRETORY COIL CELL

Figure 14.18 Intracellular recording made from a cultured secretory tubule cell showing a hyperpolarization of 19 mV in response to superfusion with methacholine (METH) (10^{-6} mol L^{-1}). A hyperpolarizing response to amiloride (AMIL) (10^{-5} mol/L^{-1}) is also shown. No change in membrane potential was observed when isoproterenol (IPR) (10^{-5} mol L^{-1}) was administered to this cell. (From reference 10, with permission)

327

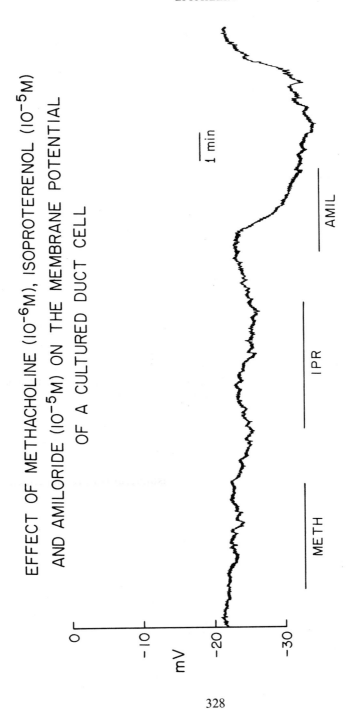

Figure 14.19 Intracellular recording from a cultured reabsorptive duct cell showing that methacholine (METH) (10^{-6} mol L^{-1}) has no effect on the membrane potential. The cell shows a hyperpolarizing response to amiloride (AML) (10^{-5} mol L^{-1}) but no response to isoproterenol (IPR) (10^{-5} mol L^{-1}). (From reference 10, with permission)

conductance in the cell membrane. Isoprenaline produced a transient hyperpolarization of 5–10 mV in three out of six cultured secretory cells tested but had no effect on cultured duct cells.

Administration of methacholine provided a means of distinguishing unequivocally between primary cultures of secretory and ductal origin. This result is in accord with the behaviour of cells in the native epithelium discussed earlier and therefore provides a most promising signature with which to distinguish cultures of secretory and ductal origin. The same cannot be said of the sodium conductance antagonist, amiloride, which produced responses in a proportion of cultured secretory and duct cells. The finding that almost half of the cultured secretory cells responded to amiloride was unexpected in view of the report that serosal amiloride had no detectable effect on methacholine-induced fluid secretion in the intact gland[20]. The possibility that amiloride-sensitive sodium channels may fulfil another role in the native secretory epithelium cannot be excluded, nor can the possibility that a proportion of secretory cells may show altered phenotypic expression when grown in culture.

SUMMARY AND CONCLUSIONS

In the secretory segment of the human eccrine sweat gland, three cell types are present. Myoepithelial cells are exclusively dye-coupled to neighbouring myoepithelial cells, dark cells to dark cells, and the remaining clear cells are either exclusively dye-coupled to each other or are uncoupled. It has been suggested in work with smooth muscle that one of the functions of cell–cell coupling is to co-ordinate the cellular response to packets of neurotransmitter released at discrete sites[66]. In this way, a signal reaching only a few cells could be transmitted to coupled neighbours by an intracellular messenger. Separate coupling compartments for clear and dark cells would mean that they could be independently regulated by hormones and neurotransmitters and so make distinct contributions to the final secretory product. In the reabsorptive duct, complete coupling throughout the epithelium exposes the maximum number of basolateral Na^+/K^+-dependent ATPase sites to intracellular sodium and so establishes a most effective gradient favouring sodium reabsorption from the lumen.

The electrophysiological properties and hormonal responsiveness of some of these cell types and coupled compartments has been discussed together with other reported findings. The function of each cell type has been considered and signatures proposed which could be used to test the veracity of cell cultures raised from secretory and reabsorptive segments. It has been shown that primary cell cultures raised from these regions retain significant differences in hormonal sensitivity appropriate to their progenitors.

Acknowledgements

The author wishes to thank the Cystic Fibrosis Foundation (USA), Cystic Fibrosis Research Trust (UK), Smith Kline Foundation (UK) and the Wellcome Trust (UK) for supporting this work.

References

1. Hayflick, L. and Moorhead, P. S. (1961). The serial cultivation of human dieploid cell strains. *Exp. Cell Res., 25*, 585–621
2. Bettger, W. J., Boyce, S. T., Walthall, B. J. and Ham, R. G. (1981). Rapid clonal growth and serial passage of human diploid fibroblasts in a lipid-enriched synthetic medium supplemented with epidermal growth factor, insulin and dexamethasone. *Proc. Natl. Acad. Sci. (USA), 78*, 5588–5592
3. Jones, C. J., Hyde, D., Lee, C. M. and Kealey, T. (1986). Electrophysiological studies on isolated human eccrine sweat glands. *Q. J. Exp. Physiol., 71*, 123–132
4. Kealey, T. (1983). The metabolism and hormonal responses of human eccrine sweat glands isolated by collagenase digestion. *Biochem. J., 212*, 143–148
5. Kealey, T., Lee, C. M., Thody, A. J. and Coaker, T. (1986). The isolation of human sebaceous glands and apocrine sweat glands by shearing. *Br. J. Dermatol., 114*, 181–188
6. Sato, K. (1977). The physiology, pharmacology and biochemistry of the eccrine sweat gland. *Rev. Physiol. Biochem. Pharmacol., 79*, 51–131
7. Lee, C. M., Jones, C. J. and Kealey, T. (1984). Biochemical and ultrastructural studies of human eccrine sweat glands isolated by shearing and maintained for seven days. *J. Cell Sci., 72*, 259–274
8. Collie, G., Buchwald, M., Harper, P. and Riordan, J. R. (1985). Culture of sweat gland epithelial cells from normal individuals and patients with cystic fibrosis. *In Vitro Cell. Devel. Biol., 21*, 517–602
9. Hazen-Martin, D. J., Spicer, S. S., Sens, M. A., Jenkins, M. Q., Westphal, M. C. and Sens, D. A. (1987). Tissue culture of normal and cystic fibrosis sweat gland duct cells: I. Alterations in dome formation. *Ped. Res., 21*, 72–78
10. Jones, C. J., Bell, C. L. and Quinton, P. M. (1989). Different physiological signatures of sweat gland secretory and duct cells in culture. *Am. J. Physiol., 255 (Cell Physiol., 24)*, C102–C111
11. Lee, C. M., Carpenter, F., Coaker, T. and Kealey, T. (1986). The primary culture of epithelia from the secretory coil and collecting duct of normal human and cystic fibrotic eccrine sweat glands. *J. Cell Sci., 83*, 103–118
12. Lee, C. M. (1988). Human eccrine sweat gland cell culture as a model for cystic fibrosis. In Mastella, G. and Quinton, P. M. (eds.) *Cellular and Molecular Basis of Cystic Fibrosis*, pp. 404–415. (San Francisco: San Francisco Press)
13. Pedersen, P. S. (1984). Primary cultures of epithelial cells derived from the reabsorptive coiled duct of human sweat glands. *IRCS Med. Sci., 12*, 752–753
14. Pedersen, P. S. (1988). The study of epithelial function by in vitro culture of sweat duct cells. In Mastella, G. and Quinton, P. M. (eds.) *Cellular and Molecular Basis of Cystic Fibrosis*, pp. 395–401. (San Francisco: San Francisco Press)
15. Lee, C. M. (1990). Cell culture systems for the study of human skin and skin glands. This volume, Chapter 15
16. Lee, C. M. and Dessi, J. (1989). NCL-SG3: A human eccrine sweat gland cell line that retains the capacity for transepithelial ion transport. *J. Cell Sci., 92*, 241–249
17. Kealey, T. (1988). Phosphorylation studies on the human eccrine sweat gland. In Mastella, G. and Quinton, P. M. (eds.) *Cellular and Molecular Basis of Cystic Fibrosis*, pp., 150–154. (San Francisco: San Francisco Press)
18. Schulz, I., Ullrich, K. J., Fromter, E., Emrich, H. M., Frick, A., Hegel, U. and Holzgreve, H. (1964). Micropuncture experiments on human sweat gland. In Di Sant'Agnese, P. A. (ed.) *Research on Pathogenesis of Cystic Fibrosis*, pp. 136–146. (Bethesda, Maryland, USA: NIAMD)
19. Sato, K. (1982). Mechanism of eccrine sweat secretion. In Quinton, P. M., Martinez, J. R. and Hopfer, U. (eds.) *Fluid and Electrolyte Abnormalities in Exocrine Glands in Cystic Fibrosis*, pp. 35–52. (San Francisco: San Francisco Press)
20. Quinton, P. M. (1981). Effects of some ion transport inhibitors on secretion and reabsorption in intact perfused single human sweat glands. *Pfleugers Arch., 391*, 309–313
21. Quinton, P. M. (1978). Techniques for microdrop analysis of fluids (sweat, saliva, urine) with an energy dispersive X-ray spectrometer on a scanning electron microscope. *Am. J. Physiol., 234*, F255–F259

22. di Sant'Agnese, P. A., Darling, R. C., Perera, G. A. and Shea, E. (1953). Abnormal electrolyte composition of sweat in cystic fibrosis of the pancreas. *Pediatrics*, **12**, 549–563
23. Sato, K. and Sato, F. (1984). Defective beta adrenergic response of cystic fibrosis sweat glands in vivo and in vitro. *J. Clin. Invest.*, **73**, 1763–1771
24. Quinton, P. M. (1983). Chloride impermeability in cystic fibrosis. *Nature (London)*, **301**, 421–422
25. Bijman, J. and Quinton, P. M. (1984). Influence of abnormal Cl impermeability on sweating in cystic fibrosis. *Am. J. Physiol.*, **247** (*Cell Physiol.*, 16), C3–C9
26. Frizzell, R. A., Rechkemmer, R. and Shoemaker, R. L. (1986). Altered regulation of airway epithelial cell chloride channels in cystic fibrosis. *Science*, **233**, 558–560
27. Welsh, M. J. and Liedtke, C. M. (1986). Chloride and potassium channels in cystic fibrosis airway epithelia. *Nature (London)*, **322**, 467–470
28. Boucher, R. C., Stutts, M. J., Knowles, M. R., Cantley, L. and Gatzy, J. T. (1986). Na⁺ transport in cystic fibrosis respiratory epithelia: Abnormal basal rate and response to adenylate cyclase activation. *J. Clin. Invest.*, **78**, 1245–1252
29. Rommens, J. M., Iannuzzi, M. C., Kerem, B., Drumm, M. L., Melmer, G., Dean, M., Rozmahel, R., Cole, J. L., Kennedy, D., Hidaka, N., Zsiga, M., Buchwald, M., Riordan, J. R., Tsui, L-C. and Collins, F. S. (1989). Identification of the cystic fibrosis gene: Chromosome walking and jumping. *Science*, **245**, 1059–1065
30. Riordan, J. R., Rommens, J. M., Kerem, B., Alon, N., Rozmahel, R., Grzelczak, Z., Zielenski, J., Lok, S., Plavsik, N., Chou, J-L., Drumm, M. L., Iannuzzi, M. C., Collins, F. S. and Tsui, L-C. (1989). Identification of the cystic fibrosis gene: Cloning and characterization of complementary DNA. *Science*, **245**, 1066–1073
31. Munger, B. L. (1961). The ultrastructure and histophysiology of human eccrine sweat glands. *J. Biochem. Biophys. Cytol.*, **11**, 385–402
32. Briggman, J. V., Bank, H. L., Bigelow, J. B., Graves, J. S. and Spicer, S. S. (1981). Structure of the tight junctions of the human eccrine sweat gland. *Am. J. Anat.*, **162**, 357–368
33. Jamieson, J. D. (1983). The exocrine pancreas and salivary glands. In Weiss, L. (ed.) *Histology, Cell and Tissue Biology*, 5th Edn., pp. 749–773. (New York: Elsevier)
34. Sorokin, S. P. (1983). The respiratory system. In Weiss, L. (ed.) *Histology, Cell and Tissue Biology*, 5th Edn., pp. 788–868. (New York: Elsevier)
35. Petersen, O. H. (1980). *The Electrophysiology of Gland Cells* (*Physiological Society Monograph no. 36*). (London: Academic Press)
36. Young, J. A. and van Lennep, E. W. (1978). *The Morphology of Salivary Glands*. (London: Academic Press)
37. Montagna, W., Chase, H. B. and Lobitz, W. C. (1953). Histology and cytochemistry of human skin: IV. The eccrine sweat glands. *J. Invest. Dermatol.*, **20**, 415
38. Montgomery, I., Jenkinson, D. McE., Elder, H. Y., Czarnecki, D. and Mackie, R. M. (1984). The effects of thermal stimulation on the ultrastructure of the human atrichial sweat gland. I. The Fundus. *Br. J. Dermatol.*, **110**, 385–397
39. Ellis, R. A. (1965). Fine structure of the myoepithelium of the eccrine sweat glands of man. *J. Cell Biol.*, **27**, 551–563
40. Sato, K. (1977). Pharmacology and function of the myoepithelial cell in the eccrine sweat gland. *Experientia*, **33**, 631–633
41. Sato, K., Nishiyama, A. and Kobayashi, M. (1979). Mechanical properties and functions of the myoepithelium in the eccrine sweat gland. *Am. J. Physiol.*, **237**, C177–184
42. Kurosumi, K., Kurosumi, U. and Tosaka, H. (1982). Ultrastructure of human eccrine sweat glands with special reference to the transitional portion. *Arch. Hist. Jpn.*, **45**, 213
43. Quinton, P. M. and Tormey, J. McD. (1976). Localization of Na/K ATPase sites in the secretory and reabsorptive epithelia of perfused eccrine sweat glands: A question to the role of the enzyme in secretion. *J. Memb. Biol.*, **29**, 383–399
44. Bisbee, C. A. (1981a). Prolactin effects on ion transport across cultured mammary epithelium. *Am. J. Physiol.*, **240** (*Cell Physiol.*), C110–C115
45. Bisbee, C. A. (1981b). Transepithelial electrophysiology of cultured mouse mammary epithelium: sensitivity to prolactins. *Am. J. Physiol.*, **241** (*Endocrinol. Metab.*, 4), E410–413

331

46. Bisbee, C. A., Machen, T. E. and Bern, H. A. (1979). Mouse mammary epithelial cells on floating collagen gels: Trans-epithelial ion transport and effects of prolactin. *Proc. Natl. Acad. Sci. USA*, **76**, 536–540

47. Falconer, I. R. and Rowe, J. M. (1975). Possible mechanism for action of prolactin on mammary cell sodium transport. *Nature (London)*, **256**, 327–328

48. Falconer, I. R. and Rowe, J. M. (1977). Effect of prolactin on sodium and potassium concentrations in mammary alveolar tissue. *Endocrinology*, **101**, 181–186

49. Bliss, D. J. and Lote, C. J. (1982). Effect of prolactin on urinary excretion and renal haemodynamics in conscious rats. *J. Physiol. London*, **322**, 399–407

50. Loretz, C. A. and Bern, H. A. (1982). Prolactin and osmoregulation in vertebrates. Progress in neuroendocrinology. *Neuroendocrinology*, **35**, 292–304

51. Stier, C. T., Cowden, E. A., Friesen, H. G. and Allison, E. M. (1984). Prolactin and the rat kidney: A clearance and micropuncture study. *Endocrinology*, **115**, 362–367

52. Mainoya, J. R. (1978). Possible influence of prolactin on intestinal hypertrophy in pregnant and lactating rats. *Experientia*, **34**, 1230–1231

53. Mainoya, J. R. (1981). Colon absorption of water and NaCl in the rat during lactation and the possible involvement of prolactin. *Experientia*, **37**, 1083–1084

54. Ramsey, D. H. and Bern, H. A. (1972). Stimulation by ovine prolactin of fluid transfer in everted sacs of rat small intestine. *J. Endocrinol.*, **53**, 453–459

55. Walker, A. M., Robertson, M. T. and Jones, C. J. (1989). Distribution of a prolactin-like material in human eccrine sweat glands. *J. Invest. Dermatol.*, **93**, 50–53

56. Stewart, W. W. (1978). Functional connections between cells as revealed by dye-coupling with a highly fluorescent naphthalimide tracer. *Cell*, **14**, 741–759

57. Stewart, W. W. (1981). Lucifer dyes: Highly fluorescent dyes for biological tracing. *Nature (London)*, **292**, 17–21

58. Jones, C. J. (1986). A simple method for embedding small specimens for photomicrography and sectioning following intracellular microiontophoresis of Lucifer Yellow CH. *Histochem. J.*, **18**, 105–108

59. Jones, C. J. and Kealey, T. (1987). Electrophysiological and dye-coupling studies on secretory, myoepithelial and duct cells in human eccrine sweat glands, with an appendix on eccrine sweat glands isolated from two patients with cystic fibrosis. *J. Physiol. (London)*, **389**, 461–481

60. Jones, C. J. and Quinton, P. M. (1989). Dye-coupling compartments in the human eccrine sweat gland. *Am. J. Physiol.*, **256** (*Cell Physiol.*, 25), C478–C485

61. Jones, C. J. and Quinton, P. M. (1987). Intracellular and transepithelial electrical measurements on microperfused ducts from human eccrine sweat glands. *J. Physiol. (London)*, **386**, 81P

62. Jones, C. J. and Quinton, P. M. (1987). Intracellular electrical measurements on perifused ducts isolated from eccrine sweat glands of normal subjects and patients with cystic fibrosis. *J. Physiol. (London)*, **394**, 95P

63. Jones, C. J. (1988). Electrophysiological and dye-coupling studies on cell types in eccrine sweat glands isolated from normal subjects and patients with cystic fibrosis: The pursuit of cell signatures. In Mastella, G. and Quinton, P. M. (eds.) *Cellular and Molecular Basis of Cystic Fibrosis*, pp. 115–123. (San Francisco: San Francisco Press)

64. Bijman, J. and Quinton, P. M. (1987). Permeability properties of cell membranes and tight junctions of normal and cystic fibrosis sweat ducts. *Pfleugers Arch.*, **408**, 505–510

65. Reddy, M. M. and Quinton, P. M. (1989). Localization of Cl conductance in normal and Cl impermeability in cystic fibrosis sweat duct epithelium. *Am. J. Physiol.*, **257** (*Cell Physiol.*, 26), C727–C735)

66. Burnstock, G. and Iwayama, T. (1971). Fine structural identification of autonomic nerves and their relation to smooth muscle. In Eranko, O. (ed.) *Progress in Brain Research, Vol. 34, Histochemistry of Nervous Transmission*, p. 389–404. (Amsterdam: Elsevier)

15
Cell Culture Systems for the Study of Human Skin and Skin Glands

C. M. LEE

Cell culture – the maintenance of a population *in vitro* through at least one cycle of division – has a long, somewhat tortuous, history. Although successful attempts to grow animal cells *in vitro* were made as early as the beginning of the century (e.g. References 1 and 2), there were major problems with microbial contamination and irreproducibility of experiments, the latter of which arose from the use of undefined natural fluids as culture media. Despite these problems, the new technique was extended to a variety of cell types via an imaginative range of biological supplements, including lymphatic, amniotic and spinal fluids, plasma, serum and various tissue and embryo extracts. The minimization of contamination and irreproducibility, however, continued to require a vigorous approach, and technical progress over the next 35 years was limited to minor refinements of very demanding experimental protocols.

The introduction of antibiotics around 1945 heralded the first major advance in cell culture. This was followed shortly afterwards by the demonstration that cultures could be used to assay and support growth of viruses[3]. Although the implications for vaccine production immediately focused attention on the requirements of cells *in vitro*, it was not until the 1960s that any significant progress was made towards eliminating the need for natural supplements in culture media, and, 30 years later, there are still cell types which cannot be grown under defined conditions. The most commonly used supplement in culture media then, as now, is serum, which preferentially supports the growth of mesenchymally-derived cells, such as fibroblasts, so that standard culture media and techniques were originally devised, and subsequently optimized, for fibroblastic cell populations. For this reason, experiments involving the culture of normal epithelial cells were, until the 1980s, largely curtailed by fibroblastic cell overgrowth in much the same way that early culture experiments were prematurely terminated by microbial overgrowth. Although the potential value of epithelial culture systems prompted persistent attempts to develop means of preventing or removing contaminating fibroblasts (Table 15.1 and Figure 15.1), none of these have proved as successful as the design of epithelia-specific culture media, which usually contain defined constituents.

Therefore, the second major advance in the science of normal cell culture, the elimination of ill-defined and variable media supplements, is coincident with the advent of satisfactory culture systems for the study of epithelia.

THE CULTURE SYSTEM

Primary epithelial cultures are initiated by outgrowth from a tissue fragment placed onto a suitable surface or by dissociating the cells which comprise the tissue and 'seeding' the resulting cell suspension into a culture chamber. After growth of the primary culture to an appropriate density in a suitable selective medium, the cells may be enzymically or mechanically raised from the substrate and reseeded onto a larger culture surface to allow further cell proliferation. This procedure is called passage: primary cultures having undergone passage are referred to as primary lines. These lines die out after a consistent and empirically determined number of passages, which is characteristic of the tissue and species, unless a specific genetic event occurs to enable their prolonged survival. If such an event does occur, the resultant cell lines are referred to as 'established' or 'immortalized', depending on the optimism of the experimenter.

METHODS OF PRIMARY EPITHELIAL CELL CULTURE

The basic laboratory and technical requirements for culture are similar for all types of cell, and an excellent manual describing these has been produced by Freshney[22]. Assuming that the requisite equipment and skills are available, the first procedure in a culture protocol is isolation of the tissue. Most epithelia form a demarcation between the body cavities and the external environment, and such surface tissues are particularly prone to contamination. Superficial contamination of many epithelial tissues, particularly skin, can be reduced by swabbing with alcohol or specific anti-microbial agents before isolation, and repeated washing with large quantities of buffer during tissue processing serves to further dilute the concentration of microbes. Subsequent incorporation of antibiotics, such as gentomycin, penicillin with streptomycin, amphotericin B or kanamycin, into the culture medium is usually sufficient to prevent growth of bacteria, fungi and mycoplasma, and eradication of these agents can be confirmed at a later stage of the culture by omitting the antibiotics. A less tractable problem, and less well recognized as a potential pitfall, is intracellular contamination by mycoplasma or viruses. The latter present particularly serious problems of detection and usually cannot be eradicated. As a consequence, all mammalian cultures need to be handled as if they are known to be contaminated with transmissible mammalian viruses.

Normal epithelial cells, whether in explant culture or seeded from

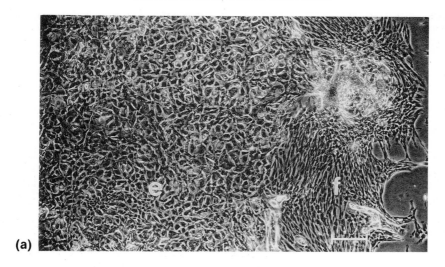

(a)

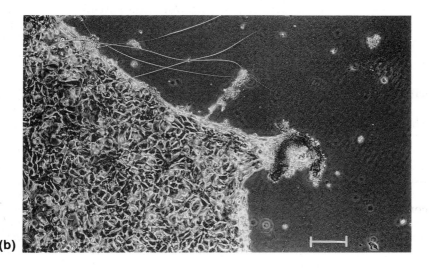

(b)

Figure 15.1   Phase-contrast micrograph of a primary culture from an explanted human sweat gland reabsorptive duct in medium containing 1% fetal calf serum (Lee *et al.*, 1986)[59] Epithelial cells (e), initially identified by their classically polygonal morphology, outgrow first; fibroblast-like cells (f), initially identified by their spindle shape and streamed growth pattern, become apparent beyond the epithelial cell sheet approximately 7 to 10 days after explantation, and thenceforth divide more rapidly than the epithelial population. Figure 15.1a shows the culture at 28 days.

Differences in the degree of epithelial and fibroblastic cell-to-plastic adhesion can be utilised to separate the populations – Figure 15.1b shows the same area of the primary culture after preferential removal of fibroblasts by trypsin treatment. Such differential removal is incomplete, necessitating frequent repetition of treatment.

Bar equals 300 μm

Table 15.1 Methods used to promote epithelial cell growth in potentially mixed primary cultures

Selection of epithelial cells	Reference
Physical and enzymic methods	
Removal of fibroblasts by mechanical means	
Collagenase in culture medium	Lasfargues and Moore, 1971[4]
Differential trypsinization	Owens *et al.*, 1974[5]
Reduced incubation temperature (32–33°C)	Jensen and Therkelsen, 1981[6]
Use of biological matrices	
Feeder cell layer	Rheinwald and Green, 1975[7]
Pigskin dermis	Freeman *et al.*, 1976[8]
Collagen	Liu and Karasek, 1978[9]
Immunological methods	
Antibody *v* fibroblastic surface marker	Gusterson *et al.*, 1981[10]
Monoclonal antibody *v* fibroblasts	Singer *et al.*, 1989[11]
Chemical methods	
Sodium ethylmercurithiosalicylate	Braaten *et al.*, 1974[12]
Use of D-valine instead of L-valine (for epithelia containing D-amino acid oxidase)	Gilbert and Migeon, 1975[13]
Proline analogue	Kao and Prockop, 1977[14]
Ammonium acetate (10 mmol/L^{-1})	Berman *et al.*, 1979[15]
Phenobarbitone	Fry and Bridges, 1979[16]
Cholera toxin	Taylor-Papadimitriou *et al.*, 1980[17]
Spermine	Webber and Chaproniere-Rickenberg, 1980[18]
Putrescine	Stoner *et al.*, 1980[19]
Reduced Ca^{2+} in culture medium	Kulesz-Martin *et al.*, 1984[20]
EDTA	Singer *et al.*, 1985[21]

suspension, require a substrate which will promote cell attachment and spreading[23]. The use of a permeable, rather than a solid, substrate is increasingly favoured since this allows nutrient uptake from the basolateral side of the cell sheet and enables measurement of secretions, transepithelial fluxes and biophysical phenomena from both apical and basolateral aspects.

Culture media are designed to contain sufficient nutrients (including dissolved gases), hormones and mitogenic factors to support metabolic function and cell division, provide an isotonic environment (usually in the range 260–320 mOsmol kg^{-1}), and maintain a pH of between 7.2 and 7.4. Although cells *in vivo* are known to be in contact with a large number of nutrients, growth factors and other potential effectors, the absolute requirements for growth and normal differentiation *in vivo* have not been fully defined for any tissue, so it is currently impossible to theoretically devise a perfect culture medium. The most practical strategy for setting up a new primary culture system is to test media which are known to support the proliferation of similar cells and then to introduce variations of these based on consideration of the influences known to affect the particular cells *in vivo.* Barnes *et al.*[24,25] and Taub[26] provide detailed methods for the culture of some epithelial cell types and a useful list of references for the culture of others.

PRIMARY CULTURE SYSTEMS IN THE STUDY OF EPITHELIAL PHYSIOLOGY

The most accurate insights into the normal functioning of an epithelium will obviously be gained by observation or measurement of the *in situ* state. Where problems of complexity or accessibility preclude such an approach, it becomes necessary to consider models further removed from the physiological state but with the virtue of increased simplicity. As such, isolated intact epithelial organs, tissues and individual cells have advantages similar to those of corresponding non-epithelial models. For isolated cells, these are generally considered to be: increased control of environmental parameters, greater quantities of experimental material and the rapid production of easily interpreted data. However, isolated epithelial cells have the unique capacity to reform a polarized cell sheet, which confers on the culture system a degree of physiological relevance normally associated with an undisturbed tissue, while retaining the simplicity of a cellular system. Where such a sheet is made from cells which normally form tubules of narrow diameter, such as sweat glands, access to the apical surface in culture may permit previously impossible biochemical, immunological and electrophysiological studies.

Furthermore, a judicious choice of culture system may permit the population development to be manipulated. Many types of epithelial cells in culture exhibit an initial proliferative phase, with minimal or no expression of differentiated function, followed by a more complete expression of differentiated function, which may occur after proliferation has slowed or ceased, and it is often possible to specifically tailor conditions at each stage of the culture in order to influence the relative duration of these phases. Expression of differentiated functions may be induced by stimuli, such as: specific hormones; alteration of nutrients in the medium; the attainment of confluence (where each epithelial cell is closely apposed to neighbouring cells); or passage onto an extracellular matrix or permeable support. Since some of the epithelia in the original population are likely to be multipotent stem cells, manipulation of the cell lineage, which determines the particular differentiated function expressed, may also be possible. Stem cell theory is discussed in more detail for colorectal carcinoma cell lines in Chapter 2 by Susan Kirkland.

PRIMARY CULTURE OF HUMAN KERATINOCYTES

The capacity of the epidermis for self-renewal, both under normal conditions and during wound healing, has long posed intriguing questions regarding the regulation of keratinocyte growth. Early attempts to answer some of these questions by developing models for mammalian keratinocyte culture did have some success[27]. However, the routine growth of human keratinocytes *in vitro* become a feasible proposition only in the mid-1970s when Rheinwald and Green developed the now-classic method for their culture on a substrate of irradiated 3T3 cells with serum and appropriate growth factors in the medium[7,28,29]. Slightly later, systems which supported growth of keratinocytes in defined serum-free media became available[30,31]. These formulae and adaptations arising from them, permitted studies on the influence of factors, such as substrate, pH and media composition, on epidermal cell proliferation for example, References 9, 32–34) and differentiation (reviewed in Reference 35).

Differentiation of keratinocytes *in vivo* involves the conversion of proliferative stem cells into non-dividing keratin-containing cells which eventually die, forming a protective cornified layer. The differentiation process is accompanied by an upward migration from the basal lamina, at the bottom of the epidermis, through the stratum spinosum and stratum granulosum to the stratum corneum, and can be followed by the sequential expression of different sets of keratin peptides[36,37], involucrin[38-41] and components of the cornified envelope[42]. However, epidermal cells cultured on solid substrates tend to proliferate, with incomplete maturation and poor formation of the stratum granulosum and stratum corneum[43]. Attachment, spreading, cell growth and maturation can be improved by use of a collagen gel[43,44], and normal development can be promoted by use of a mixed culture, including fibroblasts[45]. As knowledge of the factors influencing differentiation has accumulated it has become possible to consider keratinocyte cultures as realistic models for living epidermis[46-49] and cultured epidermal cells can be used to replace skin in the treatment of certain burns[50,51] and leg ulcers[52,53]. Such epidermal cell replacement in wound healing is recognized as being more efficient in the presence of dermis or some cultured 'dermal equivalent'[54].

PRIMARY CULTURE OF HUMAN ECCRINE SWEAT GLAND EPITHELIA

The mechanisms and regulation of ion transport across sweat gland epithelia are topics of more than inherent interest because of the characteristic malfunction of sweat glands in cystic fibrosis, the most common lethal genetic disease of Caucasian populations. Normal human eccrine sweat glands are simple tubular structures composed of two functionally distinct segments: a coiled secretory epithelium, which produces an isotonic fluid, and a reabsorptive epithelium, which recovers sodium and chloride from the sweat before it reaches the skin surface. In cystic fibrosis, the genetic defect is manifest in the secretory segments as an inability to produce fluid in response to stimulation by β-adrenergic agents[55] and in the reabsorptive segment as a chloride impermeability[56] which apparently causes an electro-chemical gradient unfavourable to the normal uptake of either sodium or chloride.

Although intact sweat glands are easily isolated, the difficulties of handling such small organs and of studying ion transport across tubules of narrow luminal diameter prompted the development of culture systems from normal reabsorptive duct[57] and normal and cystic fibrotic reabsorptive and secretory epithelia[58-61]. These culture systems exhibit the cellular polarity unique to epithelia and retain the capacity for transepithelial ion transport[61-63]. However, cells of secretory origin are capable of expressing a reabsorptive phenotype *in vitro*[63], and it is not clear whether this reflects a normal progression from secretory to reabsorptive function *in vivo* or whether an uncommitted population in the secretory coil has the capacity to enter either a secretory or a reabsorptive lineage. The second possibility offers more potential in terms of culture models since identification of the factors controlling stem cell differentiation would allow selection of either a secretory or a reabsorptive phenotype. At present, the unclear relationships between the different sweat gland cell types *in vivo*, and the absence of specific cell type markers, limits the value of culture systems in elucidating transport mechanisms. A further consequence of the lack of cell type markers is that comparative analysis of normal and cystic fibrotic populations[64] must be based on the assumption that equivalent populations result from the culture of normal and abnormal glands[65].

Such difficulties are common to many culture systems. They do not invalidate *in vitro* models, but emphasize the need to relate data from culture systems to that obtained in other experimental models.

PRIMARY CULTURE OF HUMAN SEBACEOUS GLANDS

Human sebaceous cell culture systems have not, historically speaking, had a particularly high profile, since measurement of proliferative activity, cell transition times and lipogenesis in sebaceous glands are popularly measured by radiolabelling[66] and this technique is not thwarted by the integrity of the tissue to the same extent as is the measurement of ion transport across the narrow tubes of the sweat gland epithelia. In addition, many of the most pressing questions regarding sebaceous gland function and malfunction have, in the past, related to morphology and functional organization, for which the intact gland provides the most useful model.

However, sebaceous glands can be easily isolated from surrounding skin by a variety of methods[67,68] and can either be kept on permeable supports in the organ maintenance system described for sweat glands[69] or explanted for culture. Rosenfield[70] grew human sebocytes on a 3T3 feeder layer with 20% fetal calf serum in order to study the relationship of sebaceous cell stage to growth in culture. In our laboratory, sebaceous gland cells of epithelioid morphology outgrow readily from collagenase-treated sebaceous glands explanted in serum-free media (Figure 15.2a) developed for the culture of sweat glands[59], and confluence is rapidly achieved (Figure 15.2b).

Cultured sebaceous cells are likely to originate mainly from the undifferentiated cells around the sebaceous gland acini where there is a concentration of mitotic activity[71], but will probably also contain cells in the early stages of differentiation since these retain proliferative capacity[72]. With

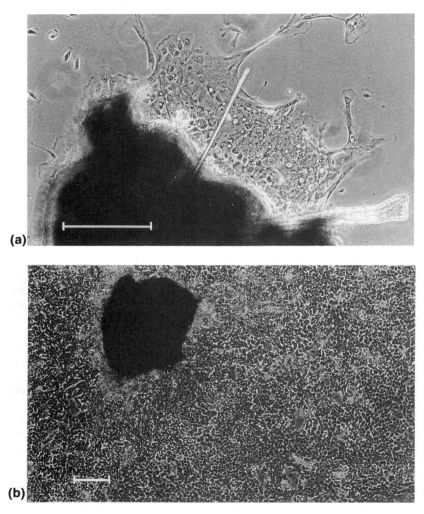

Figure 15.2 Primary culture showing (a) initial and (b) confluent outgrowth from an explanted human sebaceous gland grown in medium originally devised for primary human eccrine sweat glands (Lee and Dessi, 1989)[85]: William's E medium supplemented with penicillin (100 iu ml^{-1}), streptomycin (100 µg ml^{-1}), L-glutamine (2 mmol L^{-1}), insulin (10 µg ml^{-1}), transferrin (10 µg ml^{-1}), hydrocortisone (10 ng ml^{-1}), epidermal growth factor (10 ng ml^{-1}) and sodium selenite (10 ng ml^{-1}).
Bar equals 300 µm

Table 15.2   Intermediate filaments present in various cell types

Origin	Common name	Sequence type	Size (kDa)
All epithelia	Acidic cytokeratins	I	40–60
All epithelia	Neutral–basic cytokeratins	II	50–70
Mesenchymal cells	Vimentin	III	53
Myogenic cells	Desmin	III	52
Glial cells and astrocytes	Glial fibrilliary acid proteins	III	51
Most neurons	Neurofilaments	IV	57–150
Nuclear lamina of all eukaryocytes	Lamins	V	60–70

Adapted from Steinert and Roop (1988), Reference 89

the advent of defined media, culture systems may have an important future role in defining the parameters which control sebocyte maturation and in screening for new lipogenesis-inhibiting compounds which may have therapeutic potential in diseases such as acne, where overproduction of lipid contributes to the pathogenesis.

EPITHELIAL CELL CHARACTERIZATION

It is relatively easy to establish the epithelial origin of cells *in vivo*. Morphology is used as a general indication of cell type because different cell types have 'classical' appearances in culture. Epithelia grow as polygonal cells which become closely apposed to each other in a manner resembling a cobblestone pavement (see Figure 15.1a), which is very similar to the appearance of a magnified intact epithelial surface. However, morphology alone is an inadequate criterion of classification since this may be influenced by the culture environment and since some morphological features are common to different cell types. Cultured endothelial cells in particular, look very similar to epithelia.

A more precise criterion is the identification of intermediate filaments which are present in all cells but which vary in class according to cell type (Table 15.2). Intermediate filaments in epithelial cells are made up of cytokeratins, a family of closely related proteins, about 20 of which have been identified to date. Epithelial cells in culture retain expression of cytokeratins and this is not detectable in other cell types using standard immunocytochemical techniques.

It is, in contrast, difficult to identify the particular epithelial lineage to which cultured cells belong. Although cultures usually retain certain ultrastructural features characteristic of native epithelia, such as desmosomal cell-to-cell connections, gap junctions, tight junctions and cell polarity (often indicated by the deposition of a basal lamina between the cell substrate and the basolateral membrane and the formation of microvilli or rudimentary microvillar-like structures on the apical surface), they rarely retain other features consistently enough to enable a specific lineage to be identified.

341

Different cell types in an intact tissue will express a different range of cytokeratin filaments, but identification of individual cell types in culture is complicated by the fact that they are cell cycle specific as well as cell type specific. A degree of characterization is possible, but, in general, this system is too complex to be a reliable means of characterization.

In the absence of other markers, measuring retention of specific functions, such as ion transport or hormone production, is, in practical terms, the best indication of the origin of a cultured population.

ESTABLISHED CELL LINES AS MODEL SYSTEMS

The most serious limitation to primary culture of human cells is that there remains a frequent need for access to freshly isolated human tissue and, in addition, preparation of the tissues for culture may be very labour-intensive. These problems can be overcome by the use of established cell lines, which offer the further advantage that they are more easily stored in liquid nitrogen than the equivalent primary lines, and may also be cloned and genetically manipulated. Established lines require careful characterization because the effects of selection and adaptation to culture conditions will be more pronounced than in primary lines. The most characteristic of these effects is alteration of the karyotype after establishment. Aneuploidy seems to be an inevitable, perhaps even necessary, adjunct to the acquisition of an increased life span[73].

The 'spontaneous' development of established human cell lines *in vitro* is a very rare occurence, even though such lines readily emerge from primary cultures of other mammalian species. However, the genetic alterations required for the acquisition of an extended life span can be readily induced in human cells by physical, chemical or radiation-induced chromosomal disruption or by specific expression of inserted DNA sequences. It is well known that transforming viruses, such as the RNA-containing retroviruses or the DNA-containing papovaviruses, herpesviruses and human adenoviruses, are capable of *in vitro* immortalization of cells which are nonpermissive or semipermissive for viral growth. Where immortalizing genes for a particular cell type have been identified, viral vectors can be used to incorporate these into the host genome. Alternatively, cloned immortalizing genes can be introduced by transfection, microinjection, liposome carriers and various means of cell permeabilization.

The acquisition of immortality may be accompanied by other phenotypic alterations, such as the ability (of attachment-dependent cells) to grow in soft agar, the loss of high-density inhibition of growth, tumorigenicity etc., which are features indicative of a 'fully transformed' state. Transformation, both *in vivo* and *in vitro*, typically results from multiple genetic events[74], one of which may be immortalization. This is reflected in culture by cells which attain an increased life span without concomitant expression of other features of the fully transformed phenotype[75–77] or by those which can be transformed but fail to sustain their growth[78,79].

With respect to use as physiological models, lines which are immortalized, and which will therefore have an aneuploid karyotype, but retain a normal phenotype (at least for the characteristic to be studied), have considerable practical application. Such lines are initially selected by cloning after infection or transfection and screening the clones for expression of the required characteristics. The fact that most established human skin lines are currently of fibroblastic origin reflects the previously discussed limitations of primary epithelial culture rather than any known difference in the propensity of different cell types to become established.

ESTABLISHED LINES FROM HUMAN EPIDERMAL KERATINOCYTES

Human epidermal keratinocytes with a greatly extended life-span were obtained by Steinberg and Defendi[80] after infection of a primary keratinocyte culture with the papovavirus SV40. The proliferation phase of the cells was lengthened after virus treatment although the transformants appeared to retain the capacity to differentiate even when the usual feeder layer of fibroblasts was absent, supporting early evidence[81] that fibroblasts do not directly influence epithelial differentiation. In addition, the clonability of the immortalized cells made this an early model for genetic analysis of epithelial differentiation. However, later studies showed changes in the expression of cytokeratin filaments following keratinocyte immortalization[82,83], loss of the envelope precursor, involucrin, and a reduction in normal stratification[84], indicating that differentiation may be incomplete, at least in some established keratinocyte lines.

ESTABLISHED LINES FROM HUMAN ECCRINE SWEAT GLANDS

A similar method of immortalization was used to produce the ion-transporting human sweat gland cell line, NCL-SG3[85]. The line transports, in response to lysylbradykinin and to β-adrenergic stimuli, in a manner similar to that of primary cultures[63]. However, unlike primary cultures, the cell line showed no response to carbachol, an analogue of the cholinergic agonist acetylcholine, at a concentration known to stimulate the phosphatidyl–inositol pathway in intact sweat glands. It is not yet clear whether NCL-SG3 has lost the capacity to respond to carbachol, either as a result of extended time in culture or of chromosomal rearrangements consequent to immortalization, or whether different cell types present in the intact sweat gland respond to different stimuli[86] so that the absence of response is a function of selection during culture.

Although viral infection is a simple means of immortalization, the resulting cell line may contain complete copies of the viral genome, either incorporated into the host DNA or present as episomes., Many SV40-treated cells, such as NCL-SG3, are virus producers at early passage numbers and later cease production; this latter stage can be reached very quickly if antiserum against the transforming virus is added to the culture medium. However, all virus-immortalized lines remain potential virus producers – some transformed lines may be induced to form infectious virus on co-culture with (or fusion to) cells which permit a lytic virus cycle, or after mitomycin C, proflavine or hydrogen peroxide treatment[87]. The use of virus mutants which are defective in their origin of replication avoids the possibility of virus production and has the added advantage that transformation occurs at a higher efficiency. Several human eccrine sweat gland cell lines have now been produced in our laboratory from primary cultures of normal and cystic fibrotic origin, using an SV40 ori⁻/adenovirus hybrid[88] which contains an SV40 early region, with a defective origin of replication cloned into an adenovirus vector in place of adenovirus early region one.

Established sweat gland lines occasionally fail to retain a classically epithelial appearance (Figure 15.3). It is not clear whether this is an epithelial population exhibiting an atypical morphology as a consequence of establishment or whether selection of a fibroblastoid subpopulation originally present in low numbers has occurred. However, where lines retain their epithelial morphology, they also seem to retain the capacity for transepithelial ion transport and should therefore prove to be valuable tools for studying the mechanisms of sweat secretion in physiological and pathological states and useful screening systems for potential modulators of ion transport.

Certain of the sweat gland lines in our laboratory can be induced to differentiate morphologically as well as functionally, forming structures which imperfectly mimic epithelial tubules (Figure 15.4). If the stimuli for formation of these *in vitro* structures are similar to *in vitro* inducers of tubule formation, these lines may additionally prove useful in the study of epithelial differentiation.

In conclusion, the recent development of low-serum and serum-free culture systems for the growth of human skin and skin gland epithelia has permitted the study, at an increasingly rapid rate, of factors influencing cell growth, differentiation pathways and the expression of differentiated functions. The screening of pharmacological compounds for the capacity to modulate expression of differentiated functions has recently become simpler, both because of the advent of more sophisticated culture systems and the availability of appropriate immortal cell lines. Finally, the last few years have seen the first direct clinical applications for cultured cells, in the replacement of damaged epidermis. The practical importance of such skin grafting has yet to be established but it seems likely that these experiments herald attempts to treat other types of damage or deficiency by means of cell replacement. If so, we are likely to see even more rapid advances in cell culture in the next few years.

Figure 15.3 Cell lines established from primary cultures of human eccrine sweat gland epithelia. (a) Cell line, NCL-SG6 has retained the epithelial appearance of the parent culture, with closely opposed polygonal cells, whereas (b) cell line NCL-SG13, demonstrates a more elongated morphology with less contact between adjacent cells. (c) Cell line NCL-SG5 has a fibroblast-like morphology with bipolar cells forming parallel arrays as the culture approaches confluence.
Bar equals 300 μm

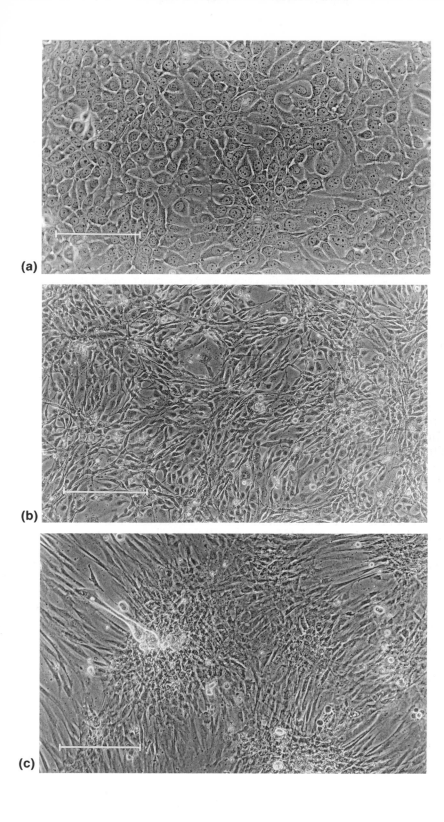

(a)

(b)

(c)

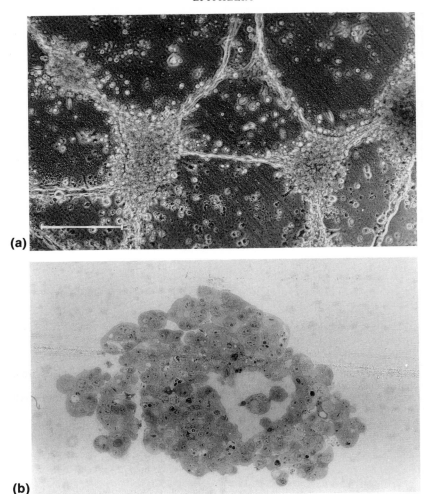

Figure 15.4 Tubule formation by established lines from human eccrine sweat gland epithelia. (a) Phase-contrast micrograph of NCL-SG5 cells forming a tubule-like network and (b) a toluidine-blue-stained cross-section through a similar tubule formed by cells of line NCL-SG3. Bar equals 300 μm

Acknowledgements

The author wishes to acknowledge financial support from the Cystic Fibrosis Research Trust, UK and the Cystic Fibrosis Foundation, USA.

References

1. Beebe, S. P. and Ewing, J. (1906). A study of the so-called infectious lymphosarcoma of dogs. *J. Med. Res., Boston,* **15**, 209–227

2. Harrison, R. G. (1907). Observations of the living, developing nerve fibre. *Proc. Soc. Exp. Biol. Med.,* **4**, 140–143

3. Enders, J. F., Weller, T. H. and Robbins, F. C. (1949). Cultivation of the Lansing strain of poliomyelitis virus in cultures of various human embryonic tissues. *Science,* **109**, 85–87

4. Lasfargues, E. Y. and Moore, D. H. (1971). A method for the continuous cultivation of mammary epithelium. *In Vitro,* **7**, 21–25

5. Owens, R. B., Smith, H. S. and Hackett, A. J. (1974). Epithelial cell cultures from normal glandular tissue of mice. Mouse epithelial cultures enriched by selective trypsinisation. *J. Nat. Cancer Inst.,* **53**, 261–269

6. Jensen, P. K. A. and Therkelsen, A. J. (1981). Cultivation at low temperature as a measure to prevent contamination with fibroblasts in epithelial cultures from human skin. *J. Invest. Dermatol.,* **77**, 210–212

7. Rheinwald, J. G. and Green, H. (1975). Serial cultivation of strains of human keratinocytes: The formation of keratinising colonies from single cells. *Cell,* **6**, 331–343

8. Freeman, A. E., Igel, H. J., Herrman, B. J. and Kleinfeld, K. L. (1976). Growth and characterization of human skin epithelial cell cultues. *In Vitro,* **12**, 352–362

9. Liu, S-C and Karasek, M. (1978). Isolation and growth of adult human epidermal keratinocytes in cell culture. *J. Invest. Dermatol.,* **71**, 157–162

10. Gusterson, B. A., Edwards, P. A. W., Foster, C. S. and Neville, A. M. (1981). The selective culture of keratinocytes using a cytotoxic antifibroblast monoclonal antibody. *Br. J. Dermatol.,* **105**, 273–277

11. Singer, K. H., Scearce, R. M., Tuck, D. T., Whichard, L. P., Denning, S. M. and Haynes, B. F. (1989). Removal of fibroblasts from human epithelial cell cultures with use of a complement fixing monoclonal antibody reactive with human fibroblasts and monocytes/macrophages. *J. Invest. Dermatol.,* **92**, 166–170

12. Braaten, J. T., Lee, M. J., Schewk, A. and Mintz, D. H. (1974). Removal of fibroblastoid cells from primary monolayer cultures of rat neonatal endocrine pancreas by sodium ethylmercurithiosalicylate. *BBRC,* **61**, 476–482

13. Gilbert, S. F. and Migeon, B. R. (1975). D-valine as a selective agent for normal human and rodent epithelial cells in culture. *Cell,* **5**, 11–17

14. Kao, W. W. and Prockop, D. J. (1977). Proline analogue removes fibroblasts from cultured mixed cell populations. *Nature (London),* **266**, 63–64

15. Berman, J., Perantoni, A., Jackson, H. A. and Kingsbury, E. (1979). Primary epithelial cell culture of adult rat kidney: enhancement of cell growth by ammonium acetate. *Exp. Cell Res.,* **121**, 47–54

16. Fry, J. and Bridges, J. W. (1979). The effect of phenobarbitone on adult rat liver cells and primary cell lines. *Toxicol. Lett.,* **4**, 295–301

17. Taylor-Papadimitriou, J., Purkis, P. and Fentiman, I. S. (1980). Choleratoxin and analogues of cyclic AMP stimulate the growth of cultured human epithelial cells. *J. Cell Physiol.,* **102** 317–322

18. Webber, M. M. and Chaproniere-Rickenburg, D. (1980). Spermine oxidation products are selectively toxic to fibroblasts in cultures of normal prostatic epithelium. *Cell Biol. Int. Rep.,* **4**, 185–193

19. Stoner, G. D., Harris, C. C., Myers, G. A., Trump, B. F. and Connor, R. D. (1980). Putrescine stimulates growth of human bronchial epithelial cells in primary culture. *In Vitro,* **16**, 399–406

20. Kulesz-Martin, M. F., Fabian, D. and Bertram, J. S. (1984). Differential calcium requirements for growth of mouse skin epithelial and fibroblast cells. *Cell Tiss. Kinet.,* **17**, 525–533

21. Singer, K. H., Harden, E. A., Robertson, A. L., Lobach, D. F. and Haynes, B. F. (1985). In vitro growth and phenotypic characterization of mesodermal-derived and epithelial components of normal and abnormal human thymus. *Hum. Immunol.,* **13**, 161–176

22. Freshney, I. (1987). *Culture of Animal Cells. A Manual of Basic Technique,* 2nd edn. (New York: A. R. Liss)

23. Grinnell, F. (1978). Cellular adhesiveness and extracellular substrata. *Int. Rev. Cytol.,* **53**, 65–144

24. Barnes, D. W., Sirbasku, D. A. and Sato, G. H. (eds.) (1984a). Methods for preparation of

media, supplements, and substrata for serum-free animal cell culture. *Cell Culture Methods for Molecular and Cell Biology*, Volume 1. (New York: A. R. Liss)

25. Barnes, D. W., Sirbasku, D. A. and Sato, G. H. (eds.) (1984b). Methods for serum-free culture of epithelial and fibroblastic cells. *Cell Culture Methods for Molecular and Cell Biology*, Volume 3. (New York: A. R. Liss)

26. Taub, M. (ed.) (1985). *Tissue Culture of Epithelial Cells*. (New York: Plenum Press)

27. Cruickshank, C. N. D., Cooper, J. R. and Hooper, C. (1960). The cultivation of cells from adult epidermis. *J. Invest. Dermatol.*, **34**, 339–342

28. Rheinwald, J. G. and Green, H. (1977). Epidermal growth factor and the multiplication of cultured human epidermal keratinocytes. *Nature (London)*, **265**, 421–424

29. Green, H. (1978). Cyclic AMP in relation to proliferation of the epidermal cell: A new view. *Cell*, **15**, 801–815

30. Macaig, T., Nemore, R. E., Weinstein, R. and Gilchrest, B. A. (1981). An endocrine approach to the control of epidermal growth: serum-free cultivation of human keratinocytes. *Science*, **211**, 1452–1454

31. Tsao, M. C., Walthall, B. J. and Ham, R. G. (1982). Clonal growth of normal human epidermal keratinocytes in a defined medium. *J. Cell Physiol.*, **110**, 219–229

32. Liu, S-C, Eaton, M. F. and Karasek, M. A. (1979). Growth characteristics of human epidermal keratinocytes from newborn foreskin in primary and serial culture. *In Vitro*, **15**, 813–822

33. Eisenger, M., Lee, J. S., Hefton, J. M., Darzynkiewicz, Z., Chiao, J. W. and deHarven, E. (1979). Human epidermal cell cultures: Growth and differentiation in the absence of dermal components or medium supplements. *Proc. Natl. Acad. Sci. USA*, **76**, 5340–5344

34. Price, F. M., Taylor, W. G., Camalier, R. F. and Sanford, K. (1983). Approaches to enhance proliferation of human epidermal keratinocytes in mass culture. *J. Nat. Cancer Inst.*, **70**, 853–861

35. Eckert, R. L. (1989). Structure, function and differentiation of the keratinocyte. *Physiol. Rev.*, **69**(4), 1316–1346

36. Fuchs, E. and Green, H. (1978). The expression of keratin genes in epidermal and cultured human epidermal cells. *Cell*, **15**, 887–897

37. Sun, T-T, Eichner, R., Nelson, W. G., Tseng, S. C. G., Weiss, R. A., Jarvinen, M. and Woodcock-Mitchell, J. (1983). Keratin classes: molecular markers for different types of epithelial differentiation. *J. Invest. Dermatol.*, **81** (Suppl.), 109s–115s

38. Rice, R. H. and Green, H. (1977). The cornified envelope of terminally differentiated human epidermal keratinocytes consists of cross-linked protein. *Cell*, **11**, 417–422

39. Banks-Schlegel, S. and Green, H. (1981). Involucrin synthesis and tissue assembly by keratinocytes in natural and cultured human epithelia. *J. Cell. Biol.*, **90**, 732–737

40. Watt, F. and Green, H. (1981). Involucrin synthesis is correlated with cell size in human epidermal cell cultures. *J. Cell. Biol.*, **90**, 738–742

41. Watt, F. M. (1983). Involucrin and other markers of keratinocyte terminal differentiation. *J. Invest. Dermatol.*, **81**, 100–103

42. Simon, M. and Green, H. (1984). Participation of membrane-associated proteins in formation of the cross-linked envelope of the keratinocyte. *Cell*, **36**, 827–834

43. Holbrook, K. A. and Hemmings, H. (1983). Phenotypic expression of epidermal cells *in vitro*: A review. *J. Invest. Dermatol.*, **81**, 11–24

44. Voigt, W-H, and Fusenig, N. E. (1979). Organotypic differentiation of mouse keratinocytes in cell culture: A light and electron microscopic study. *Biol. Chem.*, **34**, 111–118

45. Woodley, D., Didierjean, L., Regnier, M., Saurat, J. and Prunieras, M. (1980). Bullous pemphigoid antigen synthesized in vitro by human epidermal cells. *J. Invest. Dermatol.*, **75**, 148–151

46. Prunieras, M. (1979). Epidermal cell cultures as models for living epidermis. *J. Invest. Dermatol.*, **73**, 135–137

47. Green, H., Kehinde, O. and Thomas, J. (1979). Growth of cultured human epidermal cells into multiple epithelia suitable for grafting. *Proc. Natl. Acad. Sci.*, **11**, 5665–5668

48. Bell, E., Ehrlich, H. P., Sher, S., Merrill, C., Sarber, R., Hull, B., Nakatsuji, T., Church, D. and Buttle, D. (1981). Development and use of a living skin equivalent. *J. Plast. Reconst. Surg.*, **67**, 386–392

49. Bell, E., Ehrlich, H. P., Buttle, D. and Nakatsuji, T. (1981). Living tissue formed in vitro

and accepted as a skin-equivalent tissue of full thickness. *Science*, **211**, 1052–1054

50. O'Connor, N. E., Mulliken, J. B., Banks-Schlegel, S., Kehinde, O. and Green, H. (1981). Grafting of burns with cultured epithelium prepared from autologous epidermal cells. *Lancet*, **1**, 75–78

51. Phillips, T. J. (1988). Cultured skin-grafts: Past, present, future. *Arch. Dermatol.*, **124**, 1035–1038

52. Hefton, J. M., Weksler, M., Parris, A., Caldwell, D., Balin, A. K. and Carter, D. M. (1983). Grafting of skin ulcers with cultured autologous epidermal cells. *J. Invest. Dermatol.*, **80**, 322

53. Leigh, I. M. and Purkis, P. E. (1986). Culture grafted leg ulcers. *Clin. Exp. Dermatol.*, **11**, 650–652

54. Cuono, C., Langdon, R. and McGuire, J. (1986). Use of cultured epidermal autografts and dermal allografts as skin replacement after burn injury. *Lancet*, **1**, 1123–1124

55. Sato, K. and Sato, F. (1984). Defective β-adrenergic response of cystic fibrosis sweat glands *in vivo* and *in vitro*. *J. Clin. Invest.*, **73**, 1763–1771

56. Quinton, P. M. (1983). Chloride impermeability in cystic fibrosis. *Nature (London)*, **301**, 421–422

57. Pedersen, P. S. (1984). Primary culture of epithelia cells derived from the reabsorptive coiled duct of human sweat glands. *IRCS Med. Sci.*, **12**, 752–753

58. Collie, G., Buchwald, M., Harper, P. and Riordan, J. R. (1985). Culture of sweat gland epithelial cells from normal individuals and patients with cystic fibrosis. *In Vitro Cell. Devel. Biol.*, **21**, 597–602

59. Lee, C. M., Carpenter, F., Coaker, T. and Kealey, T. (1986). The primary culture of epithelia from the secretory coil and collecting duct of normal human and cystic fibrotic eccrine sweat glands. *J. Cell Sci.*, **83**, 103–118

60. Hazen-Martin, D. J., Spicer, S., Sens, M. A., Jenkins, M. Q., Westphal, M. C. and Sens, D. (1987). Tissue culture of normal and cystic fibrosis sweat gland duct cells. I. Alterations in dome formation. *Pediatr. Res.*, **21**, 72–78

61. Jones, C. J., Bell, C. L. and Quinton, P. M. (1988). Different physiological signatures of sweat gland secretory and duct cells in culture. *Am. J. Physiol.*, **255**, C102–111

62. Pedersen, P. S., Brandt, N. J. and Hainau, B. (1985). Differentiated function in primary epithelial culture derived from the coiled reabsorptive segment of human sweat glands. *IRCS Med. Sci.*, **13**, 875–876

63. Brayden, D. J., Cuthbert, A. W. and Lee, C. M. (1988). Human eccrine sweat gland epithelial cultures express ductal characteristics. *J. Physiol.*, **405**, 657–675

64. Pedersen, P. S., Larsen, E. H. and Brandt, N. J. (1987). Restitution of chloride permeability in cystic fibrosis. *Med. Sci. Res.*, **15**, 151–152

65. Lee, C. M. (1988). Human eccrine sweat gland cell culture as a model for cystic fibrosis. In Mastella, G. and Quinton, P. M. (eds.) *The Cellular and Moleular Basis of Cystic Fibrois*, pp. 404–415. (San Francisco, CA: San Francisco)

66. Thody, A. J. and Schuster, S. (1989). Control and function of sebaceous glands. *Physiol. Rev.*, **69**(2), 383–408

67. Kellum, R. E. (1966). Isolation of human sebaceous glands. *Arch. Dermatol.*, **93**, 610–612

68. Kealey, T., Lee, C. M., Thody, A. J. and Coaker, T. (1986). The isolation of human sebaceous glands and apocrine sweat glands by shearing. *Br. J. Dermatol.*, **114**, 181–188

69. Lee, C. M., Jones, C. J. and Kealey, T. (1984). Biochemical and ultrastructural studies of human eccrine sweat glands isolated by shearing and maintained for seven days. *J. Cell. Sci.*, **72**, 259–274

70. Rosenfield, R. L. (1989). Relationship of sebaceous cell stage to growth in culture. *J. Invest. Dermatol.*, **92**(5), 751–754

71. Jenkinson, D. M., Elder, H. Y., Montgomery, I. and Moss, V. A. (1985). Comparative studies of the ultrastructure of the sebaceous gland. *Tiss. Cell.*, **17**, 683–698

72. Plewig, G., Christopher, E. and Braun-Falco, O. (1971). Proliferative cells in the human sebaceous gland. *Acta Dermatovener*, **51**, 413–422

73. Terzi, M. and Hawkins, T. S. C. (1975). Chromosomal variations and the establishment of somatic cell lines *in vitro*. *Nature (London)*, **253**, 361–362

74. Bishop, J. M. (1987). The molecular genetics of cancer. *Science*, **235**, 305–311

75. Land, H., Parada, L. F. and Weinberg, R. A. (1983a). Tumorigenic conversion of primary embryo fibroblasts requires at least two cooperating oncogenes. *Nature (London)*, **304**, 596–601

76. Land, H., Parada, L. F. and Weinberg, R. A. (1983b). Cellular oncogenes and multistep carcinogenesis. *Science*, **222**, 771–778

77. Keath, E. J., Caimi, P. G. and Cole, M. D. (1984). Fibroblast lines expressing activated c-myc oncogenes are tumorigenic in nude mice and syngeneic animals. *Cell*, **39**, 339–348

78. Spandidos, D. A. and Wilkie, N. M. (1984). Malignant transformation of early passage rodent cells by a single mutated human oncogene. *Nature (London)*, **310**, 469–475

79. Land, H., Chen, A. C., Morgenstern, J. P., Parada, L. F. and Weinberg, R. A. (1986). Behaviour of myc and ras oncogenes in transformation of rat embryo fibroblasts. *Mol. Cell Biol.*, **6**, 1917–1925

80. Steinberg, M. L. and Defendi, V. (1979). Altered pattern of growth and differentiation in human keratinocytes infected by simian virus 40. *Proc. Natl. Acad. Sci.*, **76**(2), 801–805

81. Green, H. (1977). Terminal differentiation of cultured human epidermal cells. *Cell*, **11**, 405–416

82. Moll, R., Moll, I. and Wiest, W. (1982). Changes in the pattern of cytokeratin polypeptides in epidermis and hair follicles during skin development in human fetuses. *Differentiation*, **23**(2), 170–178

83. Taylor-Papadimitriou, J., Purkis, P., Lane, E. B., McKay, I. A. and Chang, S. E. (1982). Effects of SV40 transformation on the cytoskeleton and behavioural properties of human keratinocytes. *Cell Differentiation*, **11**, 169–180

84. Hronis, T. S., Steinberg, M. L., Defendi, V. and Sun, T. T. (1984). Simple epithelial nature of some simian virus-40-transformed human epidermal keratinocytes. *Cancer Res.*, **44**, 5759–5804

85. Lee, C. M. and Dessi, J. (1989). NCL-SG3: a human eccrine sweat gland cell line which retains the capacity for transepithelial ion transport. *J. Cell Sci.*, **92**, 241–249

86. Specht, N. L. and Quinton, P. M. (1989). The electrophysiological signatures of distinct cells in the secretory coil of the normal eccrine sweat glands. *Pulmonology* (Suppl.), **4**, 128

87. Pennington, T. H. and Ritchie, D. A. (1975). *Molecular Virology*. (New York: John Wiley & Sons).

88. Van Doren, K. and Gluzman, Y. (1984). Efficient transformation of human fibroblasts by adenovirus–simian virus 40 recombinants. *Mol. Cell Biol.*, **4**, 1653–1656

89. Steinert, P. M. and Roop, D. R. (1988). Molecular and cellular biology of intermediate filaments. *Ann. Rev. Biochem.*, **57**, 593–625

Index